Understanding Digital Signal Processing

Dec 1997

Understanding Digital Signal Processing

Richard G. Lyons

ADDISON-WESLEY PUBLISHING COMPANY

An imprint of Addison Wesley Longman, Inc.

Reading, Massachusetts · Harlow, England · Menlo Park, California
Berkeley, California · Don Mills, Ontario · Sydney
Bonn · Amsterdam · Tokyo · Mexico City

Many of the designations used by manufacturers and sellers to distinguish their products are claimed as trademarks. Where those designations appear in this book and Addison-Wesley was aware of a trademark claim, the designations have been printed with initial capital letters.

The publisher offers discounts on this book when ordered in quantity for special sales.

For more information, please contact:
Corporate & Professional Publishing Group
Addison Wesley Longman, Inc.
One Jacob Way
Reading, Massachusetts 02867

Library of Congress Cataloging-in-Publication Data

Lyons, Richard G., 1948–
    Understanding digital signal processing / Richard G. Lyons.
        p.   cm.
    Includes bibliographical references and index.
    ISBN 0-201-63467-8 (hc)
    1. Signal processing--Digital techniques.   I. Title.
TK5102.9.L96   1997
621.382'2--dc20   96-28818
                                                            CIP

0-201-63467-8

1 2 3 4 5 6 7 8 9—MA—00999897

First Printing, October 1996

DEDICATION

I dedicate this book to my two daughters Julie and Meredith, I wish I could go with you; to my mother Ruth for making me finish my homework; to my father Grady who didn't know what he started when he built that workbench in the basement; to my brother Ray for improving us all; to my brother Ken who succeeded where I failed; to my sister Nancy for running interference for us; to John Lennon for not giving up; to Dr. Laura Schlessinger for keeping us honest; to my advisor Glenn Caldwell and to the Iron Riders Motorcycle Club (Niles, CA) who keep me alive.

CONTENTS

Learning Digital Signal Processing

Learning the fundamentals, and how to speak the language, of digital signal processing does not require profound analytical skills or an extensive background in mathematics. All you need is a little experience with elementary algebra, knowledge of what a sinewave is, this book, and enthusiasm. This may sound hard to believe, particularly if you've just flipped through the pages of this book and seen figures and equations that appear rather complicated. The content here, you say, looks suspiciously like the material in technical journals and textbooks, material that is difficult to understand. Well, this is not just another book on digital signal processing.

This book's goal is to gently provide explanation followed by illustration, not so that you may understand the material, but that you must understand the material.[†] Remember the first time you saw two people playing chess? The game probably appeared to be mysterious and confusing. As you now know, no individual chess move is complicated. Given a little patience, the various chess moves are easy to learn. The game's complexity comes from deciding what combinations of moves to make and when to make them. So it is with understanding digital signal processing. First we learn the fundamental rules and processes and, then, practice using them in combination.

If learning digital signal processing is so easy, then why does the subject have the reputation of being difficult to understand? The answer lies partially in how the material is typically presented in the literature. It's difficult to convey technical information, with its mathematical subtleties, in written form. It's one thing to write equations, but it's another matter altogether to explain what those equations really mean from a practical standpoint, and that's the goal of this book.

Too often, written explanation of digital signal processing theory appears in one of two forms: either mathematical miracles occur and you are simply given a short and sweet equation without further explanation, or you are engulfed in a flood of complex variable equations and phrases

---

[†] "Here we have the opportunity of expounding more clearly what has already been said" (Rene Descartes).

such as "it is obvious that," "such that $W(f) \geq 0 \; \forall \; f$," and "with judicious application of the homogeneity property." Authors usually do provide the needed information, but, too often, the reader must grab a pick and shovel, put on a miner's helmet, and try to dig the information out of a mountain of mathematical expressions. (This book presents the results of several fruitful mining expeditions.) How many times have you followed the derivation of an equation, after which the author states that he or she is going to illustrate that equation with a physical example—and this turns out to be another equation? Although mathematics is necessary to describe digital signal processing, I've tried to avoid overwhelming the reader because a recipe for technical writing that's too rich in equations is hard for the beginner to digest.[†]

The intent of this book is expressed in a popular quote from E. B. White in the introduction of his *Elements of Style* (New York: Macmillan Publishing, 1959):

> Will (Strunk) felt that the reader was in serious trouble most of the time, a man floundering in a swamp, and that it was the duty of anyone attempting to write English to drain the swamp quickly and get his man up on dry ground, or at least throw him a rope.

I've attempted to avoid the traditional instructor-student relationship, but, rather, to make reading this book like talking to a friend while walking in the park. I've used just enough mathematics to develop a fundamental understanding of the theory, and, then, illustrate that theory with examples.

The Journey

Learning digital signal processing is not something you accomplish; it's a journey you take. When you gain an understanding of some topic, questions arise that cause you to investigate some other facet of digital signal processing. Armed with more knowledge, you're likely to begin exploring further aspects of digital signal processing much like those shown in the following diagram. This book is your tour guide during the first steps of your journey.

You don't need a computer to learn the material in this book, but it would help. Digital signal processing software allows the beginner to ver-

---

[†] "We need elucidation of the obvious more than investigation of the obscure" (Oliver Wendell Holmes).

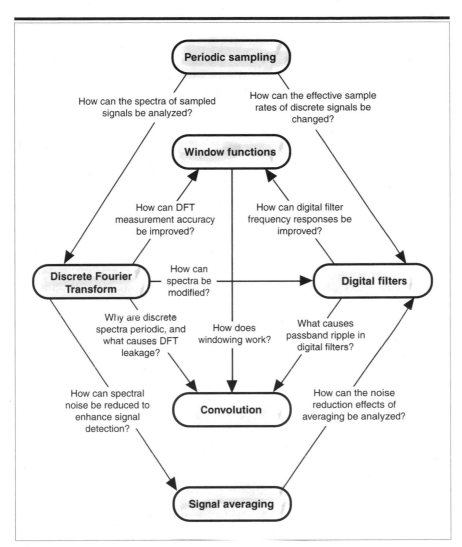

Figure P-1

ify signal processing theory through trial and error.[†] In particular, software routines that plot signal data, perform the fast Fourier transform, and analyze digital filters would be very useful.

As you go through the material in this book, don't be discouraged if your understanding comes slowly. As the Greek mathematician Menaechmus

---

[†] "One must learn by doing the thing; for though you think you know it, you have no certainty until you try it" (Sophocles).

curtly remarked to Alexander the Great, when asked for a quick explanation of mathematics, "There is no royal road to mathematics." Menaechmus was confident in telling Alexander that the only way to learn mathematics is through careful study. The same applies to digital signal processing. Also, don't worry if you have to read some of the material twice. While the concepts in this book are not as complicated as quantum physics, as mysterious as the lyrics of the song "Louie Louie," or as puzzling as the assembly instructions of a metal shed, they do get a little involved. They deserve your attention and thought. So go slow and read the material twice if you have to; you'll be glad you did. If you show persistence, to quote a phrase from Susan B. Anthony, "Failure is impossible."

Coming Attractions

Chapter 1 of this book begins by establishing the notation used throughout the remainder of our study. In that chapter, we introduce the concept of discrete signal sequences, show how they relate to continuous signals, and illustrate how those sequences can be depicted in both the time and frequency domains. In addition, Chapter 1 defines the operational symbols we'll use to build our signal processing system block diagrams. We conclude that chapter with a brief introduction to the idea of linear systems and see why linearity enables us to use a number of powerful mathematical tools in our analysis.

Chapter 2 introduces the most frequently misunderstood process in digital signal processing, periodic sampling. Although it's straightforward to grasp the concept of sampling a continuous signal, there are mathematical subtleties in the process that require thoughtful attention. Beginning gradually with simple examples of low-pass sampling and progressing to the interesting subject of bandpass sampling, Chapter 2 explains and quantifies the frequency-domain ambiguity (aliasing) associated with these important topics. The discussion there highlights the power and pitfalls of periodic sampling.

Chapter 3 is devoted to one of the foremost topics in digital signal processing, the discrete Fourier transform (DFT). Coverage begins with detailed examples illustrating the important properties of the DFT and how to interpret DFT spectral results, progresses to the topic of windows used to reduce DFT leakage, and discusses the processing gain afforded by the DFT. The chapter concludes with a detailed discussion of the various forms of the transform of rectangular functions that the beginner is likely to encounter in the literature. That last topic is included there to clarify and illustrate the DFT of both real and complex sinusoids.

Chapter 4 covers the innovation that made the most profound impact on the field of digital signal processing, the fast Fourier transform (FFT). There we show the relationship of the popular radix-2 FFT to the DFT, quantify the powerful processing advantages gained by using the FFT, demonstrate why the FFT functions as it does, and present various FFT implementation structures. Chapter 4 also includes a list of recommendations to help us when we use the FFT in practice.

Chapter 5 ushers in the subject of digital filtering. Beginning with a simple low-pass finite impulse response (FIR) filter example, we carefully progress through the analysis of that filter's frequency-domain magnitude and phase response. Next we learn how window functions affect and can be used to design FIR filters. The methods for converting low-pass FIR filter designs to bandpass and highpass digital filters are presented, and the popular Remez Exchange (Parks McClellan) FIR filter design technique is introduced and illustrated by example. In that chapter we acquaint the reader with, and take the mystery out of, the process called convolution. Proceeding through several simple convolution examples, we conclude Chapter 5 with a discussion of the powerful convolution theorem and show why it's so useful as a qualitative tool in understanding digital signal processing.

Chapter 6 introduces a second class of digital filters, infinite impulse response (IIR) filters. In discussing several methods for the design of IIR filters, the reader is introduced to the powerful digital signal processing analysis tool called the z-transform. Because the z-transform is so closely related to the continuous Laplace transform, Chapter 6 starts by gently guiding the reader from the origin, through the properties, and on to the utility of the Laplace transform in preparation for learning the z-transform. We'll see how IIR filters are designed and implemented, and why their performance is so different from FIR filters. To indicate under what conditions these filters should be used, the chapter concludes with a qualitative comparison of the key properties of FIR and IIR filters.

Chapter 7 discusses two important advanced sampling techniques prominent in digital signal processing, quadrature sampling and digital resampling. In the chapter we discover why quadrature sampling is so useful when signal phase must be analyzed and preserved, and how this special sampling process can circumvent some of the limitations of traditional periodic sampling techniques. Our introduction to digital resampling shows how we can, and when we should, change the effective sample rate of discrete data after the data has already been digitized. We've delayed the discussion of digital resampling to this chapter because some knowledge of low-pass digital filters is necessary to understand how resampling schemes operate.

Chapter 8 covers the important topic of signal averaging. There we learn how averaging increases the accuracy of signal measurement schemes by reducing measurement background noise. This accuracy enhancement is called processing gain, and the chapter shows how to predict the processing gain associated with averaging signals in both the time and frequency domains. In addition, the key differences between coherent and incoherent averaging techniques are explained and demonstrated with examples. To complete the chapter, the popular scheme known as exponential averaging is covered in some detail.

Chapter 9 presents an introduction to the various binary number formats that the reader is likely to encounter in modern digital signal processing. We establish the precision and dynamic range afforded by these formats along with the inherent pitfalls associated with their use. Our exploration of the critical subject of binary data word width (in bits) naturally leads us to a discussion of the numerical resolution limitations of analog to digital (A/D) converters and how to determine the optimum A/D converter word size for a given application. The problems of data value overflow roundoff errors are covered along with a statistical introduction to the two most popular remedies for overflow, truncation and rounding. We end the chapter by covering the interesting subject of floating-point binary formats that allow us to overcome most of the limitations induced by fixed-point binary formats, particularly in reducing the ill effects of data overflow.

Chapter 10 provides a collection of *tricks of the trade* that the professionals often use to make their digital signal processing algorithms more efficient. Those techniques are compiled into a chapter at the end of the book for two reasons. First, it seems wise to keep our collection of tricks in one chapter so that we'll know where to find them in the future. Second, many of these schemes require an understanding of the material from the previous chapters, so the last chapter is an appropriate place to keep our collection of clever tricks. Exploring these techniques in detail verifies and reiterates many of the important ideas covered in previous chapters.

The appendices include a number of topics to help the beginner understand the mathematics of digital signal processing. A comprehensive description of the arithmetic of complex numbers is covered in Appendix A, while Appendix B derives the often used, but seldom explained, closed form of a geometric series. Appendix C strives to clarify the troubling topics of complex signals and negative frequency. The statistical concepts of mean, variance, and standard deviation are introduced and illustrated in Appendix D, and Appendix E provides a discussion of the origin and utility of the logarithmic decibel scale used to improve the magnitude resolution of spectral representations. In a slightly different vein, Appendix F provides a glossary of the terminology used in the field of digital filters.

Acknowledgments

How do I sufficiently thank the people who helped me write this book? I do this by stating that any quality existing herein is due to the following talented people: for their patient efforts in the unpleasant task of reviewing early versions of the manuscript, I am grateful to Sean McCrory, Paul Chestnut, Paul Kane, John Winter, Terry Daubek, and Robin Wiprud. Special thanks go to Nancy Silva for her technical and literary guidance, and encouragement, without which this book would not have been written. For taking time to help me understand digital signal processing, I thank Frank Festini, Harry Glaze, and Dick Sanborn. I owe you people.

Gratitude goes to the reviewers, under the auspices of Addison-Wesley, whose suggestions improved much of the material. They are Mark Sullivan, David Goodman, Satyanarayan Namdhari, James Kresse, Ryerson Gewalt, David Cullen, Richard Herbert, Maggie Carr, and anonymous at Alcatel Bell. Finally, I acknowledge my good fortune in being able to work with those talented folks at Addison-Wesley: Rosa Aimée González, Simon Yates, and Tara Herries.

If you're still with me this far into the Preface, I end by saying that I had a ball writing this book and hope you get some value out of reading it.

CHAPTER ONE

Discrete Sequences and Systems

Digital signal processing has never been more prevalent or easier to perform. It wasn't that long ago when the fast Fourier transform (FFT), a topic we'll discuss in Chapter 4, was a mysterious mathematical process used only in industrial research centers and universities. Now, amazingly, the FFT is readily available to us all. It's even a built-in function provided by inexpensive spreadsheet software for home computers. The availability of more sophisticated commercial signal processing software now allows us to analyze and develop complicated signal processing applications rapidly and reliably. We can now perform spectral analysis, design digital filters, develop voice recognition, data communication, and image compression processes using software that's interactive in both the way algorithms are defined and how the resulting data are graphically displayed. Since the mid-1980s the same integrated circuit technology that led to affordable home computers has produced powerful and inexpensive hardware development systems on which to implement our digital signal processing designs.[†] Regardless, though, of the ease with which these new digital signal processing development systems and software can be applied, we still need a solid foundation in understanding the basics of digital signal processing. The purpose of this book is to build that foundation.

In this chapter we'll set the stage for the topics we'll study throughout the remainder of this book by defining the terminology used in digital signal

---

[†] During a television interview in the early 1990s, a leading computer scientist stated that had automobile technology made the same strides as the computer industry, we'd all have a car that would go a half million miles per hour and get a half million miles per gallon. The cost of that car would be so low that it would be cheaper to throw it away than pay for one day's parking in San Francisco.

1

processing, illustrating the various ways of graphically representing discrete signals, establishing the notation used to describe sequences of data values, presenting the symbols used to depict signal processing operations, and briefly introducing the concept of a linear discrete system.

1.1 Discrete Sequences and Their Notation

In general, the term *signal processing* refers to the science of analyzing time-varying physical processes. As such, signal processing is divided into two categories, analog signal processing and digital signal processing. The term *analog* is used to describe a waveform that's continuous in time and can take on a continuous range of amplitude values. An example of an analog signal is some voltage that can be applied to an oscilloscope resulting in a continuous display as a function of time. Analog signals can also be applied to a conventional spectrum analyzer to determine their frequency content. The term *analog* appears to have stemmed from the analog computers used prior to 1980. These computers solved linear differential equations by means of connecting physical (electronic) differentiators and integrators using old-style telephone operator patch cords. That way, a continuous voltage or current in the actual circuit was *analogous* to some variable in a differential equation, such as speed, temperature, air pressure, etc. (Although the flexibility and speed of modern-day digital computers have since made analog computers obsolete, a good description of the short-lived utility of analog computers can be found in reference [1].) Because present-day signal processing of continuous radio-type signals using resistors, capacitors, operational amplifiers, etc., has nothing to do with analogies, the term *analog* is actually a misnomer. The more correct term is *continuous signal processing* for what is today so commonly called analog signal processing. As such, in this book we'll minimize the use of the term *analog signals* and substitute the phrase *continuous signals* whenever appropriate.

The term *discrete-time signal* is used to describe a signal whose independent time variable is quantized so that we know only the value of the signal at discrete instants in time. Thus a discrete-time signal is not represented by a continuous waveform but, instead, a sequence of values. In addition to quantizing time, a discrete-time signal quantizes the signal amplitude. We can illustrate this concept with an example. Think of a continuous sinewave with a peak amplitude of 1 at a frequency f_o described by the equation

$$x(t) = \sin(2\pi f_o t) \, . \tag{1-1}$$

The frequency f_o is measured in hertz (Hz). (In physical systems, we usually measure frequency in units of hertz. One Hz is a single oscillation, or cycle, per second. One kilohertz (kHz) is a thousand Hz, and a megahertz (MHz) is one million Hz.[†] With t in Eq. 1-1 representing time in seconds, the $f_o t$ factor has dimensions of cycles, and the complete $2\pi f_o t$ term is an angle measured in radians.

Plotting Eq. (1-1), we get the venerable continuous sinewave curve shown in Figure 1-1(a). If our continuous sinewave represents a physical voltage, we could *sample* it once every t_s seconds using an analog-to-digital converter and represent the sinewave as a sequence of discrete values. Plotting those individual values as dots would give us the discrete waveform in Figure 1-1(b). We say that Figure 1-1(b) is the "discrete-time" version of the continuous signal in Figure 1-1(a). The independent variable t in Eq. (1-1) and Figure 1-1(a) is continuous. The independent *index* variable n in Figure 1-1(b) is discrete and can have only integer values. That is, index n is used to identify the individual elements of the discrete sequence in Figure 1-1(b).

Do not be tempted to draw lines between the dots in Figure 1-1(b). For some reason, people (particularly those engineers experienced in working with continuous signals) want to connect the dots with straight lines, or the stairstep lines shown in Figure 1-1(c). Don't fall into this innocent-looking trap. Connecting the dots can mislead the beginner into forgetting that the $x(n)$ sequence is nothing more than a list of numbers. Remember, $x(n)$ is a discrete-time sequence of individual values, and each value in that sequence plots as a single dot. It's not that we're ignorant of what lies between the dots of $x(n)$; there *is* nothing between those dots.

We can reinforce this discrete-time sequence concept by listing those Figure 1-1(b) sampled values as follows:

$$x(0) = 0 \qquad \text{(1st sequence value, index } n = 0)$$
$$x(1) = 0.31 \qquad \text{(2nd sequence value, index } n = 1)$$
$$x(2) = 0.59 \qquad \text{(3rd sequence value, index } n = 2)$$
$$x(3) = 0.81 \qquad \text{(4th sequence value, index } n = 3)$$

$$\cdots \qquad\qquad \cdots$$

$$\text{and so on,} \qquad\qquad\qquad (1\text{-}2)$$

[†] The dimension for frequency used to be *cycles/second*; that's why the tuning dials of old radios indicate frequency as kilocycles/second (kcps) or megacycles/second (Mcps). In 1960 the scientific community adopted hertz as the unit of measure for frequency in honor of the German physicist, Heinrich Hertz, who first demonstrated radio wave transmission and reception in 1887.

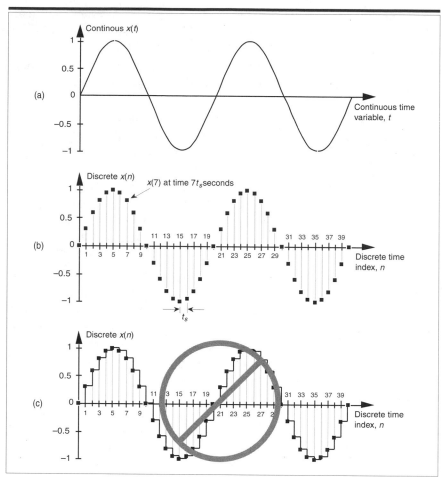

Figure 1-1 A time-domain sinewave: (a) continuous waveform representation; (b) discrete sample representation; (c) discrete samples with connecting lines.

where n represents the time index integer sequence 0, 1, 2, 3, etc., and t_s is some constant time period. Those sample values can be represented collectively, and concisely, by the discrete-time expression

$$x(n) = \sin(2\pi f_o n t_s) . \qquad (1\text{-}3)$$

(Here again, the $2\pi f_o n t_s$ term is an angle measured in radians.) Notice that the index n in Eq. (1-2) started with a value of 0, instead of 1. There's nothing sacred about this; the first value of n could just as well have been 1, but we start the index n at zero out of habit because doing so allows us to

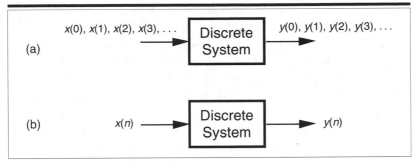

Figure 1-2 With an input applied, a discrete system provides an output: (a) the input and output are sequences of individual values; (b) input and output using the abbreviated notation of $x(n)$ and $y(n)$.

describe the sinewave starting at time zero. The variable $x(n)$ in Eq. (1-3) is read as "the sequence x of n." Equations (1-1) and (1-3) describe what are also referred to as *time-domain* signals because the independent variables, the continuous time t in Eq. (1-1), and the discrete-time nt_s values used in Eq. (1-3) are measures of time.

With this notion of a discrete-time signal in mind, let's say that a discrete system is a collection of hardware components, or software routines, that operate on a discrete-time signal sequence. For example, a discrete system could be a process that gives us a discrete output sequence $y(0)$, $y(1)$, $y(2)$, etc., when a discrete input sequence of $x(0)$, $x(1)$, $x(2)$, etc., is applied to the system input as shown in Figure 1-2(a). Again, to keep the notation concise and still keep track of individual elements of the input and output sequences, an abbreviated notation is used as shown in Figure 1-2(b) where n represents the integer sequence 0, 1, 2, 3, etc. Thus, $x(n)$ and $y(n)$ are general variables that represent two separate sequences of numbers. Figure 1-2(b) allows us to describe a system's output with a simple expression such as

$$y(n) = 2x(n) - 1 . \tag{1-4}$$

Illustrating Eq. (1-4), if $x(n)$ is the five-element sequence: $x(0) = 1$, $x(1) = 3$, $x(2) = 5$, $x(3) = 7$, and $x(4) = 9$, then $y(n)$ is the five-element sequence $y(0) = 1$, $y(1) = 5$, $y(2) = 9$, $y(3) = 13$, and $y(4) = 17$.

The fundamental difference between the way time is represented in continuous and discrete systems leads to a very important difference in how we characterize frequency in continuous and discrete systems. To illustrate, let's reconsider the continuous sinewave in Figure 1-1(a). If it represented a voltage at the end of a cable, we could measure its frequency by applying it to an oscilloscope, a spectrum analyzer, or a frequency counter. We'd have a problem, however, if we were merely given the list of values from Eq. (1-2) and asked to determine the frequency of the waveform they represent. We'd graph those discrete values, and, sure enough, we'd recognize a single sinewave as in Figure 1-1(b). We can say that the sinewave repeats every 20 samples, but there's no way to determine the exact sinewave frequency from the discrete sequence values alone. You can probably see the point we're leading to here. If we knew the time between samples—the sample period t_s—we'd be able to determine the absolute frequency of the discrete sinewave. Given that the t_s sample period is, say, 0.05 milliseconds/sample, the period of the sinewave is

$$\text{sinewave period} = \frac{20 \text{ samples}}{\text{period}} \cdot \frac{0.05 \text{ milliseconds}}{\text{sample}} = 1 \text{ millisecond.} \quad (1\text{-}5)$$

Because the frequency of a sinewave is the reciprocal of its period, we now know that the sinewave's absolute frequency is 1/(1 ms), or 1 kHz. On the other hand, if we found that the sample period was, in fact, 2 milliseconds, the discrete samples in Figure 1-1(b) would represent a sinewave whose period is 40 milliseconds and whose frequency is 25 Hz. The point here is that, in discrete systems, absolute frequency determination in Hz is dependent on the sample frequency $f_s = 1/t_s$. We'll be reminded of this dependence throughout the rest of this book.

In digital signal processing, we often find it necessary to characterize the frequency content of discrete-time domain signals. When we do so, this frequency representation takes place in what's called the *frequency domain*. By way of example, let's say we have a discrete sinewave sequence $x_1(n)$ with an arbitrary frequency f_o Hz as shown on the left side of Figure 1-3(a). We can also describe $x_1(n)$ as shown on the right side of Figure 1-3(a) by indicating that it has a frequency of 1, measured in units of f_o, and no other frequency content. Although we won't dwell in it just now, notice that the frequency-domain representations in Figure 1-3 are themselves discrete.

To illustrate our time- and frequency-domain representations further, Figure 1-3(b) shows another discrete sinewave $x_2(n)$, whose peak amplitude is 0.4, with a frequency of $2f_o$. The discrete sample values of $x_2(n)$ are expressed by the equation

$$x_2(n) = 0.4 \cdot \sin(2\pi 2 f_o n t_s) . \quad (1\text{-}6)$$

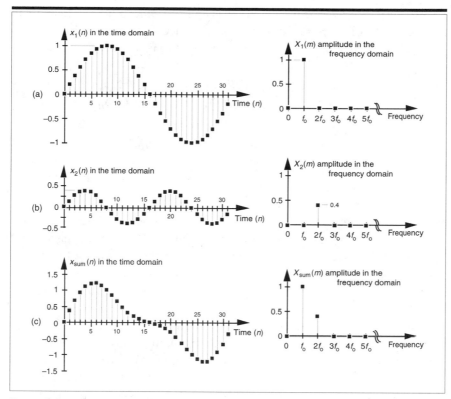

Figure 1-3 Time- and frequency-domain graphical representations: (a) sinewave of frequency f_o; (b) reduced amplitude sinewave of frequency $2f_o$; (c) sum of the two sinewaves.

When the two sinewaves, $x_1(n)$ and $x_2(n)$, are added to produce a new waveform $x_{sum}(n)$, its time-domain equation is

$$x_{sum}(n) = x_1(n) + x_2(n) = \sin(2\pi f_o n t_s) + 0.4 \cdot \sin(2\pi 2 f_o n t_s) , \qquad (1\text{-}7)$$

and its time- and frequency-domain representations are those given in Figure 1-3(c). We interpret the $X_{sum}(m)$ frequency-domain depiction, the *spectrum*, in Figure 1-3(c) to indicate that $X_{sum}(n)$ has a frequency component of f_o Hz and a reduced-amplitude frequency component of $2f_o$ Hz.

Notice three things in Figure 1-3. First, time sequences use lowercase variable names like the "x" in $x_1(n)$, and uppercase symbols for frequency-domain variables such as the "X" in $X_1(m)$. The term $X_1(m)$ is read as "the spectral sequence X sub one of m." Second, because the $X_1(m)$ frequency-domain representation of the $x_1(n)$ time sequence is itself a sequence (a list of numbers), we use the index "m" to keep track of individual elements in $X_1(m)$. We can list frequency-domain sequences just as we did with the time sequence in Eq. (1-2). For example $X_{sum}(m)$ is listed as

$$X_{sum}(0) = 0 \qquad \text{(1st } X_{sum}(m) \text{ value, index } m = 0)$$
$$X_{sum}(1) = 1.0 \qquad \text{(2nd } X_{sum}(m) \text{ value, index } m = 1)$$
$$X_{sum}(2) = 0.4 \qquad \text{(3rd } X_{sum}(m) \text{ value, index } m = 2)$$
$$X_{sum}(3) = 0 \qquad \text{(4th } X_{sum}(m) \text{ value, index } m = 3)$$
$$\cdots \qquad\qquad \cdots$$

and so on,

where the frequency index m is the integer sequence 0, 1, 2, 3, etc. Third, because the $x_1(n) + x_2(n)$ sinewaves have a phase shift of zero degrees relative to each other, we didn't really need to bother depicting this phase relationship in $X_{sum}(m)$ in Figure 1-3(c). In general, however, phase relationships in frequency-domain sequences are important, and we'll cover that subject in Chapters 3 and 5.

A key point to keep in mind here is that we now know three equivalent ways to describe a discrete-time waveform. Mathematically, we can use a time-domain equation like Eq. (1-6) for example. We can also represent a time-domain waveform graphically as we did on the left side of Figure 1-3, and we can depict its corresponding, discrete, frequency-domain equivalent as that on the right side of Figure 1-3.

As it turns out, the discrete-time domain signals we're concerned with are not only quantized in time; their amplitude values are also quantized. Because we represent all digital quantities with binary numbers, there's a limit to the resolution, or granularity, that we have in representing the values of discrete numbers. Although signal amplitude quantization can be an important consideration—we cover that particular topic in Chapter 9—we won't worry about it just now.

1.2 Signal Amplitude, Magnitude, Power

Let's define two important terms that we'll be using throughout this book: amplitude and magnitude. It's not surprising that, to the layman, these terms are typically used interchangeably. When we check our thesaurus, we find that they are synonymous.[†] In engineering, however, they mean

---

[†] Of course, layman are "other people." To the engineer, the brain surgeon is the layman. To the brain surgeon, the engineer is the layman.

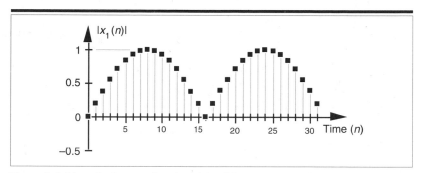

Figure 1-4 Magnitude samples, $|x_1(n)|$, of the time waveform in Figure 1-3(a).

two different things, and we must keep that difference clear in our discussions. The amplitude of a variable is the measure of how far, and in what direction, that variable differs from zero. Thus, signal amplitudes can be either positive or negative. The time-domain sequences in Figure 1-3 presented the sample value amplitudes of three different waveforms. Notice how some of the individual discrete amplitude values were positive and others were negative.

The magnitude of a variable, on the other hand, is the measure of how far, regardless of direction, its quantity differs from zero. So magnitudes are always positive values. Figure 1-4 illustrates how the magnitude of the $x_1(n)$ time sequence in Figure 1-3(a) is equal to the amplitude, but with the sign always being positive for the magnitude. We use the modulus symbol (| |) to represent the magnitude of $x_1(n)$. Occasionally, in the literature of digital signal processing, we'll find the term *magnitude* referred to as the *absolute value*.

When we examine signals in the frequency domain, we'll often be interested in the power level of those signals. The power of a signal is proportional to its amplitude (or magnitude) squared. If we assume that the proportionality constant is one, we can express the power of a sequence in the time or frequency domains as

$$x_{\text{pwr}}(n) = x(n)^2 = |x(n)|^2 , \tag{1-8}$$

or

$$X_{\text{pwr}}(m) = X(m)^2 = |X(m)|^2 . \tag{1-8'}$$

Very often we'll want to know the difference in power levels of two signals in the frequency domain. Because of the squared nature of power,

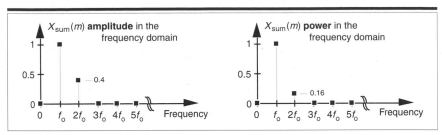

Figure 1-5 Frequency-domain amplitude and frequency-domain power of the $X_{sum}(n)$ time waveform in Figure 1-3(c).

two signals with moderately different amplitudes will have a much larger difference in their relative powers. In Figure 1-3, for example, signal $x_1(n)$'s amplitude is 2.5 times the amplitude of signal $x_2(n)$, but its power level is 6.25 that of $x_2(n)$'s power level. This is illustrated in Figure 1-5 where both the amplitude and power of $X_{sum}(m)$ are shown.

Because of their squared nature, plots of power values often involve showing both very large and very small values on the same graph. To make these plots easier to generate and evaluate, practitioners usually employ the decibel scale as described in Appendix E.

1.3 Signal Processing Operational Symbols

We'll be using block diagrams to graphically depict the way digital signal-processing operations are implemented. Those block diagrams will comprise an assortment of fundamental processing symbols, the most common of which are illustrated and mathematically defined in Figure 1-6.

Figure 1-6(a) shows the addition, element for element, of two discrete sequences to provide a new sequence. If our sequence index n begins at 0, we say that the first output sequence value is equal to the sum of the first element of the b sequence and the first element of the c sequence, or $a(0) = b(0) + c(0)$. Likewise, the second output sequence value is equal to the sum of the second element of the b sequence and the second element of the c sequence, or $a(1) = b(1) + c(1)$. Equation (1-7) is an example of adding two sequences. The subtraction process in Figure 1-6(b) generates an output sequence that's the element-for-element difference of the two input sequences. There are times when we must calculate a sequence whose elements are the sum of more than two values. This operation, illustrated in Figure 1-6(c), is called summation and is very common in

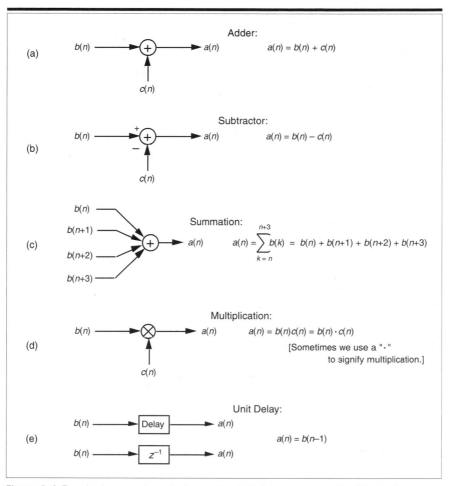

Figure 1-6 Terminology and symbols used in digital signal processing block diagrams.

digital signal processing. Notice how the lower and upper limits of the summation index k in the expression in Figure 1 6(c) tell us exactly which elements of the b sequence to sum to obtain a given $a(n)$ value. Because we'll encounter summation operations so often, let's make sure we understand their notation. If we repeat the summation equation from Figure 1-6(c) here we have

$$a(n) = \sum_{k=n}^{n+3} b(k) \ . \tag{1-9}$$

This means that

when $n = 0$, index k goes from 0 to 3, so $a(0) = b(0) + b(1) + b(2) + b(3)$

when $n = 1$, index k goes from 1 to 4, so $a(1) = b(1) + b(2) + b(3) + b(4)$

when $n = 2$, index k goes from 2 to 5, so $a(2) = b(2) + b(3) + b(4) + b(5)$

when $n = 3$, index k goes from 3 to 6, so $a(3) = b(3) + b(4) + b(5) + b(6)$

. . .   . . .

and so on.   (1-10)

We'll begin using summation operations in earnest when we discuss digital filters in Chapter 5.

The multiplication of two sequences is symbolized in Figure 1-6(d). Multiplication generates an output sequence that's the element-for-element product of two input sequences: $a(0) = b(0)c(0)$, $a(1) = b(1)c(1)$, and so on. The last fundamental operation that we'll be using is called the *unit delay* in Figure 1-6(e). While we don't need to appreciate its importance at this point, we'll merely state that the unit delay symbol signifies an operation where the output sequence $a(n)$ is equal to a delayed version of the $b(n)$ sequence. For example, $a(5) = b(4)$, $a(6) = b(5)$, $a(7) = b(6)$, etc. As we'll see in Chapter 6, due to the mathematical techniques used to analyze digital filters, the unit delay is very often depicted using the term z^{-1}.

The symbols in Figure 1-6 remind us of two important aspects of digital signal processing. First, our processing operations are always performed on sequences of individual discrete values, and second, the elementary operations themselves are very simple. It's interesting that, regardless of how complicated they appear to be, the vast majority of digital signal processing algorithms can be performed using combinations of these simple operations. If we think of a digital signal processing algorithm as a recipe, then the symbols in Figure 1-6 are the ingredients.

1.4 Introduction to Discrete Linear Time-Invariant Systems

In keeping with tradition, we'll introduce the subject of linear time-invariant (LTI) systems at this early point in our text. Although an appreciation for LTI systems is not essential in studying the next three chapters of this book, when we begin exploring digital filters, we'll build on the strict definitions of linearity and time invariance. We need to recognize and understand the notions of linearity and time invariance not just because the vast majority of discrete systems used in practice are LTI systems, but also because LTI systems are very accommodating when it comes to their analysis. That's good news for us because we can use straightforward methods to predict the performance of any digital signal processing scheme as long as it's linear and time invariant. Because linearity and time invariance are two important system characteristics having very special properties, we'll discuss them now.

1.5 Discrete Linear Systems

The term *linear* defines a special class of systems where the output is the superposition, or sum, of the individual outputs had the individual inputs been applied separately to the system. For example, we can say that the application of an input $x_1(n)$ to a system results in an output $y_1(n)$. We symbolize this situation with the following expression:

$$x_1(n) \xrightarrow{\text{results in}} y_1(n) \tag{1-11}$$

Given a different input $x_2(n)$, the system has a $y_2(n)$ output as

$$x_2(n) \xrightarrow{\text{results in}} y_2(n) \ . \tag{1-12}$$

For the system to be linear, when its input is the sum $x_1(n) + x_2(n)$, its output must be the sum of the individual outputs so that

$$x_1(n) + x_2(n) \xrightarrow{\text{results in}} y_1(n) + y_2(n) \ . \tag{1-13}$$

One way to paraphrase expression (1-13) is to state that a linear system's output is the sum of the outputs of its parts. Also, part of this description of linearity is a proportionality characteristic. This means that if the inputs are scaled by constant factors c_1 and c_2 then the output sequence parts are also scaled by those factors as

$$c_1 x_1(n) + c_2 x_2(n) \xrightarrow{\text{results in}} c_1 y_1(n) + c_2 y_2(n) \ . \tag{1-14}$$

In the literature, this proportionality attribute of linear systems in expression (1-14) is sometimes called the *homogeneity property*. With these thoughts in mind, then, let's demonstrate the concept of system linearity.

1.5.1 Example of a Linear System

To illustrate system linearity, let's say we have the discrete system shown in Figure 1-7(a) whose output is defined as

$$y(n) = \frac{-x(n)}{2} \; , \tag{1-15}$$

that is, the output sequence is equal to the negative of the input sequence with the amplitude reduced by a factor of two. If we apply an $x_1(n)$ input sequence representing a 1-Hz sinewave sampled at a rate of 32 samples

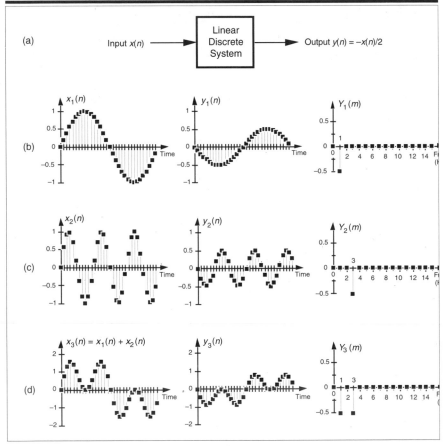

Figure 1-7 Linear system input-to-output relationships: (a) system block diagram where $y(n) = -x(n)/2$; (b) system input and output with a 1-Hz sinewave applied; (c) with a 3-Hz sinewave applied; (d) with the sum of 1-Hz and 3-Hz sinewaves applied.

per cycle, we'll have a $y_1(n)$ output as shown in the center of Figure 1-7(b). The frequency-domain spectral amplitude of the $y_1(n)$ output is the plot on the right side of Figure 1-7(b) indicating that the output comprises a single tone of peak amplitude equal to −0.5 whose frequency is 1 Hz. Next, applying an $x_2(n)$ input sequence representing a 3-Hz sinewave, the system provides a $y_2(n)$ output sequence, as shown in the center of Figure 1-7(c). The spectrum of the $y_2(n)$ output, $Y_2(m)$, confirming a single 3-Hz sinewave output is shown on the right side of Figure 1-7(c). Finally— here's where the linearity comes in—if we apply an $x_3(n)$ input sequence that's the sum of a 1-Hz sinewave and a 3-Hz sinewave, the $y_3(n)$ output is as shown in the center of Figure 1-7(d). Notice how $y_3(n)$ is the sample-for-sample sum of $y_1(n)$ and $y_2(n)$. Figure 1-7(d) also shows that the output spectrum $Y_3(m)$ is the sum of $Y_1(m)$ and $Y_2(m)$. That's linearity.

1.5.2 Example of a Nonlinear System

It's easy to demonstrate how a nonlinear system yields an output that is not equal to the sum of $y_1(n)$ and $y_2(n)$ when its input is $x_1(n) + x_2(n)$. A simple example of a nonlinear discrete system is that in Figure 1-8(a) where the output is the square of the input described by

$$y(n) = [x(n)]^2 . \tag{1-16}$$

We'll use a well-known trigonometric identity and a little algebra to predict the output of this nonlinear system when the input comprises simple sinewaves. Following the form of Eq. (1-3), let's describe a sinusoidal sequence, whose frequency $f_o = 1$ Hz, by

$$x_1(n) = \sin(2\pi f_o n t_s) = \sin(2\pi \cdot 1 \cdot n t_s) . \tag{1-17}$$

Equation (1-17) describes the $x_1(n)$ sequence on the left side of Figure 1-8(b). Given this $x_1(n)$ input sequence, the $y_1(n)$ output of the nonlinear system is the square of a 1-Hz sinewave, or

$$y_1(n) = [x_1(n)]^2 = \sin(2\pi \cdot 1 \cdot n t_s) \cdot \sin(2\pi \cdot 1 \cdot n t_s) . \tag{1-18}$$

We can simplify our expression for $y_1(n)$ in Eq. (1-18) by using the following trigonometric identity:

$$\sin(\alpha) \cdot \sin(\beta) = \frac{\cos(\alpha - \beta)}{2} - \frac{\cos(\alpha + \beta)}{2} . \tag{1-19}$$

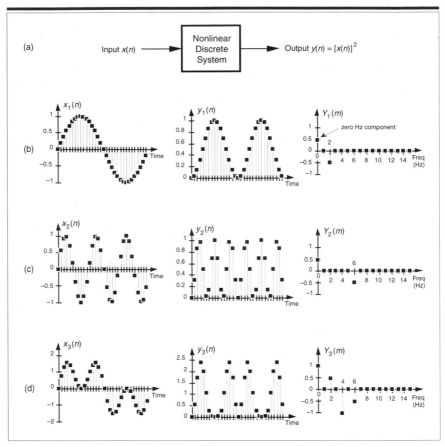

Figure 1-8 Nonlinear system input-to-output relationships: (a) system block diagram where $y(n) = (x(n))^2$; (b) system input and output with a 1-Hz sinewave applied; (c) with a 3-Hz sinewave applied; (d) with the sum of 1-Hz and 3-Hz sinewaves applied.

Using Eq. (1-19), we can express $y_1(n)$ as

$$y_1(n) = \frac{\cos(2\pi \cdot 1 \cdot nt_s - 2\pi \cdot 1 \cdot nt_s)}{2} - \frac{\cos(2\pi \cdot 1 \cdot nt_s + 2\pi \cdot 1 \cdot nt_s)}{2}$$

$$= \frac{\cos(0)}{2} - \frac{\cos(4\pi \cdot 1 \cdot nt_s)}{2} = \frac{1}{2} - \frac{\cos(2\pi \cdot 2 \cdot nt_s)}{2}, \tag{1-20}$$

which is shown as the all positive sequence in the center of Figure 1-8(b). Because Eq. (1-19) results in a frequency sum ($\alpha + \beta$) and frequency difference ($\alpha - \beta$) effect when multiplying two sinusoids, the $y_1(n)$ output sequence will be a cosine wave of 2 Hz and a peak amplitude of −0.5, added to a constant value of 1/2. The constant value of 1/2 in Eq. (1-20) is interpreted as a zero Hz frequency component, as shown in the $Y_1(m)$ spectrum in Figure 1-8(b). We could go through the same algebraic exercise to determine that, when a 3-Hz sinewave $x_2(n)$ sequence is applied to this nonlinear system, the output $y_2(n)$ would contain a zero Hz component and a 6 Hz component, as shown in Figure 1-8(c).

System nonlinearity is evident if we apply an $x_3(n)$ sequence comprising the sum of a 1-Hz and a 3-Hz sinewave as shown in Figure 1-8(d). We can predict the frequency content of the $y_3(n)$ output sequence by using the algebraic relationship

$$(a+b)^2 = a^2 + 2ab + b^2 , \tag{1-21}$$

where a and b represent the 1-Hz and 3-Hz sinewaves, respectively. From Eq. (1-19), the a^2 term in Eq. (1-21) generates the zero-Hz and 2-Hz output sinusoids in Figure 1-8(b). Likewise, the b^2 term produces in $y_3(n)$ another zero-Hz and the 6-Hz sinusoid in Figure 1-8(c). However, the $2ab$ term yields additional 2-Hz and 4-Hz sinusoids in $y_3(n)$. We can show this algebraically by using Eq. (1-19) and expressing the $2ab$ term in Eq. (1-21) as

$$2ab = 2 \cdot \sin(2\pi \cdot 1 \cdot nt_s) \cdot \sin(2\pi \cdot 3 \cdot nt_s)$$

$$= \frac{2\cos(2\pi \cdot 1 \cdot nt_s - 2\pi \cdot 3 \cdot nt_s)}{2} - \frac{2\cos(2\pi \cdot 1 \cdot nt_s + 2\pi \cdot 3 \cdot nt_s)}{2}$$

$$= \cos(2\pi \cdot 2 \cdot nt_s) - \cos(2\pi \cdot 4 \cdot nt_s) .^\dagger \tag{1-22}$$

Equation (1-22) tells us that two additional sinusoidal components will be present in $y_3(n)$ because of the system's nonlinearity, a 2-Hz cosine wave whose amplitude is +1 and a 4-Hz cosine wave having an amplitude of −1. These spectral components are illustrated in $Y_3(m)$ on the right side of Figure 1-8(d).

Notice that, when the sum of the two sinewaves is applied to the nonlinear system, the output contained sinusoids, Eq. (1-22), that were not present in either of the outputs when the individual sinewaves alone were applied. Those extra sinusoids were generated by an interaction of the

---

† The first term in Eq. (1-22) is $\cos(2\pi \cdot nt_s - 6\pi \cdot nt_s) = \cos(-4\pi \cdot nt_s) = \cos(-2\pi \cdot 2 \cdot nt_s)$. However, because the cosine function is even, $\cos(-\alpha) = \cos(\alpha)$, we can express that first term as $\cos(2\pi \cdot 2 \cdot nt_s)$.

two input sinusoids due to the squaring operation. That's nonlinearity; expression (1-13) was not satisfied. (Electrical engineers recognize this effect of internally generated sinusoids as *intermodulation distortion*.) Although nonlinear systems are usually difficult to analyze, they are occasionally used in practice. References [2], [3], and [4], for example, describe their application in nonlinear digital filters. Again, expressions (1-13) and (1-14) state that a linear system's output resulting from a sum of individual inputs, is the superposition (sum) of the individual outputs. They also stipulate that the output sequence $y_1(n)$ depends only on $x_1(n)$ combined with the system characteristics, and not on the other input $x_2(n)$, i.e., there's no interaction between inputs $x_1(n)$ and $x_2(n)$ at the output of a linear system.

1.6 Time-Invariant Systems

A time-invariant system is one where a time delay (or shift) in the input sequence causes a equivalent time delay in the system's output sequence. Keeping in mind that n is just an indexing variable we use to keep track of our input and output samples, let's say a system provides an output $y(n)$ given an input of $x(n)$, or

$$x(n) \xrightarrow{\text{results in}} y(n) \ . \tag{1-23}$$

For a system to be time invariant, with a shifted version of the original $x(n)$ input applied, $x'(n)$, the following applies:

$$x'(n) = x(n+k) \xrightarrow{\text{results in}} y'(n) = y(n+k) \ , \tag{1-24}$$

where k is some integer representing k sample period time delays. For a system to be time invariant, expression (1-24) must hold true for any integer value of k and any input sequence.

1.6.1 Example of a Time-Invariant System

Let's look at a simple example of time invariance illustrated in Figure 1-9. Assume that our initial $x(n)$ input is a unity-amplitude 1-Hz sinewave sequence with a $y(n)$ output, as shown in Figure 1-9(b). Consider a different input sequence $x'(n)$, where

$$x'(n) = x(n+4) \ . \tag{1-25}$$

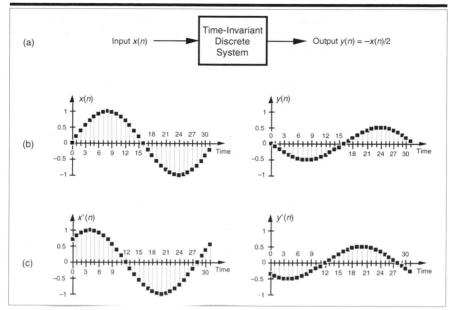

Figure 1-9 Time-invariant system input-to-output relationships: (a) system block diagram where $y(n) = -x(n)/2$; (b) system input and output with a 1-Hz sinewave applied; (c) system input and output when a 1-Hz sinewave, delayed by four samples, is applied. When $x'(n) = x(n+4)$, then, $y'(n) = y(n+4)$.

Equation (1-25) tells us that the input sequence $x'(n)$ is equal to sequence $x(n)$ shifted four samples to the left, that is, $x'(0) = x(4)$, $x'(1) = x(5)$, $x'(2) = x(6)$, and so on, as shown on the left of Figure 1-9(c). The discrete system is time invariant because the $y'(n)$ output sequence is equal to the $y(n)$ sequence shifted to the left by four samples, or $y'(n) = y(n+4)$. We can see that $y'(0) = y(4)$, $y'(1) = y(5)$, $y'(2) = y(6)$, and so on, as shown in Figure 1-9(c). For time-invariant systems, the y time shift is equal to the x time shift.

Some authors succumb to the urge to define a time-invariant system as one whose parameters do not change with time. That definition is incomplete and can get us in trouble if we're not careful. We'll just stick with the formal definition that a time-invariant system is one where a time shift in an input sequence results in an equal time shift in the output sequence. By the way, time-invariant systems in the literature are often called *shift-invariant* systems.[†]

[†] An example of a discrete process that's not time-invariant is the downsampling, or decimation, process described in Section 7.3.

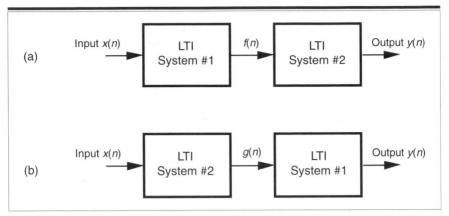

Figure 1-10 Linear time-invariant (LTI) systems in series: (a) block diagram of two LTI
systems; (b) swapping the order of the two systems does not change
the resultant output *y*(*n*).

1.7 The Commutative Property of Linear Time-Invariant Systems

Although we don't substantiate this fact until we reach Section 6.8, it's not
too early to realize that LTI systems have a useful commutative property by
which their sequential order can be rearranged with no change in their final
output. This situation is shown in Figure 1-10 where two different LTI systems are configured in series. Swapping the order of two cascaded systems
does not alter the final output. Although the intermediate data sequences
f(*n*) and *g*(*n*) will usually not be equal, the two pairs of LTI systems will have
identical *y*(*n*) output sequences. This commutative characteristic comes in
handy for designers of digital filters, as we'll see in Chapters 5 and 6.

1.8 Analyzing Linear Time-Invariant Systems

As previously stated, LTI systems can be analyzed to predict their performance. Specifically, if we know the *unit impulse response* of an LTI system,
we can calculate everything there is to know about the system; that is, the
system's unit impulse response completely characterizes the system. By
unit impulse response, we mean the system's time-domain output
sequence when the input is a single unity-valued sample (unit impulse)
preceded and followed by zero-valued samples as shown in Figure 1-11(b).

Knowing the (unit) impulse response of an LTI system, we can determine the system's output sequence for any input sequence because the
output is equal to the *convolution* of the input sequence and the system's
impulse response. Moreover, given an LTI system's time-domain impulse

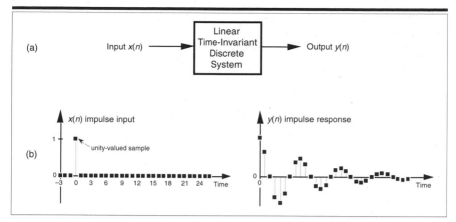

Figure 1-11 LTI system unit impulse response sequences: (a) system block diagram;
(b) impulse input sequence x(n) and impulse response output
sequence y(n).

response, we can find the system's *frequency response* by taking the Fourier
transform in the form of a *discrete Fourier transform* of that impulse
response[5].

Don't be alarmed if you're not exactly sure what is meant by convolution, frequency response, or the discrete Fourier transform. We'll introduce these subjects and define them slowly and carefully as we need them
in later chapters. The point to keep in mind here is that LTI systems can
be designed and analyzed using a number of straightforward and powerful analysis techniques. These techniques will become tools that we'll add
to our signal processing toolboxes as we journey through the subject of
digital signal processing.

References

[1] Karplus, W. J., and Soroka, W. W. *Analog Methods*, Second Edition,
McGraw-Hill, New York, 1959, p. 117.

[2] Mikami, N., Kobayashi, M., and Yokoyama, Y. "A New DSP-Oriented
Algorithm for Calculation of the Square Root Using a Nonlinear Digital
Filter," *IEEE Trans. on Signal Processing*, Vol. 40, No. 7, July 1992.

[3] Heinen, P., and Neuvo, Y. "FIR-Median Hybrid Filters," *IEEE Trans. on Acoust.
Speech, and Signal Processing*, Vol. ASSP-35, No. 6, June 1987.

[4] Oppenheim, A., Schafer, R., and Stockham, T. "Nonlinear Filtering of
Multiplied and Convolved Signals," *Proc. IEEE*, Vol. 56, August 1968.

[5] Pickerd, John. "Impulse-Response Testing Lets a Single Test Do the Work of
Thousands," *EDN*, April 27, 1995.

CHAPTER TWO

Periodic Sampling

Periodic sampling, the process of representing a continuous signal with a sequence of discrete data values, pervades the field of digital signal processing. In practice, sampling is performed by applying a continuous signal to an analog-to-digital (A/D) converter whose output is a series of digital values. Because sampling theory plays an important role in determining the accuracy and feasibility of any digital signal processing scheme, we need a solid appreciation for the often misunderstood effects of periodic sampling. With regard to sampling, the primary concern is just how fast must a given continuous signal be sampled in order to preserve its information content. We can sample a continuous signal at any sample rate we wish, and we'll get a series of discrete values—but the question is how well do these values represent the original signal? Let's learn the answer to that question and, in doing so, explore the various sampling techniques used in digital signal processing.

2.1 Aliasing: Signal Ambiguity in the Frequency Domain

There is a frequency-domain ambiguity associated with discrete-time signal samples that does not exist in the continuous signal world, and we can appreciate the effects of this uncertainty by understanding the sampled nature of discrete data. By way of example, suppose you were given the following sequence of values,

$$x(0) = 0$$
$$x(1) = 0.866$$
$$x(2) = 0.866$$
$$x(3) = 0$$
$$x(4) = -0.866$$
$$x(5) = -0.866$$
$$x(6) = 0 \ ,$$

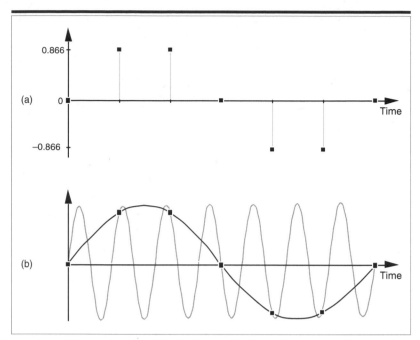

Figure 2-1 Frequency ambiguity: (a) discrete-time sequence of values; (b) two different sinewaves that pass through the points of the discrete sequence.

and were told that they represent instantaneous values of a time-domain sinewave taken at periodic intervals. Next, you were asked to draw that sinewave. You'd start by plotting the sequence of values shown by the dots in Figure 2-1(a). Next, you'd be likely to draw the sinewave, illustrated by the solid line in Figure 2-1(b), that passes through the points representing the original sequence.

Another person, however, might draw the sinewave shown by the shaded line in Figure 2-1(b). We see that the original sequence of values could, with equal validity, represent sampled values of both sinewaves. The key issue is that, if the data sequence represented periodic samples of a sinewave, we cannot unambiguously determine the frequency of the sinewave from those sample values alone.

Reviewing the mathematical origin of this frequency ambiguity enables us not only to deal with it, but to use it to our advantage. Let's derive an expression for this frequency-domain ambiguity and, then, look at a few specific examples. Consider the continuous time-domain sinusoidal signal defined as

$$x(t) = \sin(2\pi f_o t) \, . \tag{2-1}$$

This $x(t)$ signal is a garden variety sinewave whose frequency is f_o Hz. Now let's sample $x(t)$ at a rate of f_s samples/s, i.e., at regular periods of t_s seconds where $t_s = 1/f_s$. If we start sampling at time $t = 0$, we will obtain samples at times $0t_s$, $1t_s$, $2t_s$, and so on. So, from Eq. (2-1), the first n successive samples have the values

0th sample:	$x(0) = \sin(2\pi f_o 0 t_s)$
1st sample:	$x(1) = \sin(2\pi f_o 1 t_s)$
2nd sample:	$x(2) = \sin(2\pi f_o 2 t_s)$
. . .	. . .
. . .	. . .
nth sample:	$x(n) = \sin(2\pi f_o n t_s) \, .$

$$\tag{2-2}$$

Equation (2-2) defines the value of the nth sample of our $x(n)$ sequence to be equal to the original sinewave at the time instant nt_s. Because two values of a sinewave are identical if they're separated by an integer multiple of 2π radians, i.e., $\sin(\emptyset) = \sin(\emptyset + 2\pi m)$ where m is any integer, we can modify Eq. (2-2) as

$$x(n) = \sin(2\pi f_o n t_s) = \sin(2\pi f_o n t_s + 2\pi m) = \sin(2\pi(f_o + \frac{m}{nt_s})nt_s) \, . \tag{2-3}$$

If we let m be an integer multiple of n, $m = kn$, we can replace the m/n ratio in Eq. (2-3) with k so that

$$x(n) = \sin(2\pi(f_o + \frac{k}{t_s})nt_s) \, . \tag{2-4}$$

Because $f_s = 1/t_s$, we can equate the $x(n)$ sequences in Eqs. (2-2) and (2-4) as

$$x(n) = \sin(2\pi f_o n t_s) = \sin(2\pi(f_o + kf_s)nt_s) \, . \tag{2-5}$$

The f_o and $(f_o + kf_s)$ factors in Eq. (2-5) are therefore equal. The implication of Eq. (2-5) is critical. It means that an $x(n)$ sequence of digital sample values, representing a sinewave of f_o Hz, also exactly represents sinewaves at other frequencies, namely, $f_o + kf_s$. This is one of the most important relationships in the field of digital signal processing. It's the thread with which all sampling schemes are woven. In words, Eq. (2-5) states that

> **When sampling at a rate of f_s samples/s, if k is any positive or negative integer, we cannot distinguish between the sampled values of a sinewave of f_o Hz and a sinewave of (f_o+kf_s) Hz.**

It's true. No sequence of values stored in a computer, for example, can unambiguously represent one and only one sinusoid without additional information. This fact applies equally to A/D-converter output samples as well as signal samples generated by computer software routines. The sampled nature of any sequence of discrete values makes that sequence also represent an infinite number of different sinusoids.

Equation (2-5) influences all digital signal processing schemes. It's the reason that, although we've only shown it for sinewaves, we'll see in Chapter 3 that the spectrum of any discrete series of sampled values contains periodic replications of the original continuous spectrum. The period between these replicated spectra in the frequency domain will always be f_s, and the spectral replications repeat all the way from *DC to daylight* in both directions of the frequency spectrum. That's because k in Eq. (2-5) can be any positive or negative integer. (In Chapters 5 and 6, we'll learn that Eq. (2-5) is the reason that all digital filter frequency responses are periodic in the frequency domain and is crucial to analyzing and designing a popular type of digital filter known as the infinite impulse response filter.)

To illustrate the effects of Eq. (2-5), let's build on Figure 2-1 and consider the sampling of a 7-kHz sinewave at a sample rate of 6 kHz. A new sample is determined every 1/6000 seconds, or once every 167 microseconds, and their values are shown as the dots in Figure 2-2(a).

Notice that the sample values would not change at all if, instead, we were sampling a 1-kHz sinewave. In this example f_o = 7 kHz, f_s = 6 kHz, and $k = -1$ in Eq. (2-5), such that $f_o+kf_s = [7+(-1\cdot6)] = 1$ kHz. Our problem is that no processing scheme can determine if the sequence of sampled values, whose amplitudes are represented by the dots, came from a 7-kHz or a 1-kHz sinusoid. If these amplitude values are applied to a digital process that detects energy at 1 kHz, the detector output would indicate energy at 1 kHz. But we know that there is no 1-kHz tone there—our input is a spectrally pure 7-kHz tone. Equation (2-5) is causing a sinusoid, whose name is 7 kHz, to go by the *alias* of 1 kHz. Asking someone to determine which sinewave frequency accounts for the sample values in Figure 2-2(a) is like asking them "When I add two numbers I get a sum of four. What are the two numbers?" The answer is that there are an infinite number of number pairs that can add up to four.

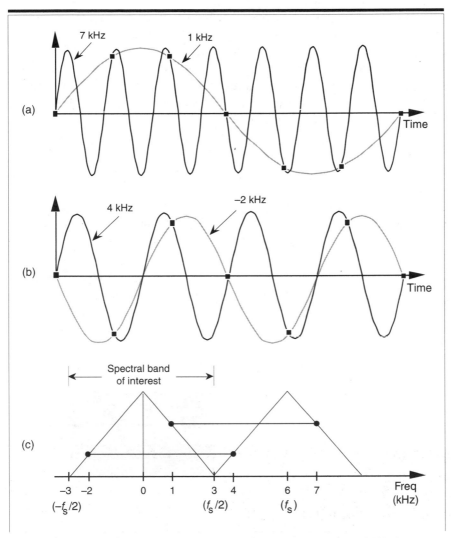

Figure 2-2 Frequency ambiguity effects of Eq. (2-5): (a) sampling a 7-kHz sinewave at a sample rate of 6 kHz; (b) sampling a 4 kHz sinewave at a sample rate of 6 kHz; (c) spectral relationships showing aliasing of the 7- and 4-kHz sinewaves.

Figure 2-2(b) shows another example of frequency ambiguity, that we'll call *aliasing*, where a 4-kHz sinewave could be mistaken for a –2-kHz sinewave. In Figure 2-2(b), f_o = 4 kHz, f_s = 6 kHz, and k = –1 in Eq. (2-5), so that f_o+kf_s = [4+(–1 · 6)] = –2 kHz. Again, if we examine a sequence of numbers representing the dots in Figure 2-2(b), we could not determine if the sampled sinewave was a 4-kHz tone or a –2-kHz tone. (Although the concept of negative frequencies might seem a bit strange, it provides a beautifully consistent methodology for predicting the spectral effects of sampling. Appendix C discusses negative frequencies and how they relate to real and complex signals.)

Now, if we restrict our spectral band of interest to the frequency range of $\pm f_s/2$ Hz, the previous two examples take on a special significance. The frequency $f_s/2$ is an important quantity in sampling theory and is referred to by different names in the literature, such as critical Nyquist, half Nyquist, and folding frequency. A graphical depiction of our two frequency aliasing examples is provided in Figure 2-2(c). We're interested in signal components that are aliased into the frequency band between $-f_s/2$ and $+f_s/2$. Notice in Figure 2-2(c) that, within the spectral band of interest (± 3 kHz, because $f_s = 6$ kHz), there is energy at -2 kHz and $+1$ kHz, aliased from 4 kHz and 7 kHz, respectively. Note also that the vertical positions of the dots in Figure 2-2(c) have no amplitude significance but that their horizontal positions indicate which frequencies are related through aliasing.

A general illustration of aliasing is provided in the *shark's tooth* pattern in Figure 2-3(a). Note how the peaks of the pattern are located at integer multiples of f_s Hz. The pattern shows how signals residing at the intersection of a horizontal line and a sloped line will be aliased to all of the intersections of that horizontal line and all other lines with like slopes. For example, the pattern in Figure 2-3(b) shows that our sampling of a 7-kHz

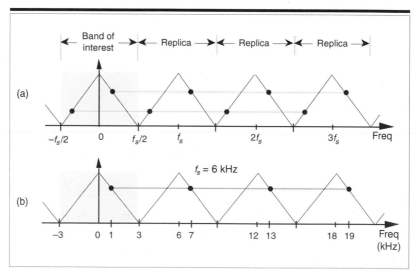

Figure 2-3 Shark's tooth pattern: (a) aliasing at multiples of the sampling frequency; (b) aliasing of the 7-kHz sinewave to 1 kHz, 13 kHz, and 19 kHz.

sinewave at a sample rate of 6 kHz will provide a discrete sequence of numbers whose spectrum ambiguously represents tones at 1 kHz, 7 kHz, 13 kHz, 19 kHz, etc. Let's pause for a moment and let these very important concepts soak in a bit. Again, discrete sequence representations of a continuous signal have unavoidable ambiguities in their frequency domains. These ambiguities must be taken into account in all practical digital signal processing algorithms.

OK, let's review the effects of sampling signals that are more interesting than just simple sinusoids.

2.2 Sampling Low-Pass Signals

Consider sampling a continuous real signal whose spectrum is shown in Figure 2-4(a). Notice that the spectrum is symmetrical about zero Hz, and the spectral amplitude is zero above $+B$ Hz and below $-B$ Hz, i.e., the signal is *band-limited*. (From a practical standpoint, the term *band-limited signal* merely implies that any signal energy outside the range of $\pm B$ Hz is below the sensitivity of our system.) Given that the signal is sampled at a rate of f_s samples/s, we can see the spectral replication effects of sampling in Figure 2-4(b), showing the original spectrum in addition to an infinite number of replications, whose period of replication is f_s Hz. (Although we stated in Section 1.1 that frequency-domain representations of discrete-time sequences are themselves discrete, the replicated spectra in Figure 2-4(b) are shown as continuous lines, instead of discrete dots, merely to keep the figure from looking too cluttered. We'll cover the full implications of discrete frequency spectra in Chapter 3.)

Let's step back a moment and understand Figure 2-4 for all it's worth. Figure 2-4(a) is the spectrum of a continuous signal, a signal that can only exist in one of two forms. Either it's a continuous signal that can be sampled, through A/D conversion, or it is merely an abstract concept such as a mathematical expression for a signal. It *cannot* be represented in a digital machine in its current band-limited form. Once the signal is represented by a sequence of discrete sample values, its spectrum takes the replicated form of Figure 2-4(b).

The replicated spectra are not just figments of the mathematics; they exist and have a profound effect on subsequent digital signal processing.[†] The replications may appear harmless, and it's natural to ask, "Why care about spectral replications? We're only interested in the frequency band

---

[†] Toward the end of Section 5.9, as an example of using the convolution theorem, another derivation of periodic sampling's replicated spectrums will be presented.

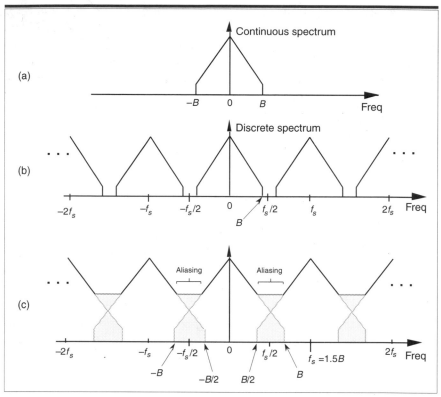

Figure 2-4 Spectral replications: (a) original continuous signal spectrum; (b) spectral replications of the sampled signal when $f_s/2 > B$; (c) frequency overlap and aliasing when the sampling rate is too low because $f_s/2 < B$.

within $\pm f_s/2$." Well, if we perform a frequency translation operation or induce a change in sampling rate through decimation or interpolation, the spectral replications will shift up or down right in the middle of the frequency range of interest $\pm f_s/2$ and could cause problems[1]. Let's see how we can control the locations of those spectral replications.

In practical A/D conversion schemes, f_s is always greater than $2B$ to separate spectral replications at the *folding frequencies* of $\pm f_s/2$. This very important relationship of $f_s \geq 2B$ is known as the Nyquist criterion. To illustrate why the term folding frequency is used, let's lower our sampling frequency to $f_s = 1.5B$ Hz. The spectral result of this *undersampling* is illustrated in Figure 2-4(c). The spectral replications are now overlapping the original baseband spectrum centered about zero Hz. Limiting our attention to the band $\pm f_s/2$ Hz, we see two very interesting effects. First, the lower edge and upper edge of the spectral replications centered at $+f_s$ and $-f_s$ now lie in our band of interest. This situation is equivalent to the original spectrum folding to the left at $+f_s/2$ and folding to the right at $-f_s/2$. Portions of the spectral replications now combine with the original spectrum, and the result is aliasing errors. The discrete sampled values associated with the spectrum of Figure 2-4(c) no longer truly represent the original input signal. The spectral information in the bands of $-B$ to $-B/2$ and $B/2$ to B Hz has been corrupted. We show the amplitude of the aliased regions in Figure 2-4(c) as dashed lines because we don't really know what the amplitudes will be if aliasing occurs.

The second effect illustrated by Figure 2-4(c) is that the entire spectral content of the original continuous signal is now residing in the band of interest between $-f_s/2$ and $+f_s/2$. This key property was true in Figure 2-4(b) and will always be true, regardless of the original signal or the sample rate. This effect is particularly important when we're digitizing (A/D converting) continuous signals. It warns us that any signal energy located above $+B$ Hz and below $-B$ Hz in the original continuous spectrum of Figure 2-4(a) will always end up in the band of interest after sampling, regardless of the sample rate. For this reason, continuous (analog) *low-pass* filters are necessary in practice.

We illustrate this notion by showing a continuous signal of bandwidth B accompanied by noise energy in Figure 2-5(a). Sampling this composite continuous signal at a rate that's greater than $2B$ prevents replications of the signal of interest from overlapping each other, but all of the noise energy still ends up in the range between $-f_s/2$ and $+f_s/2$ of our discrete spectrum shown in Figure 2-5(b). This problem is solved in practice by

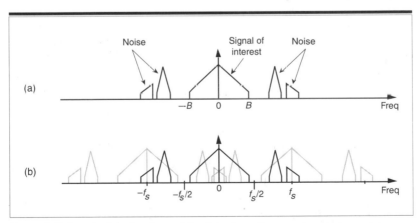

Figure 2-5 Spectral replications: (a) original continuous signal plus noise spectrum; (b) discrete spectrum with noise contaminating the signal of interest.

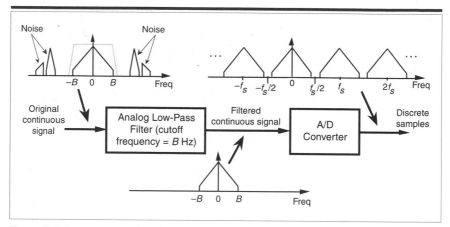

Figure 2-6 Low-pass analog filtering prior to sampling at a rate of f_s Hz.

using an analog low-pass *anti-aliasing* filter prior to A/D conversion to attenuate any unwanted signal energy above $+B$ and below $-B$ Hz as shown in Figure 2-6. An example low-pass filter response shape is shown as the dotted line superimposed on the original continuous signal spectrum in Figure 2-6. Notice how the output spectrum of the low-pass filter has been band-limited, and spectral aliasing is avoided at the output of the A/D converter.

This completes the discussion of simple low-pass sampling. Now let's go on to a more advanced sampling topic that's proven so useful in practice.

2.3 Sampling Bandpass Signals

Although satisfying the majority of sampling requirements, the sampling of low-pass signals, as in Figure 2-6, is not the only sampling scheme used in practice. We can use a technique known as *bandpass sampling* to sample a continuous bandpass signal that is centered about some frequency other than zero Hz. When a continuous input signal's bandwidth and center frequency permit us to do so, bandpass sampling not only reduces the speed requirement of A/D converters below that necessary with traditional low-pass sampling; it also reduces the amount of digital memory necessary to capture a given time interval of a continuous signal.

By way of example, consider sampling the band-limited signal shown in Figure 2-7(a) centered at $f_c = 20$ MHz, with a bandwidth $B = 5$ MHz. We use the term bandpass sampling for the process of sampling continuous signals whose center frequencies have been translated up from zero Hz. What we're calling bandpass sampling goes by various other names in the

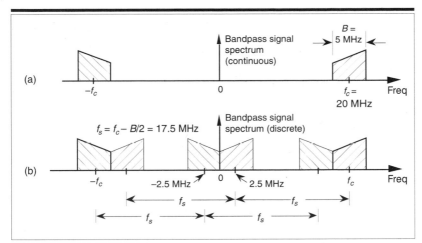

Figure 2-7 Bandpass signal sampling: (a) original continuous signal spectrum; (b) sampled signal spectrum replications when sample rate is 17.5 MHz.

literature, such as IF sampling, harmonic sampling[2], sub-Nyquist sampling, and undersampling[3]. In bandpass sampling, we're more concerned with a signal's bandwidth than its highest frequency component. Note that the negative frequency portion of the signal, centered at $-f_c$, is the mirror image of the positive frequency portion—as it must be for real signals. Our bandpass signal's highest frequency component is 22.5 MHz. Conforming to the Nyquist criterion (sampling at twice the highest frequency content of the signal) implies that the sampling frequency must be a minimum of 45 MHz. Consider the effect if the sample rate is 17.5 MHz shown in Figure 2-7(b). Note that the original spectral components remain located at $\pm f_c$, and spectral replications are located exactly at baseband, i.e., butting up against each other at zero Hz. Figure 2-7(b) shows that sampling at 45 MHz was unnecessary to avoid aliasing—instead we've used the spectral replicating effects of Eq. (2-5) to our advantage.

Bandpass sampling performs digitization and frequency translation in a single process, often called *sampling translation*. The processes of sampling and frequency translation are intimately bound together in the world of digital signal processing, and every sampling operation inherently results in spectral replications. The inquisitive reader may ask, "Can we sample at some still lower rate and avoid aliasing?" The answer is yes, but, to find out how, we have to grind through the derivation of an important bandpass sampling relationship. Our reward, however, will be worth the trouble because here's where bandpass sampling really gets interesting.

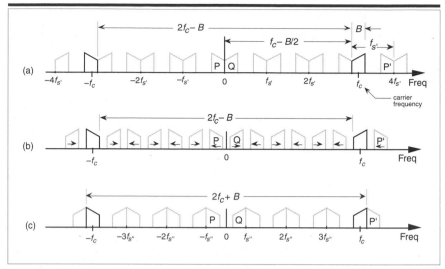

Figure 2-8 Bandpass sampling frequency limits: (a) sample rate $f_{s'} = (2f_c - B)/6$; (b) sample rate is less than $f_{s'}$; (c) minimum sample rate $f_{s''} < f_{s'}$.

Let's assume we have a continuous input bandpass signal of bandwidth *B*. Its *carrier frequency* is f_c Hz, i.e., the bandpass signal is centered at f_c Hz, and its sampled value spectrum is that shown in Figure 2-8(a). We can sample that continuous signal at a rate, say $f_{s'}$ Hz, so the spectral replications of the positive and negative bands, P and Q, just butt up against each other exactly at zero Hz. This situation, depicted in Figure 2-8(a), is reminiscent of Figure 2-7(b). With an arbitrary number of replications, say *m*, in the range of $2f_c - B$, we see that

$$mf_{s'} = 2f_c - B \quad \text{or} \quad f_{s'} = \frac{2f_c - B}{m}. \qquad (2\text{-}6)$$

In Figure 2-8(a), *m* = 6 for illustrative purposes only. Of course *m* can be any positive integer so long as $f_{s'}$ is never less than 2*B*. If the sample rate $f_{s'}$ is increased, the original spectra (bold) do not shift, but all the replications will shift. At zero Hz, the P band will shift to the right, and the Q band will shift to the left. These replications will overlap and aliasing occurs. Thus, from Eq. (2-6), for an arbitrary *m*, there is a frequency that the sample rate must not exceed, or

$$f_{s'} \leq \frac{2f_c - B}{m} \quad \text{or} \quad \frac{2f_c - B}{m} \geq f_{s'}. \qquad (2\text{-}7)$$

If we reduce the sample rate below the $f_{s'}$ value shown in Figure 2-8(a), the spacing between replications will decrease in the direction of the arrows in Figure 2-8(b). Again, the original spectra do not shift when the sample rate is changed. At some new sample rate $f_{s''}$, where $f_{s''} < f_{s'}$, the replication P' will just butt up against the positive original spectrum centered at f_c as shown in Figure 2-8(c). In this condition, we know that

$$(m+1)f_{s''} = 2f_c + B \quad \text{or} \quad f_{s''} = \frac{2f_c + B}{m+1} \,. \tag{2-8}$$

Should $f_{s''}$ be decreased in value, P' will shift further down in frequency and start to overlap with the positive original spectrum at f_c and aliasing occurs. Therefore, from Eq. (2-8) and for $m+1$, there is a frequency that the sample rate must always exceed, or

$$f_{s''} \geq \frac{2f_c + B}{m+1} \,. \tag{2-9}$$

We can now combine Eqs. (2-7) and (2-9) to say that f_s may be chosen anywhere in the range between $f_{s''}$ and $f_{s'}$ to avoid aliasing, or

$$\frac{2f_c - B}{m} \geq f_s \geq \frac{2f_c + B}{m+1} \,, \tag{2-10}$$

where m is an arbitrary, positive integer ensuring that $f_s \geq 2B$. (For this type of periodic sampling of real signals, known as real or first-order sampling, the Nyquist criterion $f_s \geq 2B$ must still be satisfied.)

To appreciate the important relationships in Eq. (2-10), let's return to our bandpass signal example, where Eq. (2-10) enables the generation of Table 2-1. This table tells us that our sample rate can be anywhere in the range of 22.5 to 35 MHz, anywhere in the range of 15 to 17.5 MHz, or anywhere in the range of 11.25 to 11.66 MHz. Any sample rate below 11.25 MHz is unacceptable because it will not satisfy Eq. (2-10) as well as $f_s \geq 2B$. The spectra resulting from several of the sampling rates from Table 2-1 are shown in Figure 2-9 for our bandpass signal example. Notice in Figure 2-9(f) that when f_s equals 7.5 MHz ($m = 5$), we have aliasing problems because neither the greater than relationships in Eq. (2-10) nor $f_s \geq 2B$ have been satisfied. The $m = 4$ condition is also unacceptable because $f_s \geq 2B$ is not satisfied. The last column in Table 2-1 gives the *optimum* sampling frequency for each acceptable m value. Optimum sampling frequency is defined here as that

Table 2-1 Equation (2-10) Applied to the Bandpass Signal Example

m	$(2f_c - B)/m$	$(2f_c + B)/(m+1)$	*Optimum Sampling Rate*
1	35.0 MHz	22.5 MHz	22.5 MHz
2	17.5 MHz	15.0 MHz	17.5 MHz
3	11.66 MHz	11.25 MHz	11.25 MHz
4	8.75 MHz	9.0 MHz	—
5	7.0 MHz	7.5 MHz	—

frequency where spectral replications do not butt up against each other except at zero Hz. For example, in the $m = 1$ range of permissible sampling frequencies, it is much easier to perform subsequent digital filtering or other processing on the signal samples whose spectrum is that of Figure 2-9(b), as opposed to the spectrum in Figure 2-9(a).

The reader may wonder, "Is the optimum sample rate always equal to the minimum permissible value for f_s using Eq. (2-10)?" The answer depends on the specific application—perhaps there are certain system constraints that must be considered. For example, in digital telephony, to simplify the follow-on processing, sample frequencies are chosen to be integer multiples of 8 kHz[4]. Another application-specific factor in choosing the optimum f_s is the shape of analog anti-aliasing filters[5]. Often, in practice, high-performance A/D converters have their hardware components *fine-tuned* during manufacture to ensure maximum linearity at high frequencies (>5 MHz). Their use at lower frequencies is not recommended.

An interesting way of illustrating the nature of Eq. (2-10) is to plot the minimum sampling rate, $(2f_c + B)/(m+1)$, for various values of m, as a function of R defined as

$$R = \frac{\text{highest signal frequency component}}{\text{bandwidth}} = \frac{f_c + B/2}{B} \quad . \tag{2-11}$$

If we normalize the minimum sample rate from Eq. (2-10) by dividing it by the bandwidth B, we get a curve whose axes are normalized to the bandwidth shown as the solid curve in Figure 2-10. This figure shows us the minimum normalized sample rate as a function of the normalized highest frequency component in the bandpass signal. Notice that, regardless of the value of R, the minimum sampling rate need never exceed $4B$ and approaches $2B$ as the carrier frequency increases. Surprisingly, the minimum acceptable sampling frequency actually decreases as the bandpass signal's carrier frequency increases. We can

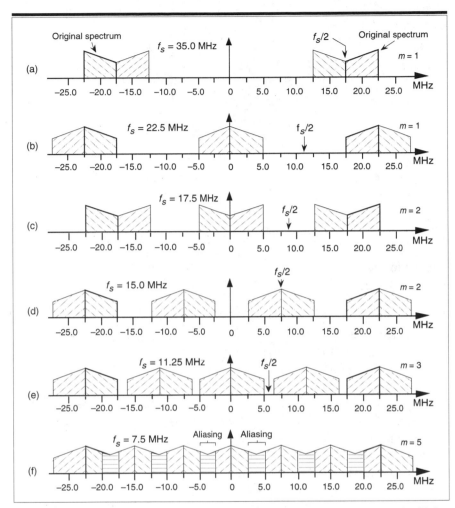

Figure 2-9 Various spectral replications from Table 2-1: (a) f_s = 35 MHz; (b) f_s = 22.5 MHz; (c) f_s = 17.5 MHz; (d) f_s = 15 MHz; (e) f_s = 11.25 MHz; (f) f_s = 7.5 MHz.

interpret Figure 2-10 by reconsidering our bandpass signal example from Figure 2-7 where R = 22.5/5 = 4.5. This R value is indicated by the dashed line in Figure 2-10 showing that m = 3 and f_s/B is 2.25. With B = 5 MHz, then, the minimum f_s = 11.25 MHz in agreement with Table 2-1. The leftmost line in Figure 2-10 shows the low-pass sampling case, where the sample rate f_s must be twice the signal's highest frequency component. So the normalized sample rate f_s/B is twice the highest frequency component over B or $2R$.

Figure 2-10 has been prominent in the literature, but its normal presentation enables the reader to jump to the false conclusion that any

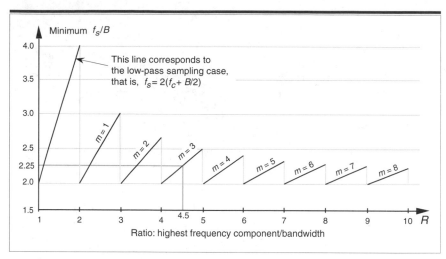

Figure 2-10 Minimum bandpass sampling rate from Eq. (2-10).

sample rate above the minimum shown in the figure will be an acceptable sample rate[6–12]. There's a clever way to avoid any misunder-
standing[13]. If we plot the acceptable ranges of bandpass sample frequencies from Eq. (2-10) as a function of R we get the depiction shown
in Figure 2-11. As we saw from Eq. (2-10), Table 2-1, and Figure 2-9,
acceptable bandpass sample rates are a series of frequency ranges separated by unacceptable ranges of sample rate frequencies, that is, an
acceptable bandpass sample frequency must be above the minimum
shown in Figure 2-10, but cannot be just any frequency above that minimum. The shaded region in Figure 2-11 shows those normalized bandpass sample rates that will lead to spectral aliasing. Sample rates within
the white regions of Figure 2-11 are acceptable. So, for bandpass sampling, we want our sample rate to be in the white wedged areas associated with some value of m from Eq. (2-10). Let's understand the significance of Figure 2-11 by again using our previous bandpass signal
example from Figure 2-7.

Figure 2-12 shows our bandpass signal example R value (highest frequency component/bandwidth) of 4.5 as the dashed vertical line. Because
that line intersects just three white wedged areas, we see that there are
only three frequency regions of acceptable sample rates, and this agrees
with our results from Table 2-1. The intersection of the $R = 4.5$ line and the
borders of the white wedged areas are those sample rate frequencies listed in Table 2-1. So Figure 2-11 gives a depiction of bandpass sampling
restrictions much more realistic than Figure 2-10.

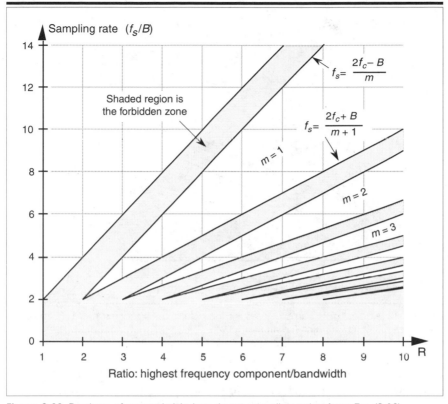

Figure 2-11 Regions of acceptable bandpass sampling rates from Eq. (2-10), normalized to the sample rate over the signal bandwidth (f_s/B).

Although Figures 2-11 and 2-12 indicate that we can use a sample rate that lies on the boundary between a white and shaded area, these sample rates should be avoided in practice. Nonideal analog bandpass filters, sample rate clock generator instabilities, and slight imperfections in available A/D converters make this *ideal* case impossible to achieve exactly. It's prudent to keep f_s somewhat separated from the boundaries. Consider the bandpass sampling scenario shown in Figure 2-13. With a typical (nonideal) analog bandpass filter, whose frequency response is indicated by the dashed line, it's prudent to consider the filter's bandwidth not as B, but as B_{gb} in our equations. That is, we create a guard band on either side of our filter so that there can be a small amount of aliasing in the discrete spectrum without distorting our desired signal, as shown at the bottom of Figure 2-13.

We can relate this idea of using guard bands to Figure 2-11 by looking more closely at one of the white wedges. As shown in Figure 2-14, we'd like

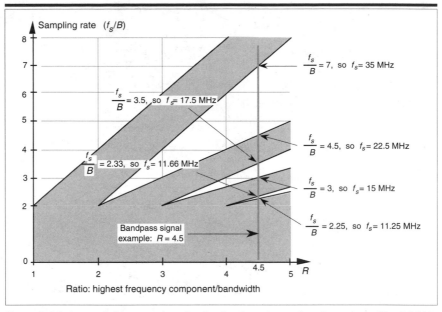

Figure 2-12 Acceptable sample rates for the bandpass signal example (B = 5 MHz) with a value of R = 4.5.

to set our sample rate as far down toward the vertex of the white area as we can—lower in the wedge means a lower sampling rate. However, the closer we operate to the boundary of a shaded area, the more narrow the guard band must be, requiring a sharper analog bandpass filter, as well as the tighter the tolerance we must impose on the stability and accuracy of our A/D clock generator. (Remember, operating on the boundary between a white and shaded area in Figure 2-11 causes spectral replications to butt up against each other.) So, to be safe, we operate at some intermediate point away from any shaded boundaries as shown in Figure 2-14. Further analysis of how guard band widths and A/D clock parameters relate to the geometry of Figure 2-14 is available in reference [13]. For this discussion, we'll just state that it's a good idea to ensure that our selected sample rate does not lie too close to the boundary between a white and shaded area in Figure 2-11.

There are a couple of ways to make sure we're not operating near a boundary. One way is to set the sample rate in the middle of a white wedge for a given value of R. We do this by taking the average between the maximum and minimum sample rate terms in Eq. (2-10) for a particular value of m, that is, to center the sample rate operating point within a wedge we use a sample rate of

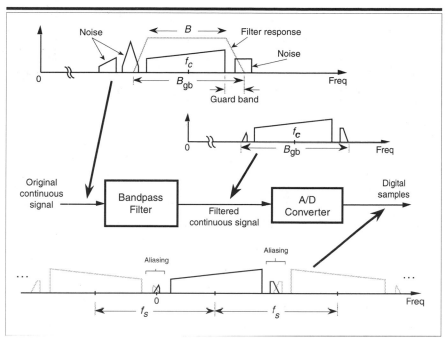

Figure 2-13 Bandpass sampling with aliasing occurring only in the filter guard bands.

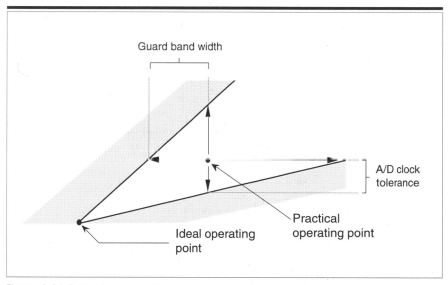

Figure 2-14 Typical operating point for f_s to compensate for nonideal hardware.

$$f_{S_{cntr}} = \frac{1}{2} \cdot \left[\frac{2f_c - B}{m} + \frac{2f_c + B}{m+1} \right] = \frac{f_c - B/2}{m} + \frac{f_c + B/2}{m+1} . \qquad (2\text{-}12)$$

Another way to avoid the boundaries of Figure 2-14 is to use the following expression to determine an intermediate f_{s_i} operating point:

$$f_{s_i} = \frac{4f_c}{m_{odd}} , \qquad (2\text{-}13)$$

where m_{odd} is an odd integer[14]. Of course the choice of m_{odd} must ensure that the Nyquist restriction of $f_{s_i} > 2B$ be satisfied. We show the results of Eqs. (2-12) and (2-13) for our bandpass signal example in Figure 2-15.

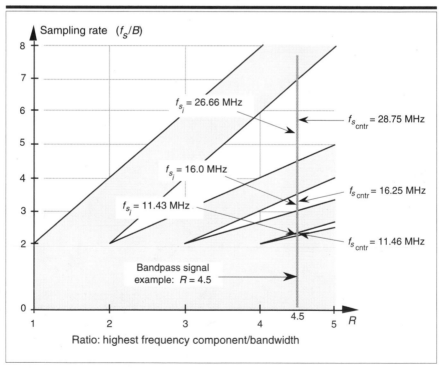

Figure 2-15 Intermediate f_{s_i} and $f_{s_{cntr}}$ operating points, from Eqs. (2-12) and (2-13), to avoid operating at the shaded boundaries for the bandpass signal example. B = 5 MHz and R = 4.5.

2.4 Spectral Inversion in Bandpass Sampling

Some of the permissible f_s values from Eq. (2-10) will, although avoiding aliasing problems, provide a sampled baseband spectrum (located near zero Hz) that is inverted from the original positive and negative spectral shapes, that is, the positive baseband will have the inverted shape of the negative half from the original spectrum. This spectral inversion happens whenever m, in Eq. (2-10), is an odd integer, as illustrated in Figures 2-9(b) and 2-9(e). When the original positive spectral bandpass components are symmetrical about the f_c frequency, spectral inversion presents no problem and any nonaliasing value for f_s from Eq. (2-10) may be chosen. However, if spectral inversion is something to be avoided, for example, when single sideband signals are being processed, the minimum applicable sample rate to avoid spectral inversion is defined by Eq. (2-10) with the restriction that m is the largest even integer such that $f_s \geq 2B$ is satisfied.[†] Using our definition of optimum sampling rate, the expression that provides the optimum noninverting sampling rates and avoids spectral replications butting up against each other, except at zero Hz, is

$$f_{s_o} = \frac{2f_c - B}{m_{even}} \, , \tag{2-14}$$

where m_{even} = 2, 4, 6, etc. For our bandpass signal example, Eq. (2-14) and $m = 2$ provide an optimum noninverting sample rate of f_{s_o} = 17.5 MHz, as shown in Figure 2-9(c). In this case, notice that the spectrum translated toward zero Hz has the same orientation as the original spectrum centered at 20 MHz.

Then again, if spectral inversion is unimportant for your application, we can determine the absolute minimum sampling rate without having to choose various values for m in Eq. (2-10) and creating a table like we did for Table 2-1. Considering Figure 2-16, the question is "How many replications of the positive and negative images of bandwidth B can we squeeze into the frequency range of $2f_c + B$ without overlap?" That number of replications is

$$R = \frac{\text{frequency span}}{\text{twice the bandwidth}} = \frac{2f_c + B}{2B} = \frac{f_c + B/2}{B} \, . \tag{2-15}$$

[†] Single sideband signals are discussed in Section C.4 of Appendix C.

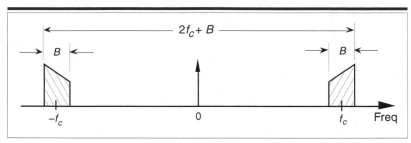

Figure 2-16 Frequency span of a continuous bandpass signal.

To avoid overlap, we have to make sure that the number of replications is an integer less than or equal to R in Eq. (2-15). So, we can define the integral number of replications to be R_{int} where

$$R_{int} \leq R < R_{int} + 1 ,$$

or

$$R_{int} \leq \frac{f_c + B/2}{B} < R_{int} + 1 . \qquad (2\text{-}16)$$

With R_{int} replications in the frequency span of $2f_c + B$, then, the spectral repetition period, or minimum sample rate $f_{s_{min}}$, is

$$f_{s_{min}} = \frac{2f_c + B}{R_{int}} . \qquad (2\text{-}17)$$

In our bandpass signal example, finding $f_{s_{min}}$ first requires the appropriate value for R_{int} in Eq. (2-13) as

$$R_{int} \leq \frac{22.5}{5} < R_{int} + 1 ,$$

so $R_{int} = 4$. Then, from Eq. (2-17), $f_{s_{min}} = (40+5)/4 = 11.25$ MHz, which is the sample rate illustrated in Figures 2-9(e) and 2-12. So, we can use Eq. (2-17) and avoid using various values for m in Eq. (2-10) and having to create a table like Table 2-1. (Be careful though. Eq. (2-17) places our sampling rate at the boundary between a white and shaded area of Figure 2-12, and we have to consider the guard band strategy discussed above.) To recap the bandpass signal example, sampling at 11.25 MHz, from Eq. (2-17), avoids aliasing and inverts the spectrum, while sampling at 17.5 MHz, from Eq. (2-14), avoids aliasing with no spectral inversion.

Now here's some good news. With a little additional digital processing, we can sample at 11.25 MHz, with its spectral inversion and easily reinvert the spectrum back to its original orientation. The discrete spectrum of any digital signal can be inverted by multiplying the signal's discrete-time samples by a sequence of alternating plus ones and minus ones (1, –1, 1, –1, etc.), indicated in the literature by the succinct expression $(-1)^n$. This scheme allows bandpass sampling at the lower rate of Eq. (2-17) while correcting for spectral inversion, thus avoiding the necessity of using the higher sample rates from Eq. (2-14). Although multiplying time samples by $(-1)^n$ is explored in detail in Section 10.1, all we need to remember at this point is the simple rule that multiplication of real signal samples by $(-1)^n$ is equivalent to multiplying by a cosine whose frequency is $f_s/2$. In the frequency domain, this multiplication flips the positive frequency band of interest, from zero to $+f_s/2$ Hz, about $f_s/4$ Hz, and flips the negative frequency band of interest, from $-f_s/2$ to zero Hz, about $-f_s/4$ Hz as shown in Figure 2-17. The $(-1)^n$ sequence is not only used for inverting the spectra of bandpass sampled sequences; it can be used to invert the spectra of low-pass sampled signals. Be aware, however, that, in the low-pass sampling case, any DC (zero Hz) component in the original continuous signal will be translated to both $+f_s/2$ and $-f_s/2$ after multiplication by $(-1)^n$.

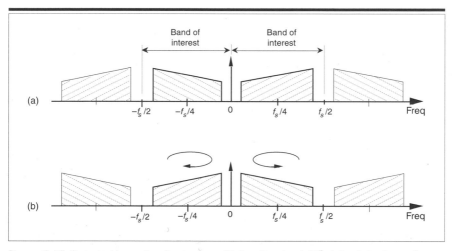

Figure 2-17 Spectral inversion through multiplication by $(-1)^n$: (a) original spectrum of a time-domain sequence; (b) new spectrum of the product of original time sequence and the $(-1)^n$ sequence.

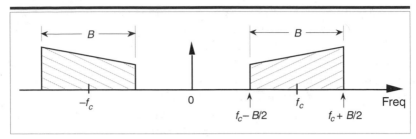

Figure 2-18 Continuous signal spectrum where bandpass sampling is not possible because $f_c - B/2 < B$.

Now that we have an understanding of bandpass sampling, we'd better remind ourselves of the situation where bandpass sampling is not possible. If a continuous bandpass signal's lowest frequency component is less than the bandwidth, we can't use bandpass sampling. This condition is shown in Figure 2-18 where $f_c - B/2 < B$. There's no way to squeeze any spectral replications between $f_c - B/2$ and zero Hz with bandpass sampling. In this case, we'd have to low-pass sample the signal in Figure 2-17 at a rate of at least twice the highest frequency component, or $f_s \geq 2(f_c + B/2)$.

We conclude this topic by consolidating in Table 2-2 what we need to know about bandpass sampling.

Table 2-2 Bandpass Sampling Relationships

Requirement	Sample Rate Expression	Conditions
Acceptable ranges of f_s for bandpass sampling: Eq. (2-10)	$\dfrac{2f_c - B}{m} \geq f_s \geq \dfrac{2f_c + B}{m+1}$	m = any positive integer so that $f_s \geq 2B$.
Sample rate in the middle of the acceptable sample rate bands: Eq. (2-12)	$f_{s_{cntr}} = \dfrac{f_c - B/2}{m} + \dfrac{f_c + B/2}{m+1}$	m = any positive integer so that $f_{s_{cntr}} \geq 2B$.
Sample rate at an intermediate point in the acceptable sample rate bands: Eq. (2-13)	$f_{s_i} = \dfrac{4f_c}{m_{odd}}$	m_{odd} = any positive odd integer so that $f_{s_i} \geq 2B$. (Spectral inversion occurs when $m_{odd} = 3, 7, 11$, etc.)
Optimum sample rate to avoid spectral inversion: Eq. (2-14)	$f_{s_o} = \dfrac{2f_c - B}{m_{even}}$	m_{even} = any even positive integer so that $f_{s_o} \geq 2B$.
Absolute minimum f_s to avoid aliasing: Eq. (2-17)	$f_{s_{min}} = \dfrac{2f_c + B}{R_{int}}$	where $R_{int} \leq \dfrac{f_c + B/2}{B} < R_{int} + 1$.

References

[1] Crochiere, R.E. and Rabiner, L.R. "Optimum FIR Digital Implementations for Decimation, Interpolation, and Narrow-band Filtering," *IEEE Trans. on Acoust. Speech, and Signal Proc.*, Vol. ASSP-23, No. 5, October 1975.

[2] Steyskal, H. "Digital Beamforming Antennas," *Microwave Journal*, January 1987.

[3] Hill, G. "The Benefits of Undersampling," *Electronic Design*, July 11, 1994.

[4] Yam, E., and Redman, M. "Development of a 60-channel FDM-TDM Trans-multiplexer," *COMSAT Technical Review*, Vol. 13, No. 1, Spring 1983.

[5] Floyd, P., and Taylor, J. "Dual-Channel Space Quadrature-Interferometer System," *Microwave System Designer's Handbook*, Fifth Edition, Microwave Systems News, 1987.

[6] Lyons, R. G. "How Fast Must You Sample," *Test and Measurement World*, November 1988.

[7] Stremler, F. *Introduction to Communication Systems*, Chapter 3, Second Edition, Addison Wesley Publishing Co., Reading, Massachusetts, p. 125.

[8] Webb, R. C. "IF Signal Sampling Improves Receiver Detection Accuracy," *Microwaves & RF*, March 1989.

[9] Haykin, S. *Communications Systems*, Chapter 7, John Wiley and Sons, New York, 1983, p. 376.

[10] Feldman, C. B., and Bennett, W. R. "Bandwidth and Transmission Performance," *Bell System Tech. Journal*, Vol. 28, 1989, p. 490.

[11] Panter, P. F. *Modulation Noise, and Spectral Analysis*, McGraw-Hill, New York, 1965, p. 527.

[12] Shanmugam, K. S. *Digital and Analogue Communications Systems*, John Wiley and Sons, New York, 1979, p. 378.

[13] Vaughan, R., Scott, N. and White, D. "The Theory of Bandpass Sampling," *IEEE Trans. on Signal Processing*, Vol. 39, No. 9, September 1991, pp. 1973–1984.

[14] Xenakis B., and Evans, A. "Vehicle Locator Uses Spread Spectrum Technology," *RF Design*, October 1992.

CHAPTER THREE

The Discrete Fourier Transform

The discrete Fourier transform (DFT) is one of the two most common, and powerful, procedures encountered in the field of digital signal processing. (Digital filtering is the other.) The DFT enables us to analyze, manipulate, and synthesize signals in ways not possible with continuous (analog) signal processing. Even though it's now used in almost every field of engineering, we'll see applications for DFT continue to flourish as its utility becomes more widely understood. Because of this, a solid understanding of the DFT is mandatory for anyone working in the field of digital signal processing.

The DFT is a mathematical procedure used to determine the harmonic, or frequency, content of a discrete signal sequence. Although, for our purposes, a discrete signal sequence is a set of values obtained by periodic sampling of a continuous signal in the time domain, we'll find that the DFT is useful in analyzing any discrete sequence regardless of what that sequence actually represents. The DFT's origin, of course, is the continuous Fourier transform $X(f)$ defined as

$$X(f) = \int_{-\infty}^{\infty} x(t)e^{-j2\pi ft}dt \ , \tag{3-1}$$

where $x(t)$ is some continuous time-domain signal.[†]

In the field of continuous signal processing, Eq. (3-1) is used to transform an expression of a continuous time-domain function $x(t)$ into a continuous frequency-domain function $X(f)$. Subsequent evaluation of the $X(f)$ expression enables us to determine the frequency content of any practical signal of interest and opens up a wide array of signal analysis and processing

---

[†] Fourier is pronounced 'for-yā. In engineering school, we called Eq. (3-1) the "four-year" transform because it took about four years to do one homework problem.

possibilities in the fields of engineering and physics. One could argue that the Fourier transform is the most dominant and widespread mathematical mechanism available for the analysis of physical systems. (A prominent quote from Lord Kelvin better states this sentiment: "Fourier's theorem is not only one of the most beautiful results of modern analysis, but it may be said to furnish an indispensable instrument in the treatment of nearly every recondite question in modern physics." By the way, the history of Fourier's original work in harmonic analysis, relating to the problem of heat conduction, is fascinating. References [1] and [2] are good places to start for those interested in the subject.)

With the advent of the digital computer, the efforts of early digital processing pioneers led to the development of the DFT defined as the discrete frequency-domain sequence $X(m)$, where

DFT equation (exponential form): →

$$X(m) = \sum_{n=0}^{N-1} x(n)e^{-j2\pi nm/N} \quad . \tag{3-2}$$

For our discussion of Eq. (3-2), $x(n)$ is a discrete sequence of time-domain sampled values of the continuous variable $x(t)$. The "e" in Eq. (3-2) is, of course, the base of natural logarithms and $j = \sqrt{-1}$.

3.1 Understanding the DFT Equation

Equation (3-2) has a tangled, almost unfriendly, look about it. Not to worry. After studying this chapter, Eq. (3-2) will become one of our most familiar and powerful tools in understanding digital signal processing. Let's get started by expressing Eq. (3-2) in a different way and examining it carefully. From *Euler's relationship* $e^{-j\phi} = \cos(\phi) - j\sin(\phi)$, Eq. (3-2) is equivalent to

DFT equation (rectangular form): → $X(m) = \sum_{n=0}^{N-1} x(n)\left[\cos(2\pi nm/N) - j\sin(2\pi nm/N)\right]$. (3-3)

We have separated the complex exponential of Eq. (3-2) into its real and imaginary components where

$X(m)$ = the mth DFT output component, i.e., $X(0)$, $X(1)$, $X(2)$, $X(3)$, etc.,

m = the index of the DFT output in the frequency domain,
$m = 0, 1, 2, 3, \ldots, N-1$,

$x(n)$ = the sequence of input samples, $x(0)$, $x(1)$, $x(2)$, $x(3)$, etc.,

n = the time-domain index of the input samples, $n = 0, 1, 2, 3, \ldots, N-1$,

j = $\sqrt{-1}$, and

N = the number of samples of the input sequence and the number of frequency points in the DFT output.

Although it looks more complicated than Eq. (3-2), Eq. (3-3) turns out to be easier to understand. (If you're not too comfortable with it, don't let the $j = \sqrt{-1}$ concept bother you too much. It's merely a convenient abstraction that helps us compare the phase relationship between various sinusoidal components of a signal. Appendix C discusses the j operator in some detail.)[†] The indices for the input samples (n) and the DFT output samples (m) always go from 0 to $N-1$ in the standard DFT notation. This means that with N input time-domain sample values, the DFT determines the spectral content of the input at N equally spaced frequency points. The value N is an important parameter because it determines how many input samples are needed, the resolution of the frequency-domain results, and the amount of processing time necessary to calculate an N-point DFT.

It's useful to see the structure of Eq. (3-3) by eliminating the summation and writing out all the terms. For example, when $N = 4$, n and m both go from 0 to 3, and Eq. (3-3) becomes

$$X(m) = \sum_{n=0}^{3} x(n)[\cos(2\pi nm / 4) - j\sin(2\pi nm / 4)] \ . \qquad \text{(3-4a)}$$

Writing out all the terms for the first DFT output term corresponding to $m = 0$,

$$\begin{aligned}
X(0) = \ &x(0)\cos(2\pi \cdot 0 \cdot 0 / 4) - jx(0)\sin(2\pi \cdot 0 \cdot 0 / 4) \\
&+ x(1)\cos(2\pi \cdot 1 \cdot 0 / 4) \quad jx(1)\sin(2\pi \cdot 1 \cdot 0 / 4) \\
&+ x(2)\cos(2\pi \cdot 2 \cdot 0 / 4) - jx(2)\sin(2\pi \cdot 2 \cdot 0 / 4) \\
&+ x(3)\cos(2\pi \cdot 3 \cdot 0 / 4) - jx(3)\sin(2\pi \cdot 3 \cdot 0 / 4). \qquad \text{(3-4b)}
\end{aligned}$$

For the second DFT output term corresponding to $m = 1$, Eq. (3-4a) becomes

---

[†] Instead of the letter j, be aware that mathematicians often use the letter i to represent the $\sqrt{-1}$ operator.

$$X(1) = x(0)\cos(2\pi \cdot 0 \cdot 1/4) - jx(0)\sin(2\pi \cdot 0 \cdot 1/4)$$
$$+ x(1)\cos(2\pi \cdot 1 \cdot 1/4) - jx(1)\sin(2\pi \cdot 1 \cdot 1/4)$$
$$+ x(2)\cos(2\pi \cdot 2 \cdot 1/4) - jx(2)\sin(2\pi \cdot 2 \cdot 1/4)$$
$$+ x(3)\cos(2\pi \cdot 3 \cdot 1/4) - jx(3)\sin(2\pi \cdot 3 \cdot 1/4). \qquad \text{(3-4c)}$$

For the third output term corresponding to $m = 2$, Eq. (3-4a) becomes

$$X(2) = x(0)\cos(2\pi \cdot 0 \cdot 2/4) - jx(0)\sin(2\pi \cdot 0 \cdot 2/4)$$
$$+ x(1)\cos(2\pi \cdot 1 \cdot 2/4) - jx(1)\sin(2\pi \cdot 1 \cdot 2/4)$$
$$+ x(2)\cos(2\pi v2 \cdot 2/4) - jx(2)\sin(2\pi \cdot 2 \cdot 2/4)$$
$$+ x(3)\cos(2\pi \cdot 3 \cdot 2/4) - jx(3)\sin(2\pi \cdot 3 \cdot 2/4). \qquad \text{(3-4d)}$$

Finally, for the fourth and last output term corresponding to $m = 3$, Eq. (3-4a) becomes

$$X(3) = x(0)\cos(2\pi \cdot 0 \cdot 3/4) - jx(0)\sin(2\pi \cdot 0 \cdot 3/4)$$
$$+ x(1)\cos(2\pi \cdot 1 \cdot 3/4) - jx(1)\sin(2\pi \cdot 1 \cdot 3/4)$$
$$+ x(2)\cos(2\pi \cdot 2 \cdot 3/4) - jx(2)\sin(2\pi \cdot 2 \cdot 3/4)$$
$$+ x(3)\cos(2\pi \cdot 3 \cdot 3/4) - jx(3)\sin(2\pi \cdot 3 \cdot 3/4). \qquad \text{(3-4e)}$$

The above multiplication symbol "·" in Eq. (3-4) is used merely to separate the factors in the sine and cosine terms. The pattern in Eq. (3-4b) through (3-4e) is apparent now, and we can certainly see why it's convenient to use the summation sign in Eq. (3-3). Each $X(m)$ DFT output term is the sum of the *point for point* product between an input sequence of signal values and a complex sinusoid of the form $\cos(\emptyset) - j\sin(\emptyset)$. The exact frequencies of the different sinusoids depend on both the sampling rate f_s at which the original signal was sampled, and the number of samples N. For example, if we are sampling a continuous signal at a rate of 500 samples/s and, then, perform a 16-point DFT on the sampled data, the fundamental frequency of the sinusoids is $f_s/N = 500/16$ or 31.25 Hz. The other $X(m)$ analysis frequencies are integral multiples of the fundamental frequency, i.e.,

$X(0)$ = 1st frequency term, with analysis frequency = $0 \cdot 31.25 = 0$ Hz,
$X(1)$ = 2nd frequency term, with analysis frequency = $1 \cdot 31.25 = 31.25$ Hz,
$X(2)$ = 3rd frequency term, with analysis frequency = $2 \cdot 31.25 = 62.5$ Hz,
$X(3)$ = 4th frequency term, with analysis frequency = $3 \cdot 31.25 = 93.75$ Hz,

. . .

. . .

$X(15)$ = 16th frequency term, with analysis frequency = $15 \cdot 31.25 = 468.75$ Hz.

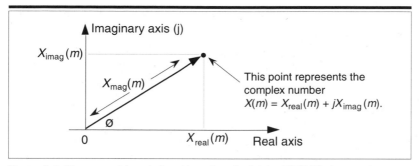

Figure 3-1 Trigonometric relationships of an individual DFT $X(m)$ complex output value.

The N separate DFT analysis frequencies are

$$f_{\text{analysis}}(m) = \frac{mf_s}{N} \ . \tag{3-5}$$

So, in this example, the $X(0)$ DFT term tells us the magnitude of any 0-Hz ("DC") component contained in the input signal, the $X(1)$ term specifies the magnitude of any 31.25-Hz component in the input signal, and the $X(2)$ term indicates the magnitude of any 62.5-Hz component in the input signal, etc. Moreover, as we'll soon show by example, the DFT output terms also determine the phase relationship between the various analysis frequencies contained in an input signal.

Quite often we're interested in both the magnitude and the power (magnitude squared) contained in each $X(m)$ term, and the standard definitions for right triangles apply here as depicted in Figure 3-1.

If we represent an arbitrary DFT output value, $X(m)$, by its real and imaginary parts

$$X(m) = X_{\text{real}}(m) + jX_{\text{imag}}(m) = X_{\text{mag}}(m) \text{ at an angle of } X_\varnothing(m) \ , \tag{3-6}$$

the magnitude of $X(m)$ is

$$X_{\text{mag}}(m) = |X(m)| = \sqrt{X_{\text{real}\,(m)}^2 + X_{\text{imag}}(m)^2} \ . \tag{3-7}$$

By definition, the phase angle of $X(m)$, $X_\varnothing(m)$, is

$$X_\varnothing(m) = \tan^{-1}\left(\frac{X_{\text{imag}}(m)}{X_{\text{real}}(m)}\right). \tag{3-8}$$

The power of $X(m)$, referred to as the power spectrum, is the magnitude squared where

$$X_{PS}(m) = X_{mag}(m)^2 = X_{real}(m)^2 + X_{imag}(m)^2 . \qquad (3-9)$$

3.1.1 DFT Example 1

The above Eqs. (3-2) and (3-3) will become more meaningful by way of an example, so, let's go through a simple one step-by-step. Let's say we want to sample and perform an 8-point DFT on a continuous input signal containing components at 1 kHz and 2 kHz, expressed as

$$x_{in}(t) = \sin(2\pi \cdot 1000 \cdot t) + 0.5\sin(2\pi \cdot 2000 \cdot t + 3\pi/4) . \qquad (3-10)$$

To make our example input signal $x_{in}(t)$ a little more interesting, we have the 2-kHz term shifted in phase by 135° ($3\pi/4$ radians) relative to the 1-kHz sinewave. With a sample rate of f_s, we sample the input every $1/f_s = t_s$ seconds. Because $N = 8$, we need 8 input sample values on which to perform the DFT. So the 8-element sequence $x(n)$ is equal to $x_{in}(t)$ sampled at the nt_s instants in time so that

$$x(n) = x_{in}(nt_s) = \sin(2\pi \cdot 1000 \cdot nt_s) + 0.5\sin(2\pi \cdot 2000 \cdot nt_s + 3\pi/4) . \qquad (3-11)$$

If we choose to sample $x_{in}(t)$ at a rate of $f_s = 8000$ samples/s from Eq. (3-5), our DFT results will indicate what signal amplitude exists in $x(n)$ at the analysis frequencies of mf_s/N, or 0 kHz, 1 kHz, 2 kHz, . . ., 7 kHz. With $f_s = 8000$ samples/s, our eight $x(n)$ samples are

$$
\begin{aligned}
&x(0) = 0.3535, \quad x(1) = 0.3535, \\
&x(2) = 0.6464, \quad x(3) = 1.0607, \\
&x(4) = 0.3535, \quad x(5) = -1.0607, \\
&x(6) = -1.3535, \quad x(7) = -0.3535 .
\end{aligned} \qquad (3\text{-}11')
$$

These $x(n)$ sample values are the dots plotted on the solid continuous $x_{in}(t)$ curve in Figure 3-2(a). (Note that the sum of the sinusoidal terms in Eq. (3-10), shown as the dashed curves in Figure 3-2(a), is equal to $x_{in}(t)$.)

Now we're ready to apply Eq. (3-3) to determine the DFT of our $x(n)$ input. We'll start with $m = 1$ because the $m = 0$ case leads to a special result that we'll discuss shortly. So, for $m = 1$, or the 1-kHz ($mf_s/N = 1 \cdot 8000/8$) DFT frequency term, Eq. (3-3) for this example becomes

$$X(1) = \sum_{n=0}^{7} x(n)\cos(2\pi n/8) - jx(n)\sin(2\pi n/8) \ . \qquad (3\text{-}12)$$

Next we multiply $x(n)$ by successive points on the cosine and sine curves of the first analysis frequency that have a single cycle over our 8 input samples. In our example, for $m = 1$, we'll sum the products of the $x(n)$ sequence with a 1-kHz cosine wave and a 1-kHz sinewave evaluated at the angular values of $2\pi n/8$. Those analysis sinusoids are shown as the dashed curves in Figure 3-2(b). Notice how the cosine and sinewaves have $m = 1$ complete cycles in our sample interval.

Substituting our $x(n)$ sample values into Eq. (3-12) and listing the cosine terms in the left column and the sine terms in the right column, we have

$X(1) = 0.3535 \cdot 1.0$	$- j(0.3535 \cdot 0.0)$	$\leftarrow$ this is the $n = 0$ term
$+ 0.3535 \cdot 0.707$	$- j(0.3535 \cdot 0.707)$	$\leftarrow$ this is the $n = 1$ term
$+ 0.6464 \cdot 0.0$	$- j(0.6464 \cdot 1.0)$	$\leftarrow$ this is the $n = 2$ term
$+ 1.0607 \cdot -0.707$	$- j(1.0607 \cdot 0.707)$	$\ldots$
$+ 0.3535 \cdot -1.0$	$- j(0.3535 \cdot 0.0)$	$\ldots$
$- 1.0607 \cdot -0.707$	$- j(-1.0607 \cdot -0.707)$	$\ldots$
$- 1.3535 \cdot 0.0$	$- j(-1.3535 \cdot -1.0)$	$\ldots$
$- 0.3535 \cdot 0.707$	$- j(-0.3535 \cdot -0.707)$	$\leftarrow$ this is the $n = 7$ term

$= 0.3535$	$+ j0.0$
$+ 0.250$	$- j0.250$
$+ 0.0$	$- j0.6464$
$- 0.750$	$- j0.750$
$+ 0.3535$	$- j0.0$
$+ 0.750$	$- j0.750$
$+ 0.0$	$- j1.3535$
$- 0.250$	$- j0.250$

$$= \ 0.0 - j4.0 = 4 \angle -90°.$$

So we now see that the input $x(n)$ contains a signal component at a frequency of 1 kHz. Using Eq. (3-7), Eq. (3-8), and Eq. (3-9) for our $X(1)$ result, $X_{mag}(1) = 4$, $X_{PS}(1) = 16$, and $X(1)$'s phase angle relative to a 1-kHz cosine is $X_\emptyset(1) = -90°$.

For the $m = 2$ frequency term, we correlate $x(n)$ with a 2-kHz cosine wave and a 2-kHz sinewave. These waves are the dashed curves in Figure 3-2(c). Notice here that the cosine and sinewaves have $m = 2$ complete cycles in our

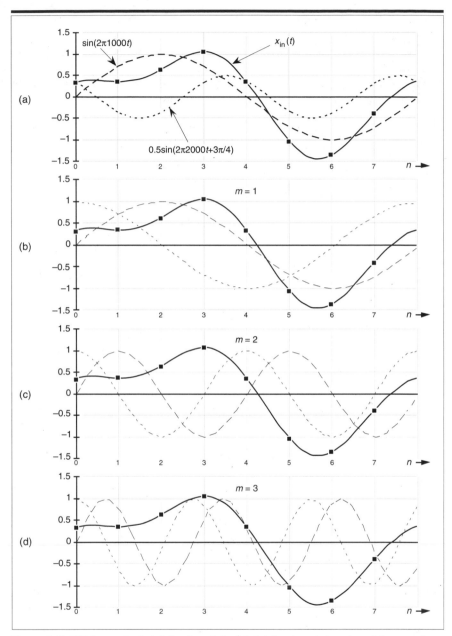

Figure 3-2 DFT Example 1: (a) the input signal; (b) the input signal and the *m* = 1 sinusoids; (c) the input signal and the *m* = 2 sinusoids; (d) the input signal and the *m* = 3 sinusoids.

sample interval in Figure 3-2(c). Substituting our $x(n)$ sample values in Eq. (3-3), for $m = 2$, gives

$$
\begin{aligned}
X(2) = \;& 0.3535 \cdot 1.0 && -j(0.3535 \cdot 0.0) \\
& + 0.3535 \cdot 0.0 && -j(0.3535 \cdot 1.0) \\
& + 0.6464 \cdot -1.0 && -j(0.6464 \cdot 0.0) \\
& + 1.0607 \cdot 0.0 && -j(1.0607 \cdot -1.0) \\
& + 0.3535 \cdot 1.0 && -j(0.3535 \cdot 0.0) \\
& -1.0607 \cdot 0.0 && -j(-1.0607 \cdot 1.0) \\
& -1.3535 \cdot -1.0 && -j(-1.3535 \cdot 0.0) \\
& -0.3535 \cdot 0.0 && -j(-0.3535 \cdot -1.0)
\end{aligned}
$$

$$
\begin{aligned}
= \;& 0.3535 && + j0.0 \\
& + 0.0 && - j0.3535 \\
& + 0.6464 && - j0.0 \\
& - 0.0 && + j1.0607 \\
& + 0.3535 && - j0.0 \\
& + 0.0 && + j1.0607 \\
& + 1.3535 && - j0.0 \\
& - 0.0 && - j0.3535
\end{aligned}
$$

$$
= \; 1.414 + j1.414 = 2 \angle 45°.
$$

Here our input $x(n)$ contains a signal at a frequency of 2 kHz whose relative amplitude is 2, and whose phase angle relative to a 2-kHz cosine is 45°. For the $m = 3$ frequency term, we correlate $x(n)$ with a 3-kHz cosine wave and a 3-kHz sinewave. These waves are the dashed curves in Figure 3-2(d). Again, see how the cosine and sinewaves have $m = 3$ complete cycles in our sample interval in Figure 3-2(d). Substituting our $x(n)$ sample values in Eq. (3-3) for $m = 3$, gives

$$
\begin{aligned}
X(3) = \;& 0.3535 \cdot 1.0 && -j(0.3535 \cdot 0.0) \\
& + 0.3535 \cdot -0.707 && -j(0.3535 \cdot 0.707) \\
& + 0.6464 \cdot 0.0 && -j(0.6464 \cdot -1.0) \\
& + 1.0607 \cdot 0.707 && -j(1.0607 \cdot 0.707) \\
& + 0.3535 \cdot -1.0 && -j(0.3535 \cdot 0.0) \\
& - 1.0607 \cdot 0.707 && -j(-1.0607 \cdot -0.707) \\
& - 1.3535 \cdot 0.0 && -j(-1.3535 \cdot 1.0) \\
& - 0.3535 \cdot -0.707 && -j(-0.3535 \cdot -0.707)
\end{aligned}
$$

$$
\begin{array}{ll}
= \ \ 0.3535 & + \, j0.0 \\
- \ 0.250 & - \, j0.250 \\
+ \ 0.0 & + \, j0.6464 \\
+ \ 0.750 & - \, j0.750 \\
- \ 0.3535 & - \, j0.0 \\
- \ 0.750 & - \, j0.750 \\
+ \ 0.0 & + \, j1.3535 \\
+ \ 0.250 & - \, j0.250
\end{array}
$$

$$= \ 0.0 - j0.0 = 0 \angle 0°.$$

Our DFT indicates that $x(n)$ contained no signal at a frequency of 3 kHz. Let's continue our DFT for the $m = 4$ frequency term using the sinusoids in Figure 3-3(a).
So Eq. (3-3) is

$$
\begin{array}{ll}
X(4) = 0.3535 \cdot 1.0 & - \, j(0.3535 \cdot 0.0) \\
+ \ 0.3535 \cdot -1.0 & - \, j(0.3535 \cdot 0.0) \\
+ \ 0.6464 \cdot 1.0 & - \, j(0.6464 \cdot 0.0) \\
+ \ 1.0607 \cdot -1.0 & - \, j(1.0607 \cdot 0.0) \\
+ \ 0.3535 \cdot 1.0 & - \, j(0.3535 \cdot 0.0) \\
- \ 1.0607 \cdot -1.0 & - \, j(-1.0607 \cdot 0.0) \\
- \ 1.3535 \cdot 1.0 & - \, j(-1.3535 \cdot 0.0) \\
- \ 0.3535 \cdot -1.0 & - \, j(-0.3535 \cdot 0.0)
\end{array}
$$

$$
\begin{array}{ll}
= \ \ 0.3535 & - \, j0.0 \\
- \ 0.3535 & - \, j0.0 \\
+ \ 0.6464 & - \, j0.0 \\
- \ 1.0607 & - \, j0.0 \\
+ \ 0.3535 & - \, j0.0 \\
+ \ 1.0607 & - \, j0.0 \\
- \ 1.3535 & - \, j0.0 \\
+ \ 0.3535 & - \, j0.0
\end{array}
$$

$$= \ 0.0 - j0.0 = 0 \angle 0°.$$

Our DFT for the $m = 5$ frequency term using the sinusoids in Figure 3-3(b) yields

$$
\begin{array}{ll}
X(5) = 0.3535 \cdot 1.0 & - \, j(0.3535 \cdot 0.0) \\
+ \ 0.3535 \cdot -0.707 & - \, j(0.3535 \cdot -0.707) \\
+ \ 0.6464 \cdot 0.0 & - \, j(0.6464 \cdot 1.0) \\
+ \ 1.0607 \cdot 0.707 & - \, j(1.0607 \cdot -0.707)
\end{array}
$$

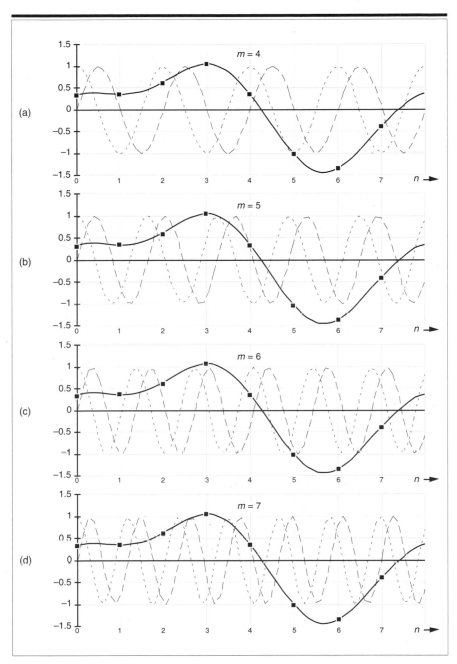

Figure 3-3 DFT Example 1: (a) the input signal and the $m = 4$ sinusoids; (b) the input and the $m = 5$ sinusoids; (c) the input and the $m = 6$ sinusoids; (d) the input and the $m = 7$ sinusoids.

$$
\begin{aligned}
&+\ 0.3535 \cdot -1.0 && -\ j(0.3535 \cdot 0.0) \\
&-\ 1.0607 \cdot 0.707 && -\ j(-1.0607 \cdot 0.707) \\
&-\ 1.3535 \cdot 0.0 && -\ j(-1.3535 \cdot -1.0) \\
&-\ 0.3535 \cdot -0.707 && -\ j(-0.3535 \cdot 0.707)
\end{aligned}
$$

$$
\begin{aligned}
=\ &0.3535 && -\ j0.0 \\
&-\ 0.250 && +\ j0.250 \\
&+\ 0.0 && -\ j0.6464 \\
&+\ 0.750 && +\ j0.750 \\
&-\ 0.3535 && -\ j0.0 \\
&-\ 0.750 && +\ j0.750 \\
&+\ 0.0 && -\ j1.3535 \\
&+\ 0.250 && +\ j0.250
\end{aligned}
$$

$$
=\ 0.0 - j.0 = 0\ \angle 0°.
$$

For the $m = 6$ frequency term using the sinusoids in Figure 3-3(c), Eq. (3-3) is

$$
\begin{aligned}
X(6) =\ &0.3535 \cdot 1.0 && -\ j(0.3535 \cdot 0.0) \\
&+\ 0.3535 \cdot 0.0 && -\ j(0.3535 \cdot -1.0) \\
&+\ 0.6464 \cdot -1.0 && -\ j(0.6464 \cdot 0.0) \\
&+\ 1.0607 \cdot 0.0 && -\ j(1.0607 \cdot 1.0) \\
&+\ 0.3535 \cdot 1.0 && -\ j(0.3535 \cdot 0.0) \\
&-\ 1.0607 \cdot 0.0 && -\ j(-1.0607 \cdot -1.0) \\
&-\ 1.3535 \cdot -1.0 && -\ j(-1.3535 \cdot 0.0) \\
&-\ 0.3535 \cdot 0.0 && -\ j(-0.3535 \cdot 1.0)
\end{aligned}
$$

$$
\begin{aligned}
=\ &0.3535 && -\ j0.0 \\
&+\ 0.0 && +\ j0.3535 \\
&-\ 0.6464 && -\ j0.0 \\
&+\ 0.0 && -\ j1.0607 \\
&+\ 0.3535 && -\ j0.0 \\
&+\ 0.0 && -\ j1.0607 \\
&+\ 1.3535 && -\ j0.0 \\
&+\ 0.0 && +\ j0.3535
\end{aligned}
$$

$$
=\ 1.414 - j1.414 = 2\ \angle -45°.
$$

For the $m = 7$ frequency term using the sinusoids in Figure 3-3(d), Eq. (3-3) is

$$
\begin{aligned}
X(7) =\ &0.3535 \cdot 1.0 && -\ j(0.3535 \cdot 0.0) \\
&+\ 0.3535 \cdot 0.707 && -\ j(0.3535 \cdot -0.707) \\
&+\ 0.6464 \cdot 0.0 && -\ j(0.6464 \cdot -1.0) \\
&+\ 1.0607 \cdot -0.707 && -\ j(1.0607 \cdot -0.707)
\end{aligned}
$$

$$
\begin{array}{ll}
+\ 0.3535 \cdot -1.0 & -\ j(0.3535 \cdot 0.0) \\
-\ 1.0607 \cdot -0.707 & -\ j(-1.0607 \cdot 0.707) \\
-\ 1.3535 \cdot 0.0 & -\ j(-1.3535 \cdot 1.0) \\
-\ 0.3535 \cdot 0.707 & -\ j(-0.3535 \cdot 0.707)
\end{array}
$$

$$
\begin{array}{ll}
=\ 0.3535 & +\ j0.0 \\
+\ 0.250 & +\ j0.250 \\
+\ 0.0 & +\ j0.6464 \\
-\ 0.750 & +\ j0.750 \\
-\ 0.3535 & -\ j0.0 \\
+\ 0.750 & +\ j0.750 \\
+\ 0.0 & +\ j1.3535 \\
-\ 0.250 & +\ j0.250
\end{array}
$$

$$
=\ 0.0 + j4.0 = 4 \angle 90°.
$$

If we plot the $X(m)$ output magnitudes as a function of frequency, we get the magnitude *spectrum* of the $x(n)$ input sequence, shown in Figure 3-4(a). The phase angles of the $X(m)$ output terms are depicted in Figure 3-4(b).

Hang in there, we're almost finished with our example. We've saved the calculation of the $m = 0$ frequency term to the end because it has a special significance. When $m = 0$, we correlate $x(n)$ with $\cos(0) - j\sin(0)$ so that Eq. (3-3) becomes

$$
X(0) = \sum_{n=0}^{N-1} x(n)[\cos(0) - j\sin(0)] . \tag{3-13}
$$

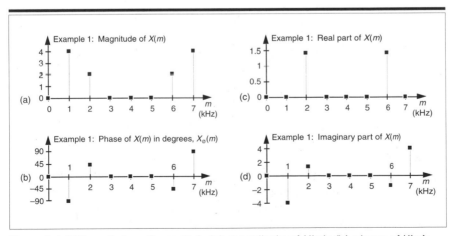

Figure 3-4 DFT results from Example 1: (a) magnitude of $X(m)$; (b) phase of $X(m)$; (c) real part of $X(m)$; (d) imaginary part of $X(m)$.

Because $\cos(0) = 1$, and $\sin(0) = 0$,

$$X(0) = \sum_{n=0}^{N-1} x(n) .$$

(3-13')

We can see that Eq. (3-13') is the sum of the $x(n)$ samples. This sum is, of course, proportional to the average of $x(n)$. (Specifically, $X(0)$ is equal to N times $x(n)$'s average value). This makes sense because the $X(0)$ frequency term is the nontime-varying (DC) component of $x(n)$. If $X(0)$ were nonzero, this would tell us that the $x(n)$ sequence is riding on a DC bias and has some nonzero average value. For our specific example input from Eq. (3-10), the sum, however, is zero. The input sequence has no DC component, so we know that $X(0)$ will be zero. But let's not be lazy—we'll calculate $X(0)$ anyway just to be sure. Evaluating Eq. (3-3) or Eq. (3-13') for $m = 0$, we see that

$$
\begin{aligned}
X(0) &= 0.3535 \cdot 1.0 && - j(0.3535 \cdot 0.0) \\
&+ 0.3535 \cdot 1.0 && - j(0.3535 \cdot 0.0) \\
&+ 0.6464 \cdot 1.0 && - j(0.6464 \cdot 0.0) \\
&+ 1.0607 \cdot 1.0 && - j(1.0607 \cdot 0.0) \\
&+ 0.3535 \cdot 1.0 && - j(0.3535 \cdot 0.0) \\
&- 1.0607 \cdot 1.0 && - j(-1.0607 \cdot 0.0) \\
&- 1.3535 \cdot 1.0 && - j(-1.3535 \cdot 0.0) \\
&- 0.3535 \cdot 1.0 && - j(-0.3535 \cdot 0.0)
\end{aligned}
$$

$$
\begin{aligned}
X(0) &= 0.3535 && - j0.0 \\
&+ 0.3535 && - j0.0 \\
&+ 0.6464 && - j0.0 \\
&+ 1.0607 && - j0.0 \\
&+ 0.3535 && - j0.0 \\
&- 1.0607 && - j0.0 \\
&- 1.3535 && - j0.0 \\
&- 0.3535 && - j0.0
\end{aligned}
$$

$$= 0.0 - j0.0 = 0 \angle 0°.$$

So our $x(n)$ had no DC component, and, thus, its average value is zero. Notice that Figure 3-4 indicates that $x_{in}(t)$, from Eq. (3-10), has signal components at 1 kHz ($m = 1$) and 2 kHz ($m = 2$). Moreover, the 1-kHz tone has a magnitude twice that of the 2-kHz tone. The DFT results depicted in Figure 3-4 tell us exactly what's the spectral content of the signal defined by Eqs. (3-10) and (3-11).

The perceptive reader should be asking two questions at this point. First, what do those nonzero magnitude values at $m = 6$ and $m = 7$ in Figure 3-4(a) mean? Also, why do the magnitudes seem four times larger than we would expect? We'll answer those good questions shortly. The above 8-point DFT example, although admittedly simple, illustrates two very important characteristics of the DFT that we should never forget. First, any individual $X(m)$ output value is nothing more than the sum of the term-by-term products, a correlation, of an input signal sample sequence with a cosine and a sinewave whose frequencies are m complete cycles in the total sample interval of N samples. This is true no matter what the f_s sample rate is and no matter how large N is in an N-point DFT. The second important characteristic of the DFT of real input samples is the symmetry of the DFT output terms.

3.2 DFT Symmetry

Looking at Figure 3-4(a) again, there is an obvious symmetry in the DFT results. Although the standard DFT is designed to accept complex input sequences, most physical DFT inputs (such as digitized values of some continuous signal) are referred to as *real*, that is, real inputs have nonzero real sample values, and the imaginary sample values are assumed to be zero. When the input sequence $x(n)$ is real, as it will be for all of our examples, the complex DFT outputs for $m = 0$ to $m = (N/2) - 1$ are redundant with frequency output values for $m > (N/2)$. The mth DFT output will have the same magnitude as the $(N-m)$th DFT output. The phase angle of the DFT's mth output is the negative of the phase angle of the $(N-m)$th DFT output. So the mth and $(N-m)$th outputs are related by the following:

$$X(m) = |X(m)| \text{ at } X_{\theta}(m) \text{ degrees}$$

$$= |X(N-m)| \text{ at } -X_{\theta}(N-m) \text{ degrees} . \tag{3-14}$$

We can state that when the DFT input sequence is real, $X(m)$ is the complex conjugate of $X(N-m)$, or

$$X(m) = X^*(N-m) ,^{\dagger} \tag{3-14'}$$

where the superscript * symbol denotes conjugation.

[†] Using our notation, the complex conjugate of $x = a + jb$ is defined as $x^* = a - jb$; that is, we merely change the sign of the imaginary part of x. In an equivalent form, if $x = e^{j\theta}$, then $x^* = e^{-j\theta}$.

In our example above, notice in Figures 3-3(b) and 3-3(d), that $X(5)$, $X(6)$, and $X(7)$ are the complex conjugates of $X(3)$, $X(2)$, and $X(1)$, respectively. Like the DFT's magnitude symmetry, the real part of $X(m)$ has what is called *even symmetry*, as shown in Figure 3-4(c), while the DFT's imaginary part has *odd symmetry*, as shown in Figure 3-4(d). This relationship is what is meant when the DFT is called conjugate symmetric in the literature. It means that, if we perform an N-point DFT on a real input sequence, we'll get N separate complex DFT output terms, but only the first $N/2$ terms are independent. So to obtain the DFT of $x(n)$, we need only compute the first $N/2$ values of $X(m)$ where $0 \le m \le (N/2) - 1$; the $X(N/2)$ to $X(N-1)$ DFT output terms provide no additional information about the spectrum of the real sequence $x(n)$.

Although Eqs. (3-2) and (3-3) are equivalent, expressing the DFT in the exponential form of Eq. (3-2) has a terrific advantage over the form of Eq. (3-3). Not only does Eq. (3-2) save pen and paper, Eq. (3-2)'s exponentials are so much easier to manipulate when we're trying to analyze DFT relationships. Using Eq. (3-2), products of terms become the addition of exponents and, with due respect to Euler, we don't have all those trigonometric relationships to memorize. Let's demonstrate this by proving Eq. (3-14) to show the symmetry of the DFT of real input sequences. Substituting $N-m$ for m in Eq. (3-2), we get the expression for the $(N-m)$th component of the DFT:

$$X(N - m) = \sum_{n=0}^{N-1} x(n) e^{-j2\pi n(N-m)/N} = \sum_{n=0}^{N-1} x(n) e^{-j2\pi nN/N} e^{-j2\pi n(-m)/N}$$

$$= \sum_{n=0}^{N-1} x(n) e^{-j2\pi n} e^{j2\pi nm/N} . \tag{3-15}$$

Because $e^{-j2\pi n} = \cos(2\pi n) - j\sin(2\pi n) = 1$ for all integer values of n,

$$X(N - m) = \sum_{n=0}^{N-1} x(n) e^{j2\pi nm/N} . \tag{3-15'}$$

We see that $X(N-m)$ in Eq. (3-15') is merely $X(m)$ in Eq. (3-2) with the sign reversed on $X(m)$'s exponent—and that's the definition of the complex conjugate. This is illustrated by the DFT output phase-angle plot in Figure 3-4(b) for our DFT Example 1. Try deriving Eq. (3-15') using the cosines and sines of Eq. (3-3), and you'll see why the exponential form of the DFT is so convenient for analytical purposes.

There's an additional symmetry property of the DFT that deserves mention at this point. In practice, we're occasionally required to determine the DFT of real input functions where the input index n is defined over both positive and negative values. If that real input function is even, then $X(m)$ is always real and even; that is, if the real $x(n) = x(-n)$, then, $X_{real}(m)$ is in general nonzero and $X_{imag}(m)$ is zero. Conversely, if the real input function is odd, $x(n) = -x(-n)$, then $X_{real}(m)$ is always zero and $X_{imag}(m)$ is, in general, nonzero. This characteristic of input function symmetry is a property that the DFT shares with the continuous Fourier transform, and (don't worry) we'll cover specific examples of it later in Section 3.13 and in Chapter 5.

3.3 DFT Linearity

The DFT has a very important property known as *linearity*. This property states that the DFT of the sum of two signals is equal to the sum of the transforms of each signal; that is, if an input sequence $x_1(n)$ has a DFT $X_1(m)$ and another input sequence $x_2(n)$ has a DFT $X_2(m)$, then the DFT of the sum of these sequences $x_{sum}(n) = x_1(n) + x_2(n)$ is

$$X_{sum}(m) = X_1(m) + X_2(m) . \tag{3-16}$$

This is certainly easy enough to prove. If we plug $x_{sum}(n)$ into Eq. (3-2) to get $X_{sum}(m)$, then

$$X_{sum}(m) = \sum_{n=0}^{N-1} [x_1(n) + x_2(n)] e^{-j2\pi nm/N}$$

$$= \sum_{n=0}^{N-1} x_1(n) e^{-j2\pi nm/N} + \sum_{n=0}^{N-1} x_2(n) e^{-j2\pi nm/N} = X_1(m) + X_2(m) .$$

Without this property of linearity, the DFT would be useless as an analytical tool because we could transform only those input signals that contain a single sinewave. The real-world signals that we want to analyze are much more complicated than a single sinewave.

3.4 DFT Magnitudes

The DFT Example 1 results of $|X(1)| = 4$ and $|X(2)| = 2$ may puzzle the reader because our input $x(n)$ signal, from Eq. (3-11), had peak amplitudes of 1.0 and 0.5, respectively. There's an important point to keep in mind regarding DFTs defined by Eq. (3-2). When a real input signal contains a sinewave component of peak amplitude A_o with an integral number of cycles over N input samples, the output magnitude of the DFT for that particular sinewave is M_r where

$$M_r = A_o N / 2 . \tag{3-17}$$

If the DFT input is a complex sinusoid of magnitude A_o (i.e., $A_o e^{j2\pi ft}$) with an integral number of cycles over N samples, the output magnitude of the DFT is M_c where

$$M_c = A_o N . \tag{3-17'}$$

As stated in relation to Eq. (3-13'), if the DFT input was riding on a DC value equal to D_o, the magnitude of the DFT's $X(0)$ output will be $D_o N$.

Looking at the real input case for the 1000 Hz component of Eq. (3-11), $A_o = 1$ and $N = 8$, so that $M_{real} = 1 \cdot 8/2 = 4$, as our example shows. Equation (3-17) may not be so important when we're using software or floating-point hardware to perform DFTs, but if we're implementing the DFT with fixed-point hardware, we have to be aware that the output can be as large as $N/2$ times the peak value of the input. This means that, for real inputs, hardware memory registers must be able to hold values as large as $N/2$ times the maximum amplitude of the input sample values. We discuss DFT output magnitudes in further detail later in this chapter. The DFT magnitude expressions in Eqs. (3-17) and (3-17') are why we occasionally see the DFT defined in the literature as

$$X'(m) = \frac{1}{N} \sum_{n=0}^{N-1} x(n) e^{-j2\pi nm / N} . \tag{3-18}$$

The $1/N$ scale factor in Eq. (3-18) makes the amplitudes of $X'(m)$ equal to half the time-domain input sinusoid's peak value at the expense of the additional division by N computation. Thus, hardware or software implementations of the DFT typically use Eq. (3-2) as opposed to Eq. (3-18). Of course, there are always exceptions. There are commercial software packages using

$$X''(m) = \frac{1}{\sqrt{N}} \sum_{n=0}^{N-1} x(n)e^{-j2\pi nm/N},$$

and

$$x(n) = \frac{1}{\sqrt{N}} \sum_{n=0}^{N-1} X''(m)e^{j2\pi nm/N} \qquad (3\text{-}18')$$

for the forward and inverse DFTs. (In Section 3.7, we discuss the meaning and significance of the inverse DFT.) The $1/\sqrt{N}$ scale factors in Eqs. (3-18') seem a little strange, but they're used so that there's no scale change when transforming in either direction. When analyzing signal spectra in practice, we're normally more interested in the relative magnitudes rather than absolute magnitudes of the individual DFT outputs, so scaling factors aren't usually that important to us.

3.5 DFT Frequency Axis

The frequency axis m of the DFT result in Figure 3-4 deserves our attention once again. Suppose we hadn't previously seen our DFT Example 1, were given the eight input sample values, from Eq. (3-11'), and asked to perform an 8-point DFT on them. We'd grind through Eq. (3-2) and get the $X(m)$ values shown in Figure 3-4. Next we ask, "What's the frequency of the highest magnitude component in $X(m)$ in Hz?" The answer is not "1." The answer depends on the original sample rate f_s. Without prior knowledge, we have no idea over what time interval the samples were taken, so we don't know the absolute scale of the $X(m)$ frequency axis. The correct answer to the question is to take f_s and plug it into Eq. (3-5) with $m = 1$. Thus, if $f_s = 8000$ samples/s, then, the frequency associated with the largest DFT magnitude term is

$$f_{\text{analysis}}(m) = \frac{mf_s}{N} = f_{\text{analysis}}(1) = \frac{1 \cdot 8000}{8} = 1000 \text{ Hz}.$$

If we said the sample rate f_s was 75 samples/s, we'd know, from Eq. (3-5), that the frequency associated with the largest magnitude term is now

$$f_{\text{analysis}}(1) = \frac{1 \cdot 75}{8} = 9.375 \text{ Hz}.$$

OK, enough of this—just remember that the DFT's frequency spacing (resolution) is f_s/N.

To recap what we've learned so far:

- each DFT output term is the sum of the term-by-term products of an input time-domain sequence with sequences representing a sine and a cosine wave,

- for real inputs, an N-point DFT's output provides only $N/2$ independent terms,

- the DFT is a linear operation,

- the magnitude of the DFT results are directly proportional to N, and

- the DFT's frequency resolution is f_s/N.

It's also important to realize, from Eq. (3-5), that $X(N/2)$, when $m = N/2$, corresponds to half the sample rate, i.e. the folding (Nyquist) frequency $f_s/2$.

3.6 DFT Shifting Theorem

There's an important property of the DFT known as the shifting theorem. It states that a shift in time of a periodic $x(n)$ input sequence manifests itself as a constant phase shift in the angles associated with the DFT results. (We won't derive the shifting theorem equation here because its derivation is included in just about every digital signal processing textbook in print.) If we decide to sample $x(n)$ starting at n equals some integer k, as opposed to $n = 0$, the DFT of those time-shifted sample values is $X_{shifted}(m)$ where

$$X_{shifted}(m) = e^{j2\pi km/N} X(m) \,. \tag{3-19}$$

Equation (3-19) tells us that, if the point where we start sampling $x(n)$ is shifted to the right by k samples, the DFT output spectrum of $X_{shifted}(m)$ is $X(m)$ with each of $X(m)$'s complex terms multiplied by the linear phase shift $e^{j2\pi km/N}$, which is merely a phase shift of $2\pi km/N$ radians or $360km/N$ degrees. Conversely, if the point where we start sampling $x(n)$ is shifted to the left by k samples, the spectrum of $X_{shifted}(m)$ is $X(m)$ multiplied by $e^{-j2\pi km/N}$. Let's illustrate Eq. (3-19) with an example.

3.6.1 DFT Example 2

Suppose we sampled our DFT Example 1 input sequence later in time by $k = 3$ samples. Figure 3-5 shows the original input time function,

$$x_{in}(t) = \sin(2\pi 1000t) + 0.5\sin(2\pi 2000t + 3\pi/4) \,.$$

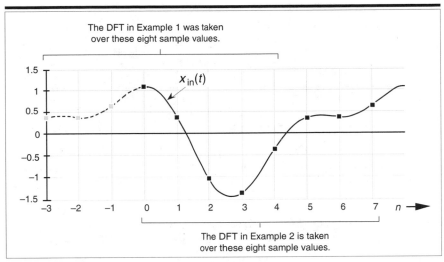

The DFT in Example 1 was taken over these eight sample values.

$x_{in}(t)$

The DFT in Example 2 is taken over these eight sample values.

Figure 3-5 Comparison of sampling times between DFT Example 1 and DFT Example 2.

We can see that Figure 3-5 is a continuation of Figure 3-2(a). Our new $x(n)$ sequence becomes the values represented by the solid black dots in Figure 3-5 whose values are

$$x(0) = 1.0607, \qquad x(1) = 0.3535,$$
$$x(2) = -1.0607, \qquad x(3) = -1.3535,$$
$$x(4) = -0.3535, \qquad x(3) = 0.3535,$$
$$x(6) = 0.3535, \qquad x(7) = 0.6464 . \qquad\qquad (3\text{-}20)$$

Performing the DFT on Eq. (3-20), $X_{shifted}(m)$ is

$X_{shifted}(m) =$	m	$X_{shifted}(m)'s$ magnitude	$X_{shifted}(m)'s$ phase	$X_{shifted}(m)'s$ real part	$X_{shifted}(m)'s$ imaginary part	
	0	0	0	0	0	
	1	4	+45	2.8284	2.8284	
	2	2	−45	1.4142	−1.414	
	3	0	0	0	0	
	4	0	0	0	0	
	5	0	0	0	0	
	6	2	+45	1.4142	1.4142	
	7	4	−45	2.8284	−2.828	(3-21)

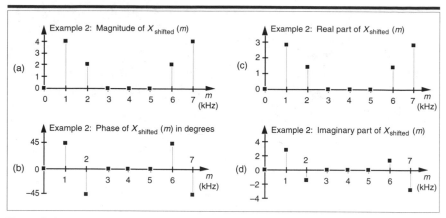

Figure 3-6 DFT results from Example 2: (a) magnitude of $X_{shifted}(m)$; (b) phase of $X_{shifted}(m)$; (c) real part of $X_{shifted}(m)$; (d) imaginary part of $X_{shifted}(m)$.

The values in Eq. (3-21) are illustrated as the dots in Figure 3-6. Notice that Figure 3-6(a) is identical to Figure 3-4(a). Equation (3-19) told us that the magnitude of $X_{shifted}(m)$ should be unchanged from that of $X(m)$. That's a comforting thought, isn't it? We wouldn't expect the DFT magnitude of our original periodic $x_{in}(t)$ to change just because we sampled it over a different time interval. The phase of the DFT result does, however, change depending on the instant at which we started to sample $x_{in}(t)$.

By looking at the $m = 1$ component of $X_{shifted}(m)$, for example, we can double-check to see that phase values in Figure 3-6(b) are correct. Using Eq. (3-19) and remembering that $X(1)$ from DFT Example 1 had a magnitude of 4 at a phase angle of –90 (or $-\pi/2$ radians), $k = 3$ and $N = 8$ so that

$$X_{shifted}(1) = e^{j2\pi km/N} \cdot X(1) = e^{j2\pi 3 \cdot 1/8} \cdot 4e^{-j\pi/2} = 4e^{j(6\pi/8 - 4\pi/8)} = 4e^{j\pi/4} . \quad (3\text{-}22)$$

So $X_{shifted}(1)$ has a magnitude of 4 and a phase angle of $\pi/4$ or +45°, which is what we set out to prove using Eq. (3-19).

3.7 Inverse DFT

Although the DFT is the major topic of this chapter, it's appropriate, now, to introduce the inverse discrete Fourier transform (IDFT). Typically we think of the DFT as transforming time-domain data into a frequency-domain representation. Well, we can reverse this process and obtain the original time-domain signal by performing the IDFT on the $X(m)$ frequency-domain values. The standard expressions for the IDFT are

$$x(n) = \frac{1}{N}\sum_{m=0}^{N-1} X(m)e^{j2\pi mn/N} \qquad (3\text{-}23)$$

and equally,

$$x(n) = \frac{1}{N}\sum_{m=0}^{N-1} X(m)[\cos(2\pi mn/N) + j\sin(2\pi mn/N)] \ . \qquad (3\text{-}23')$$

Remember the statement we made in Section 3.1 that a discrete time-domain signal can be considered the sum of various sinusoidal analytical frequencies and that the $X(m)$ outputs of the DFT are a set of N complex values indicating the magnitude and phase of each analysis frequency comprising that sum. Equations (3-23) and (3-23') are the mathematical expressions of that statement. It's very important for the reader to understand this concept. If we perform the IDFT by plugging our results from DFT Example 1 into Eq. (3-23), we'll go from the frequency-domain back to the time-domain and get our original real Eq. (3-11') $x(n)$ sample values of

$$x(0) = 0.3535 + j0.0 \quad x(1) = 0.3535 + j0.0$$
$$x(2) = 0.6464 + j0.0 \quad x(3) = 1.0607 + j0.0$$
$$x(4) = 0.3535 + j0.0 \quad x(3) = -1.0607 + j0.0$$
$$x(6) = -1.3535 + j0.0 \quad x(7) = -0.3535 + j0.0 \ .$$

Notice that Eq. (3-23)'s IDFT expression differs from the DFT's Eq. (3-2) only by a $1/N$ scale factor and a change in the sign of the exponent. Other than the magnitude of the results, every characteristic that we've covered, thus far, regarding the DFT, also applies to the IDFT.

3.8 DFT Leakage

Hold on to your seat now. Here's where the DFT starts to get really interesting. The two previous DFT examples gave us correct results because the input $x(n)$ sequences were very carefully chosen sinusoids. As it turns out, the DFT of sampled real-world signals provides frequency-domain results that can be misleading. A characteristic, known as leakage, causes our DFT results to be only an approximation of the true spectra of the original input signals prior to digital sampling. Although there are ways to minimize leakage, we can't eliminate it entirely. Thus, we need to understand exactly what effect it has on our DFT results.

Let's start from the beginning. DFTs are constrained to operate on a finite set of N input values sampled at a sample rate of f_s, to produce an N-point transform whose discrete outputs are associated with the individual analytical frequencies $f_{analysis}(m)$, with

$$f_{analysis}(m) = \frac{mf_s}{N}, \text{ where } m = 0, 1, 2, ..., N-1. \tag{3-24}$$

Equation (3-24), illustrated in DFT Example 1, may not seem like a problem, but it is. The DFT produces correct results only when the input data sequence contains energy precisely at the analysis frequencies given in Eq. (3-24), at integral multiples of our fundamental frequency f_s/N. If the input has a signal component at some intermediate frequency between our analytical frequencies of mf_s/N, say $1.5f_s/N$, this input signal will show up to some degree in *all* of the N output analysis frequencies of our DFT! (We typically say that input signal energy shows up in all of the DFT's output *bins*, and we'll see, in a moment, why the phrase "output bins" is appropriate.) Let's understand the significance of this problem with another DFT example.

Assume we're taking a 64-point DFT of the sequence indicated by the dots in Figure 3-7(a). The sequence is a sinewave with exactly three cycles contained in our $N = 64$ samples. Figure 3-7(b) shows the first half of the DFT of the input sequence and indicates that the sequence has an average value of zero ($X(0) = 0$) and no signal components at any frequency other than the $m = 3$ frequency. No surprises so far. Figure 3-7(a) also shows, for example, the $m = 4$ sinewave analysis frequency, superimposed over the input sequence, to remind us that the analytical frequencies always have an integral number of cycles over our total sample interval of 64 points. The sum of the products of the input sequence and the $m = 4$ analysis frequency is zero. (Or we can say, the correlation of the input sequence, and the $m = 4$ analysis frequency is zero.) The sum of the products of this particular three-cycle input sequence and any analysis frequency other than $m = 3$ is zero. Continuing with our leakage example, the dots in Figure 3-8(a) show an input sequence having 3.4 cycles over our $N = 64$ samples. Because the input sequence does not have an integral number of cycles over our 64-sample interval, input energy has leaked into all the other DFT output bins as shown in Figure 3-8(b). The $m = 4$ bin, for example, is not zero because the sum of the products of the input sequence and the $m = 4$ analysis frequency is no longer zero. This is leakage—it causes any input signal whose frequency is not exactly at a DFT bin center to leak into all of the other DFT output bins. Moreover,

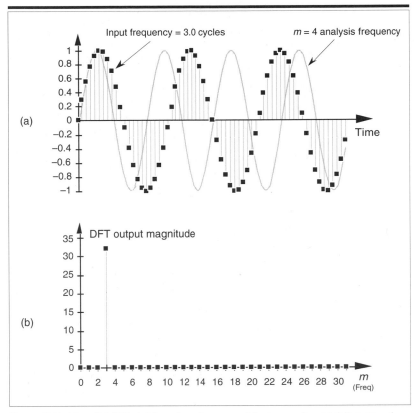

Figure 3-7 64-point DFT: (a) input sequence of three cycles and the $m = 4$ analysis frequency sinusoid; (b) DFT output magnitude.

leakage is an unavoidable fact of life when we perform the DFT on real-world finite-length time sequences.

Now, as the English philosopher Douglas Adams would say, "Don't panic." Let's take a quick look at the cause of leakage to learn how to predict and minimize its unpleasant effects. To understand the effects of leakage, we need to know the amplitude response of a DFT when the DFT's input is an arbitrary, real sinusoid. Although Sections 3.14 and 3.15 discuss this issue in detail, for our purposes, here, we'll just say that, for a real cosine input having k cycles in the N-point input time sequence, the amplitude response of an N-point DFT bin in terms of the bin index m is approximated by the sinc function

$$X(m) \approx \frac{N}{2} \cdot \frac{\sin[\pi(k - m)]}{\pi(k - m)} \ . \tag{3-25}$$

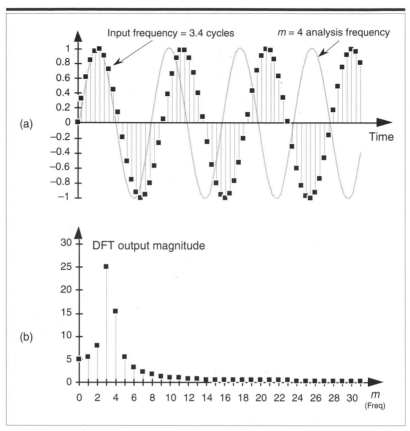

Figure 3-8 64-point DFT: (a) 3.4 cycles input sequence and the $m = 4$ analysis frequency sinusoid; (b) DFT output magnitude.

We'll use Eq. (3-25), illustrated in Figure 3-9(a), to help us determine how much leakage occurs in DFTs. We can think of the curve in Figure 3-9(a), comprising a main lobe and periodic peaks and valleys known as *sidelobes*, as the continuous positive spectrum of an N-point, real cosine time sequence having k complete cycles in the N-point input time interval. The DFT's outputs are discrete samples that reside on the curves in Figure 3-9; that is, our DFT output will be a sampled version of the continuous spectrum. (We show the DFT's magnitude response to a real input in terms of frequency (Hz) in Figure 3-9(b).) When the DFT's input sequence has exactly an integral k number of cycles (centered exactly in the $m = k$ bin), no leakage occurs, as in Figure 3-9, because when the angle in the numerator of Eq. (3-25) is a nonzero integral multiple of π, the sine of that angle is zero.

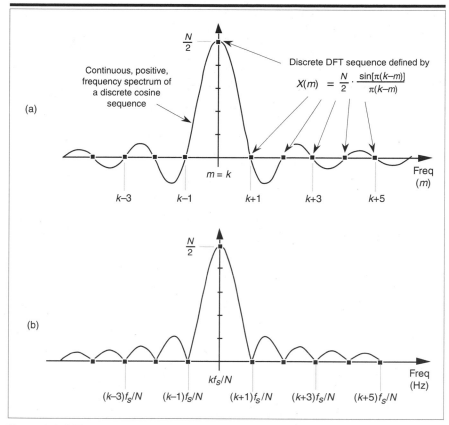

Figure 3-9 DFT positive frequency response due to an *N*-point input sequence containing *k* cycles of a real cosine: (a) amplitude response as a function of bin index *m*; (b) magnitude response as a function of frequency in Hz.

By way of example, we can illustrate again what happens when the input frequency *k* is not located at a bin center. Assume that a real 8-kHz sinusoid, having unity amplitude, has been sampled at a rate of $f_s - 32{,}000$ samples/s. If we take a 32-point DFT of the samples, the DFT's frequency resolution, or bin spacing, is $f_s/N = 32{,}000/32$ Hz = 1.0 kHz. We can predict the DFT's magnitude response by centering the input sinusoid's spectral curve at the positive frequency of 8 kHz, as shown in Figure 3-10(a). The dots show the DFT's output bin magnitudes.

Again, here's the important point to remember: the DFT output is a sampled version of the continuous spectral curve in Figure 3-10(a). Those sampled values in the frequency-domain, located at mf_s/N, are the dots in Figure 3-10(a). Because the input signal frequency is exactly at a DFT

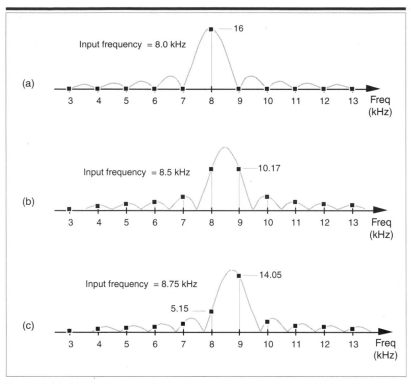

Figure 3-10 DFT bin positive frequency responses: (a) DFT input frequency = 8.0 kHz; (b) DFT input frequency = 8.5 kHz; (c) DFT input frequency = 8.75 kHz.

bin center, the DFT results have only one nonzero value. Stated in another way, when an input sinusoid has an integral number of cycles over N time-domain input sample values, the DFT outputs reside on the continuous spectrum at its peak and exactly at the curve's zero crossing points. From Eq. (3-25) we know the peak output magnitude is $32/2 = 16$. (If the real input sinusoid had an amplitude of 2, the peak of the response curve would be $2 \cdot 32/2$, or 32.) Figure 3-10(b) illustrates DFT leakage where the input frequency is 8.5 kHz, and we see that the frequency-domain sampling results in nonzero magnitudes for all DFT output bins. An 8.75-kHz input sinusoid would result in the leaky DFT output shown in Figure 3-10(c). If we're sitting at a computer studying leakage by plotting the magnitude of DFT output values, of course, we'll get the dots in Figure 3-10 and won't see the continuous spectral curves.

At this point, the attentive reader should be thinking: "If the continuous spectrums that we're sampling are symmetrical, why does the DFT output in

Figure 3-8(b) look so asymmetrical?" In Figure 3-8(b), the bins to the right of the third bin are decreasing in amplitude faster than the bins to the left of the third bin. "And another thing, evaluating the continuous spectrum's $X(mf_s)$ function at an abscissa value of 0.4 gives a magnitude scale factor of 0.75. Applying this factor to the DFT's maximum possible peak magnitude of 32, we should have a third bin magnitude of approximately $32 \cdot 0.75 = 24$—but Figure 3-8(b) shows that the third-bin magnitude is slightly greater than 25. What's going on here?" We answer this by remembering what Figure 3-8(b) really represents. When examining a DFT output, we're normally interested only in the $m = 0$ to $m = (N/2–1)$ bins. Thus, for our 3.4 cycles per sample interval example in Figure 3-8(b), only the first 32 bins are shown. Well, the DFT is periodic in the frequency domain as illustrated in Figure 3-11. Upon examining the DFT's output for higher and higher frequencies, we end up going in circles, and the spectrum repeats itself forever.

The more conventional way to view a DFT output is to *unwrap* the spectrum in Figure 3-11 to get the spectrum in Figure 3-12. Figure 3-12 shows some of the additional replications in the spectrum for the 3.4 cycles per sample interval example. Concerning our DFT output asymmetry problem, as some of the input 3.4-cycle signal amplitude leaks into the 2nd bin, the 1st bin, and the 0th bin, leakage continues into the –1st bin, the –2nd bin, the –3rd bin, etc. Remember, the 63rd bin is the –1st bin, the 62nd bin is the –2nd bin, and so on. These bin equivalencies allow

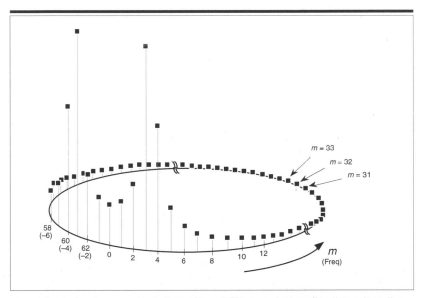

Figure 3-11 Cyclic representation of the DFT's spectral replication when the DFT input is 3.4 cycles per sample interval.

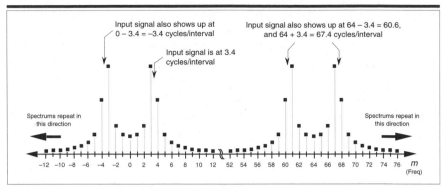

Figure 3-12 Spectral replication when the DFT input is 3.4 cycles per sample interval.

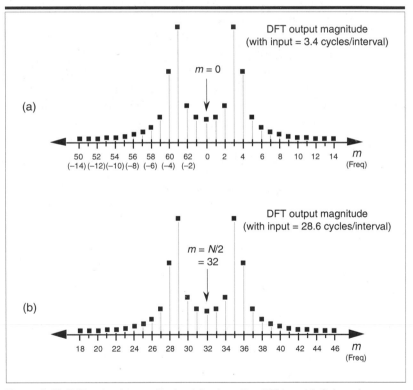

Figure 3-13 DFT output magnitude: (a) when the DFT input is 3.4 cycles per sample interval; (b) when the DFT input is 28.6 cycles per sample interval.

us to view the DFT output bins as if they extend into the negative frequency range, as shown in Figure 3-13(a). The result is that the leakage *wraps around* the $m = 0$ frequency bin, as well as around the $m = N$ frequency bin. This is not surprising because the $m = 0$ frequency *is* the $m = N$

frequency. The leakage wraparound at the $m = 0$ frequency accounts for the asymmetry about the DFT's $m = 3$ bin in Figure 3-8(b).

Recall from the DFT symmetry discussion, when a DFT input sequence $x(n)$ is real, the DFT outputs from $m = 0$ to $m = (N/2–1)$ are redundant with frequency bin values for $m > (N/2)$, where N is the DFT size. The mth DFT output will have the same magnitude as the $(N–m)$th DFT output. That is, $|X(m)| = |X(N–m)|$. What this means is that leakage wraparound also occurs about the $m = N/2$ bin. This can be illustrated using an input of 28.6 cycles per sample interval $(32 – 3.4)$ whose spectrum is shown in Figure 3-13(b). Notice the similarity between Figure 3-13(a) and Figure 3-13(b). So the DFT exhibits leakage wraparound about the $m = 0$ and $m = N/2$ bins. Minimum leakage asymmetry will occur near the $N/4$th bin as shown in Figure 3-14(a) where the full spectrum of a 16.4 cycles per sample interval input is provided. Figure 3-14(b) shows a close-up view of the first 32 bins of the 16.4 cycles per sample interval spectrum.

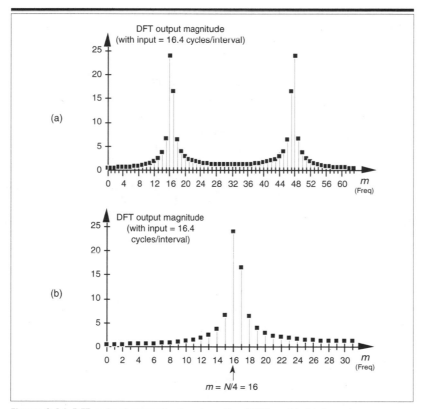

Figure 3-14 DFT output magnitude when the DFT input is 16.4 cycles per sample interval: (a) full output spectrum view; (b) close-up view showing minimized leakage asymmetry at frequency $m = N/4$.

You can read about leakage all day. However, the best way to appreciate its effects is to sit down at a computer and use a software program to take DFTs, in the form of fast Fourier transforms (FFTs), of your personally generated test signals like those in Figures 3-6 and 3-7. You can, then, experiment with different combinations of input frequencies and various DFT sizes. You'll be able to demonstrate that DFT leakage effect is troublesome because the bins containing low-level signals are corrupted by the sidelobe levels from neighboring bins containing high-amplitude signals.

Although there's no way to eliminate leakage completely, an important technique known as *windowing* is the most common remedy to reduce its unpleasant effects. Let's look at a few DFT window examples.

3.9 Windows

Windowing reduces DFT leakage by minimizing the magnitude of Eq. (3-25)'s sinc function's $\sin(x)/x$ sidelobes shown in Figure 3-9. We do this by forcing the amplitude of the input time sequence at both the beginning and the end of the sample interval to go smoothly toward a single common amplitude value. Figure 3-15 shows how this process works. If we consider the infinite-duration time signal shown in Figure 3-15(a), a DFT can only be performed over a finite-time sample interval like that shown in Figure 3-15(c). We can think of the DFT input signal in Figure 3-15(c) as the product of an input signal existing for all time, Figure 3-15(a), and the rectangular window whose magnitude is 1 over the sample interval shown in Figure 3-15(b). Anytime we take the DFT of a finite-extent input sequence we are, by default, multiplying that sequence by a window of all ones and effectively multiplying the input values outside that window by zeros. As it turns out, Eq. (3-25)'s sinc function's $\sin(x)/x$ shape, shown in Figure 3-9, is caused by this rectangular window because the continuous Fourier transform of the rectangular window in Figure 3-15(b) *is* the sinc function.

As we'll soon see, it's the rectangular window's abrupt changes between one and zero that are the cause of the sidelobes in the the $\sin(x)/x$ sinc function. To minimize the spectral leakage caused by those sidelobes, we have to reduce the sidelobe amplitudes by using window functions other than the rectangular window. Imagine if we multiplied our DFT input, Figure 3-15(c), by the triangular window function shown in Figure 3-15(d) to obtain the windowed input signal shown in Figure 3-15(e). Notice that the values of our final input signal appear to be the same at the beginning and end of the sample interval in Figure 3-15(e). The reduced discontinuity decreases the level of relatively high frequency components in our overall DFT output; that is, our DFT bin sidelobe levels are reduced in magnitude using a triangular window. There are other window functions that reduce leakage even more than the triangular window, such as the Hanning window in Figure 3-15(f). The product of the window in Figure 3-15(f) and the input sequence provides the signal shown in Figure 3-15(g) as the input to the DFT. Another common window function is the Hamming window shown in Figure 3-15(h). It's much like the Hanning window, but it's raised on a *pedestal*.

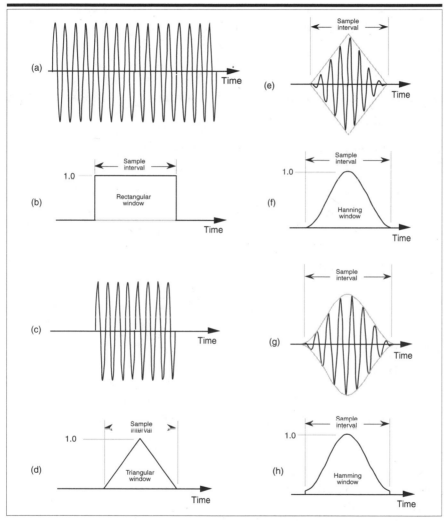

Figure 3-15 Minimizing sample interval endpoint discontinuities: (a) infinite duration input sinusoid; (b) rectangular window due to finite-time sample interval; (c) product of rectangular window and infinite-duration input sinusoid; (d) triangular window function; (e) product of triangular window and infinite-duration input sinusoid; (f) Hanning window function; (g) product of Hanning window and infinite-duration input sinusoid; (h) Hamming window function.

Before we see exactly how well these windows minimize DFT leakage, let's define them mathematically. Assuming that our original N input signal samples are indexed by n, where $0 \leq n \leq N-1$, we'll call the N time-domain window coefficients $w(n)$; that is, an input sequence $x(n)$ is multiplied by the corresponding window $w(n)$ coefficients before the DFT is performed. So the DFT of the windowed $x(n)$ input sequence, $X_w(m)$, takes the form of

$$X_\omega(m) = \sum_{n=0}^{N-1} \omega(n) \cdot x(n) e^{-j2\pi nm/N}. \tag{3-26}$$

To use window functions, we need mathematical expression of them in terms of n. The following expressions define our window function coefficients:

Rectangular window: (also called the uniform, or boxcar, window)

$w(n) = 1$, for $n = 0, 1, 2, \ldots, N-1$. $\qquad$ (3-27)

Triangular window: (very similar to the Bartlett[3], and Parzen[4,5] windows)

$$\omega(n) = \frac{n}{N/2}, \text{ for } n = 0, 1, 2, ..., N/2, \text{ and}$$

$$= 2 - \frac{n}{N/2},$$

for $n = N/2+1, N/2+2, ..., N-1$. $\qquad$ (3-28)

Hanning window: (also called the raised cosine, Hann, or von Hann window)

$w(n) = 0.5 - 0.5\cos(2\pi n/N)$, for $n = 0, 1, 2, \ldots, N-1$. $\qquad$ (3-29)

Hamming window:

$w(n) = 0.54 - 0.46\cos(2\pi n/N)$, for $n = 0, 1, 2, \ldots, N-1$. $\qquad$ (3-30)

If we plot the $w(n)$ values from Eqs. (3-27) through (3-30), we'd get the corresponding window functions like those in Figures 3-15(b), 3-15(d), 3-15(f), and 3-15(h).[†]

[†] In the literature, the equations for window functions depend on the range of the sample index n. We define n to be in the range $0 < n < N-1$. Some authors define n to be in the range $-N/2 < n < N/2$, in which case, for example, the expression for the Hanning window would have a sign change and be $w(n) = 0.5 + 0.5\cos(2\pi n/N)$.

The rectangular window's amplitude response is the yardstick we normally use to evaluate another window function's amplitude response; that is, we typically get an appreciation for a window's response by comparing it to the rectangular window that exhibits the magnitude response shown in Figure 3-9(b). The rectangular window's $\sin(x)/x$ magnitude response, $|W(m)|$, is repeated in Figure 3-16(a). Also included in Figure 3-16(a) are the Hamming, Hanning, and triangular window magnitude responses. (The frequency axis in Figure 3-16 is such that the curves show the response of a single N-point DFT bin when the various window functions are used.) We can see that the last three windows give reduced sidelobe levels relative to the rectangular window. Because the Hamming, Hanning, and triangular windows reduce the time-domain signal levels applied to the DFT, their main lobe peak values are reduced relative to the rectangular window. (Because of the near-zero $w(n)$ coefficients at the beginning and end of the sample interval, this signal level loss is called the processing gain, or loss, of a window.) Be that as it may, we're primarily interested in the windows' sidelobe levels, which are difficult to see in Figure 3-16(a)'s linear scale. We will avoid this difficulty by plotting the windows' magnitude responses on a logarithmic decibel scale, and normalize each plot so its main lobe peak values are zero dB. (Appendix E provides a discussion of the origin and utility of measuring frequency-domain responses on a logarithmic scale using decibels.) Defining the log magnitude response to be $|W_{dB}(m)|$, we get $|W_{dB}(m)|$ by using the expression

$$|W_{dB}(m)| = 20 \cdot \log_{10}\left(\frac{|W(m)|}{|W(0)|}\right). \tag{3-31}$$

(The $|W(0)|$ term in the denominator of Eq. (3-31) is the value of $W(m)$ at the peak of the main lobe when $m = 0$.) The $|W_{dB}(m)|$ curves for the various window functions are shown in Figure 3-16(b). Now we can really see how the various window sidelobe responses compare to each other.

Looking at the rectangular window's magnitude response we see that its main lobe is the most narrow, $2f_s/N$. However, unfortunately, its first sidelobe level is only −13 dB below the main lobe peak, which is not so good. (Notice that we're only showing the positive frequency portion of the window responses in Figure 3-16.) The triangular window has reduced sidelobe levels, but the price we've paid is that the triangular window's main lobe width is twice as wide as that of the rectangular window's. The various nonrectangular windows' wide main lobes degrade the windowed DFT's frequency resolution by almost a factor of two. However, as we'll see, the important benefits of leakage reduction usually outweigh the loss in DFT frequency resolution.

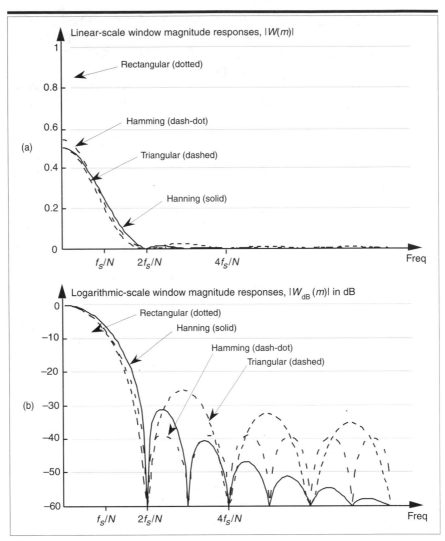

Figure 3-16 Window magnitude responses: (a) | W(m) | on a linear scale; (b) | $W_{dB}(m)$ | on a normalized logarithmic scale.

Notice the further reduction of the first sidelobe level, and the rapid sidelobe roll-off of the Hanning window. The Hamming window has even lower first sidelobe levels, but this window's sidelobes roll off slowly relative to the Hanning window. This means that leakage three or four bins away from the center bin is lower for the Hamming window than for the Hanning window, and leakage a dozen or so bins away from the center bin is lower for the Hanning window than for the Hamming window.

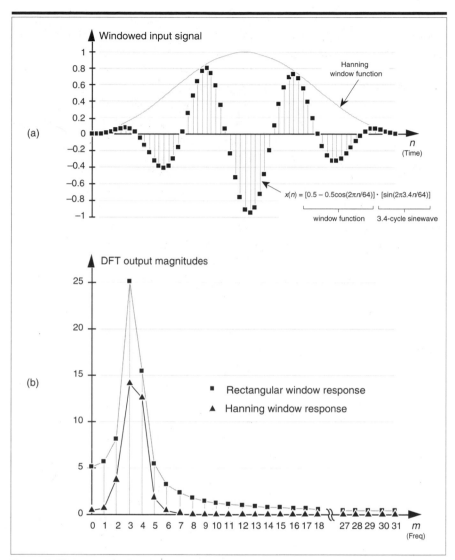

Figure 3-17 Hanning window: (a) 64-sample product of a Hanning window and a 3.4 cycles per sample interval input sinewave; (b) Hanning DFT output response vs. rectangular window DFT output response.

When we apply the Hanning window to Figure 3-8(a)'s 3.4 cycles per sample interval example, we end up with the DFT input shown in Figure 3-17(a) under the Hanning window envelope. The DFT outputs for the windowed waveform are shown in Figure 3-17(b) along with the DFT results with no windowing, i.e., the rectangular window. As we expected, the shape of the Hanning window's response looks broader and has a lower peak amplitude, but its sidelobe leakage is noticeably reduced from that of the rectangular window.

We can demonstrate the benefit of using a window function to help us detect a low-level signal in the presence of a nearby high-level signal. Let's add 64 samples of a 7 cycles per sample interval sinewave, with a peak amplitude of only 0.1, to Figure 3-8(a)'s unity-amplitude 3.4 cycles per sample sinewave. When we apply a Hanning window to the sum of these sinewaves, we get the time-domain input shown in Figure 3-18(a). Had we not windowed the input data, our DFT output would be the squares in Figure 3-18(b) where DFT leakage causes the input signal component at $m = 7$ to be barely discernible. However, the DFT of the windowed data shown as the triangles in Figure 3-18(b), makes it easier for us to detect the presence of the $m = 7$ signal component. From a practical standpoint, people who use the DFT to perform real-world signal detection have learned that their overall frequency resolution and signal sensitivity are affected much more by the size and shape of their window function than the mere size of their DFTs.

As we become more experienced using window functions on our DFT input data, we'll see how different window functions have their own individual advantages and disadvantages. Furthermore, regardless of the window function used, we've decreased the leakage in our DFT output from that of the rectangular window. There are many different window functions described in the literature of digital signal processing—so many, in fact, that they've been named after just about everyone in the digital signal processing business. It's not that clear that there's a great deal of difference among many of these window functions. What we find is that window selection is a trade-off between main lobe widening, first sidelobe levels, and how fast the sidelobes decrease with increased frequency. The use of any particular window depends on the application[5], and there are many applications.

Windows are used to improve DFT spectrum analysis accuracy[6], to design digital filters[7,8], to simulate antenna radiation patterns, and even in the hardware world to improve the performance of certain mechanical force to voltage conversion devices[9]. So there's plenty of window information available for those readers seeking further knowledge. (The Mother of all technical papers on windows is that by Harris[10]. A useful paper by Nuttall corrected and extended some portions of Harris's paper[11].) Again, the best way to appreciate windowing effects is to have access to a computer software package that contains DFT, or FFT, routines and start analyzing windowed signals. (By the way,

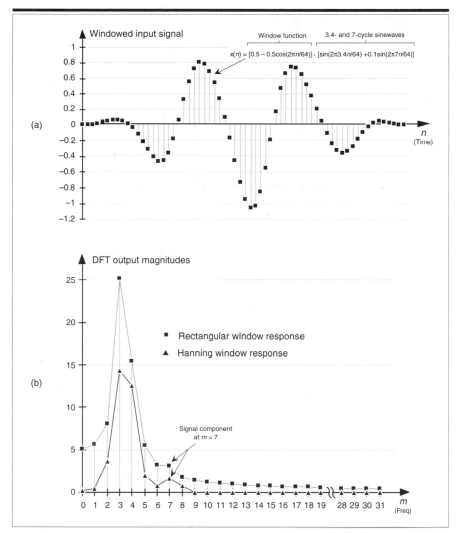

Figure 3-18 Increased signal detection sensitivity afforded using windowing: (a) 64-sample product of a Hanning window and the sum of a 3.4 cycles and a 7 cycles per sample interval sinewaves; (b) reduced leakage Hanning DFT output response vs. rectangular window DFT output response.

while we delayed their discussion until Section 5.3, there are two other commonly used window functions that can be used to reduce DFT leakage. They're the Chebyshev and Kaiser window functions, that have adjustable parameters, enabling us to strike a compromise between widening main lobe width and reducing sidelobe levels.)

3.10 DFT Scalloping Loss

Scalloping is the name used to describe fluctuations in the overall magnitude response of an N-point DFT. Although we derive this fact in Section 3.16, for now we'll just say that when no input windowing function is used, the $\sin(x)/x$ shape of the sinc function's magnitude response applies to each DFT output bin. Figure 3-19(a) shows a DFT's aggregate magnitude response by superimposing several $\sin(x)/x$ bin magnitude responses.[†] (Because the sinc function's sidelobes are not key to this discussion, we don't show them in Figure 3-19(a).) Notice from Figure 3-19(b) that the overall DFT frequency-domain response is indicated by the bold envelope curve. This rippled curve, also called the picket fence effect, illustrates the processing loss for input frequencies between the bin centers.

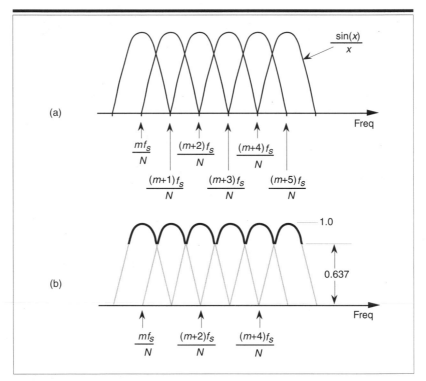

Figure 3-19 DFT bin magnitude response curves: (a) individual $\sin(x)/x$ responses for each DFT bin; (b) equivalent overall DFT magnitude response.

[†] Perhaps Figure 3-19(a) is why individual DFT outputs are called "bins." Any signal energy under a $\sin(x)/x$ curve will show up in the *enclosed storage compartment* of that DFT's output sample.

From Figure 3-19(b), we can determine that the magnitude of the DFT response fluctuates from 1.0, at bin center, to 0.637 halfway between bin centers. If we're interested in DFT output power levels, this envelope ripple exhibits a scalloping loss of almost −4 dB halfway between bin centers. Figure 3-19 illustrates a DFT output when no window (i.e., a rectangular window) is used. Because nonrectangular window functions broaden the DFT's main lobe, their use results in a scalloping loss that will not be as severe as the rectangular window[10,12]. That is, their wider main lobes overlap more and fill in the valleys of the envelope curve in Figure 3-19(b). For example, the scalloping loss of a Hanning window is approximately 0.82, or −1.45 dB, halfway between bin centers. Scalloping loss is not, however, a severe problem in practice. Real-world signals normally have bandwidths that span many frequency bins so that DFT magnitude response ripples can go almost unnoticed. Let's look at a scheme called *zero stuffing* that's used to both alleviate scalloping loss effects and to improve the DFT's frequency resolution.

3.11 DFT Resolution, Zero Stuffing, and Frequency-Domain Sampling

One popular method used to improve DFT frequency resolution is known as *zero stuffing*, or *zero padding*. This process involves the addition of zero-valued data samples to an original DFT input sequence to increase the total number of input data samples. Investigating this zero stuffing technique illustrates the DFT's important property of frequency-domain sampling alluded to in the discussion on leakage. When we sample a continuous time-domain function, having a continuous Fourier transform (CFT), and take the DFT of those samples, the DFT results in a frequency-domain sampled approximation of the CFT. The more points in our DFT, the better our DFT output approximates the CFT.

To illustrate this idea, suppose we want to approximate the CFT of the continuous $f(t)$ function in Figure 3-20(a). This $f(t)$ waveform extends to infinity in both directions but is nonzero only over the time interval of T seconds. If the nonzero portion of the time function is a sinewave of three cycles in T seconds, the magnitude of its CFT is shown in Figure 3-20(b). (Because the CFT is taken over an infinitely wide time interval, the CFT has infinitesimally small frequency resolution, resolution so fine-grained that it's continuous.) It's this CFT that we'll approximate with a DFT.

Suppose we want to use a 16-point DFT to approximate the CFT of $f(t)$ in Figure 3-20(a). The 16 discrete samples of $f(t)$, spanning the three periods of $f(t)$'s sinusoid, are those shown on the left side of Figure 3-21(a). Applying

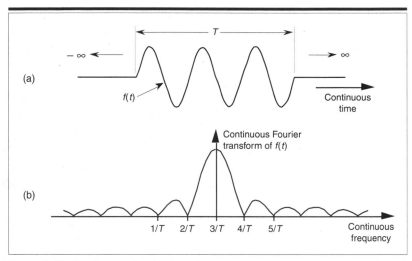

Figure 3-20 Continuous Fourier transform: (a) continuous time-domain $f(t)$ of a truncated sinusoid of frequency $3/T$; (b) continuous Fourier transform of $f(t)$.

those time samples to a 16-point DFT results in discrete frequency-domain samples, the positive frequency of which are represented by the dots on the right side of Figure 3-21(a). We can see that the DFT output samples Figure 3-20(b)'s CFT. If we append (or zero stuff) 16 zeros to the input sequence and take a 32-point DFT, we get the output shown on the right side of Figure 3-21(b), where we've increased our DFT frequency resolution by a factor of two. Our DFT is sampling the input function's CFT more often now. Adding 32 more zeros and taking a 64-point DFT, we get the output shown on the right side of Figure 3-21(c). The 64-point DFT output now begins to show the true shape of the CFT. Adding 64 more zeros and taking a 128-point DFT, we get the output shown on the right side of Figure 3-21(d). The DFT frequency-domain sampling characteristic is obvious now, but notice that the bin index for the center of the main lobe is different for each of the DFT outputs in Figure 3-21.

Does this mean we have to redefine the DFT's frequency axis when using the zero-stuffing technique? Not really. If we perform zero stuffing on L nonzero input samples to get a total of N time samples for an N-point DFT, the zero-stuffed DFT output bin center frequencies are related to the original f_s by our old friend Eq. (3-5), or

$$\text{center frequency of the } m\text{th bin} = \frac{mf_s}{N}. \qquad (3\text{-}32)$$

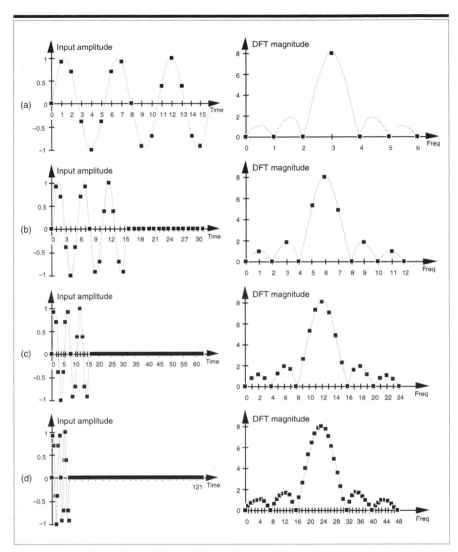

Figure 3-21 DFT frequency domain sampling: (a) 16 input data samples and
 $N = 16$; (b) 16 input data samples, 16 stuffed zeros, and $N = 32$; (c) 16
input data samples, 48 stuffed zeros, and $N = 64$; (d) 16 input data
samples, 112 stuffed zeros, and $N = 128$.

So in our Figure 3-21(a) example, we use Eq. (3-32) to show that,
although the zero-stuffed DFT output bin index of the main lobe
changes as N increases, the zero-stuffed DFT output frequency associated with the main lobe remains the same. The following list shows how
this works:

Figure No.	Main lobe peak located at m =	L =	N =	Frequency of main lobe peak relative to f_s =
Figure 3-21(a)	3	16	16	$3f_s/16$
Figure 3-21(b)	6	16	32	$6 \cdot f_s/32 = 3f_s/16$
Figure 3-21(c)	12	16	64	$12 \cdot f_s/64 = 3f_s/16$
Figure 3-21(d)	24	16	128	$24 \cdot f_s/128 = 3f_s/16$

Do we gain anything by appending more zeros to the input sequence and taking larger DFTs? Not really, because our 128-point DFT is sampling the input's CFT sufficiently now in Figure 3-21(d). Sampling it more often with a larger DFT won't improve our understanding of the input's frequency content. The issue here is that adding zeros to an input sequence will improve our DFT's output resolution, but there's a practical limit on how much we gain by adding more zeros. For our example here, a 128-point DFT shows us the detailed content of the input spectrum. We've hit a *law of diminishing returns* here. Performing a 256-point or 512-point DFT, in our case, would serve little purpose.[†] There's no reason to *oversample* this particular input sequence's CFT. Of course, there's nothing sacred about stopping at a 128-point DFT. Depending on the number of samples in some arbitrary input sequence and the sample rate, we might, in practice, need to append any number of zeros to get some desired DFT frequency resolution.

There are two final points to be made concerning zero stuffing. First, the DFT magnitude expressions in Eqs. (3-17) and (3-17') don't apply if zero stuffing is being used. If we perform zero stuffing on L nonzero samples of a sinusoid whose frequency is located at a bin center to get a total of N input samples for an N-point DFT, we must replace the N with L in Eqs. (3-17) and (3-17') to predict the DFT's output magnitude for that particular sinewave. Second, in practical situations, if we want to perform both zero stuffing and windowing on a sequence of input data samples, we must be careful not to apply the window to the entire input including the appended zero-valued samples. The window function must be applied only to the original nonzero time samples, otherwise the stuffed zeros will *zero out* and distort part of the window function,

---

[†] Notice that the DFT sizes (N) we've discussed are powers of 2 (64, 128, 256, 512). That's because we actually perform DFTs using a special algorithm known as the fast Fourier transform (FFT). As we'll see in Chapter 4, the typical implementation of the FFT requires that N be a power of 2.

leading to erroneous results. (Section 4.5 gives additional practical pointers on performing the DFT using the FFT algorithm to analyze real-world signals.)

To digress slightly, now's a good time to define the term *discrete-time Fourier transform* (DTFT) that the reader may encounter in the literature. The DTFT is the continuous Fourier transform of an L-point discrete time-domain sequence; some authors use the DTFT to describe many of the digital signal processing concepts we've covered in this chapter. We can't perform the DTFT on a computer because it has an infinitely fine frequency resolution—but we could, if we wished, approximate the DTFT by performing an N-point DFT on an L-point discrete time sequence where $N > L$. That is, in fact, what we did in Figure 3-21 when we zero-stuffed the original 16-point time sequence. While we don't emphasize the DTFT in this book, keep in mind that when $N = L$ the DTFT approximation is identical to the DFT.

3.12 DFT Processing Gain

There are two types of processing gain associated with DFTs. People who use the DFT to detect signal energy embedded in noise often speak of the DFT's *processing gain* because the DFT can *pull* signals out of background noise. This is due to the inherent correlation gain that takes place in any N-point DFT. Beyond this natural processing gain, additional *integration gain* is possible when multiple DFT outputs are averaged. Let's look at the DFT's inherent processing gain first.

3.2.1 Processing Gain of a Single DFT

The concept of the DFT having processing gain is straightforward if we think of a particular DFT bin output as the output of a narrowband filter. Because a DFT output bin has the amplitude response of the $\sin(x)/x$ function, that bin's output is primarily due to input energy residing under, or very near, the bin's main lobe. It's valid to think of a DFT bin as a kind of *bandpass* filter whose band center is located at mf_s/N. We know from Eq. (3-17) that the maximum possible DFT output magnitude increases as the number of points (N) in a DFT increases. Also, as N increases, the DFT output bin main lobes become more narrow. So a DFT output bin can be treated as a bandpass filter whose gain can be increased and whose bandwidth can be reduced by increasing the value of N. Decreasing a bandpass filter's bandwidth is

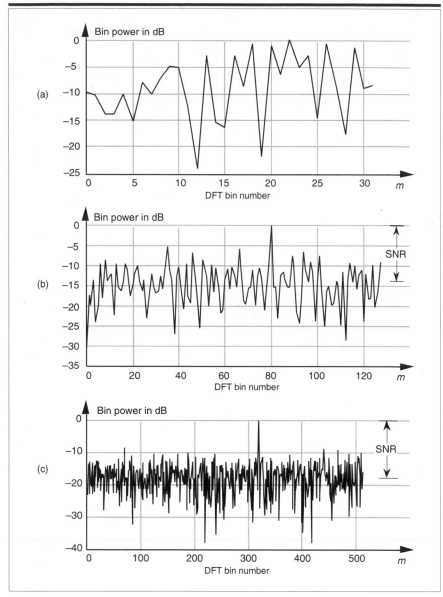

Figure 3-22 Single DFT processing gain: (a) $N = 64$; (b) $N = 256$; (c) $N = 1024$.

useful in energy detection because the frequency resolution improves in addition to the filter's ability to minimize the amount of background noise that resides within its passband. We can demonstrate this by looking at the DFT of a spectral tone (a constant frequency sinewave) added to random noise. Figure 3-22(a) is a logarithmic plot showing the first 32 outputs of a 64-point DFT when the input tone is at the center of the DFT's $m = 20$th bin. The output power levels (DFT magnitude squared) in Figure 3-22(a) are normalized so that the highest bin output power is set to 0 dB. Because the tone's original signal power is below the average noise power level, the tone is a bit difficult to detect when $N = 64$. (The time-domain noise, used to generate Figure 3-22(a), has an average value of zero, i.e., no DC bias or amplitude offset.) If we quadruple the number of input samples and increase the DFT size to $N = 256$, we can now see the tone power raised above the average background noise power as shown for $m = 80$ in Figure 3-22(b). Increasing the DFT's size to $N = 1024$ provides additional processing gain to pull the tone further up out of the noise as shown in Figure 3-22(c).

To quantify the idea of DFT processing gain, we can define a signal-to-noise ratio (SNR) as the DFT's *output signal-power level* over the *average output noise-power level*. (In practice, of course, we like to have this ratio as large as possible.) For several reasons, it's hard to say what any given single DFT output SNR will be. That's because we can't exactly predict the energy in any given N samples of random noise. Also, if the input signal frequency is not at bin center, leakage will raise the effective background noise and reduce the DFT's output SNR. In addition, any window being used will have some effect on the leakage and, thus, on the output SNR. What we'll see is that the DFT's output SNR increases as N gets larger because a DFT bin's output noise standard deviation (*rms*) value is proportional to $\sqrt{N}$, and the DFT's output magnitude for the bin containing the signal tone is proportional to N. More generally for real inputs, if $N > N'$, an N-point DFT's output SNR_N increases over the N'-point DFT $SNR_{N'}$ by the following relationship:

$$SNR_N = SNR_{N'} + 20 \cdot \log_{10}\left(\frac{N/N'}{\sqrt{N/N'}}\right). \qquad (3\text{-}33)$$

If we increase a DFT's size from N' to $N = 2N'$, from Eq. (3-33), the DFT's output SNR increases by 3 dB. So we say that a DFT's processing gain increases by 3 dB whenever N is doubled. Be aware that we may double a DFT's size and get a resultant processing gain of less than 3 dB in the presence of random noise; then again, we may gain slightly more than 3 dB. That's the nature of random noise. If we perform many DFTs, we'll see an average processing gain, shown in Figure 3-23(a), for various input signal SNRs. Because we're interested in the slope of the curves in Figure

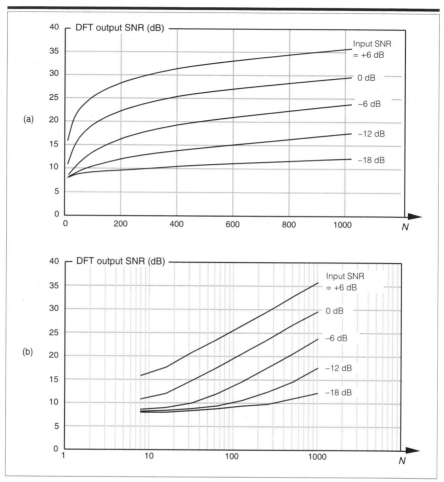

Figure 3-23 DFT processing gain vs. number of DFT points N for various input signal-to-noise ratios: (a) linear N axis; (b) logarithmic N axis.

3-23(a), we plot those curves on a logarithmic scale for N in Figure 3-23(b) where the curves straighten out and become linear. Looking at the slope of the curves in Figure 3-23(b), we can now see the 3 dB increase in processing gain as N doubles so long as N is greater than 20 or 30 and the signal is not overwhelmed by noise. There's nothing sacred about the absolute values of the curves in Figures 3-23(a) and 3-23(b). They were generated through a simulation of noise and a tone whose frequency was at a DFT bin center. Had the tone's frequency been between bin centers, the processing gain curves would have been shifted downward, but their shapes would still be the same,[†] that is, Eq. (3-33) is still valid regardless of the input tone's frequency.

3.12.2 Integration Gain Due to Averaging Multiple DFTs

Theoretically, we could get very large DFT processing gains by increasing the DFT size arbitrarily. The problem is that the number of necessary DFT multiplications increases proportionally to N^2, and larger DFTs become very computationally intensive. Because addition is easier and faster to perform than multiplication, we can average the outputs of multiple DFTs to obtain further processing gain and signal detection sensitivity. The subject of averaging multiple DFT outputs is covered in Section 8.3.

3.13 The DFT of Rectangular Functions

We conclude this chapter by providing the mathematical details of two important aspects of the DFT. First, we obtain the expressions for the DFT of a rectangular function (rectangular window), and then we'll use these results to illustrate the magnitude response of the DFT. We're interested in the DFT's magnitude response because it provides an alternate viewpoint to understand the leakage that occurs when we use the DFT as a signal analysis tool.

One of the most prevalent and important computations encountered in digital signal processing is the DFT of a rectangular function. We see it in sampling theory, window functions, discussions of convolution, spectral analysis, and in the design of digital filters. As common as it is, however, the literature covering the DFT of rectangular functions can be confusing for several reasons for the digital signal processing beginner. The standard mathematical notation is a bit hard to follow at first, and sometimes the equations are presented with too little explanation. Compounding the problem, for the beginner, are the various expressions of this particular DFT. In the literature, we're likely to find any one of the following forms for the DFT of a rectangular function:

$$\text{DFT}_{\text{rect.function}} = \frac{\sin(x)}{\sin(x/N)}, \text{ or } \frac{\sin(x)}{x}, \text{ or } \frac{\sin(Nx/2)}{\sin(x/2)}. \qquad (3\text{-}34)$$

[†] The curves would be shifted downward, indicating a lower SNR, because leakage would raise the average noise-power level, and scalloping loss would reduce the DFT bin's output power level.

In this section we'll show how the forms in Eq. (3-34) were obtained, see how they're related, and create a kind of *Rosetta stone* table allowing us to move back and forth between the various DFT expressions. Take a deep breath and let's begin our discussion with the definition of a rectangular function.

3.13.1 DFT of a General Rectangular Function

A general rectangular function $x(n)$ can be defined as N samples containing K unity-valued samples as shown in Figure 3-24. The full N-point sequence, $x(n)$, is the rectangular function that we want to transform. We call this the general form of a rectangular function because the K unity samples begin at a arbitrary index value of $-n_o$. Let's take the DFT of $x(n)$ in Figure 3-24 to get our desired $X(m)$. Using m as our frequency-domain sample index, the expression for an N-point DFT is

$$X(m) = \sum_{n=-(N/2)+1}^{N/2} x(n)e^{-j2\pi nm/N}. \tag{3-35}$$

With $x(n)$ being nonzero only over the range of $-n_o \leq n \leq -n_o+(K-1)$, we can modify the summation limits of Eq. (3-35) to express $X(m)$ as

$$X(m) = \sum_{n=-n_o}^{-n_o+(K-1)} 1 \cdot e^{-j2\pi nm/N}, \tag{3-36}$$

because only the K samples contribute to $X(m)$. That last step is important because it allows us to eliminate the $x(n)$ terms and make Eq. (3-36) easier to handle. To keep the following equations from being too messy, let's use the dummy variable $q = 2\pi m/N$.

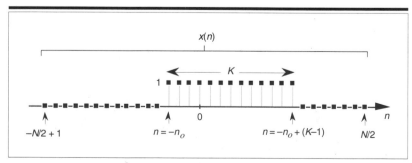

Figure 3-24 Rectangular function of width K samples defined over N samples where $K < N$.

OK, here's where the algebra comes in. Over our new limits of summation, we eliminate the factor of 1 and Eq. (3-36) becomes

$$X(q) = \sum_{n=-n_o}^{-n_o+(K-1)} e^{-jqn}$$

$$= e^{-jq(-n_o)} + e^{-jq(-n_o+1)} + e^{-jq(-n_o+2)} + ... + e^{-jq(-n_o+(K-1))}$$

$$= e^{-jq(-n_o)}e^{-j0q} + e^{-jq(-n_o)}e^{-j1q} + e^{-jq(-n_o)}e^{-j2q} + ... + e^{-jq(-n_o)}e^{-jq(K-1)}$$

$$= e^{jq(n_o)} \cdot \left[e^{-j0q} + e^{-j1q} + e^{-j2q} + ... + e^{jq(K-1)} \right] . \tag{3-37}$$

The series inside the brackets of Eq. (3-37), allows the use of a summation, such as

$$X(q) = e^{jq(n_o)} \sum_{p=0}^{K-1} e^{-jpq} . \tag{3-38}$$

Equation (3-38) certainly doesn't look any simpler than Eq. (3-36), but it is. Equation (3-38) is a *geometric series* and, from the discussion in Appendix B, it can be evaluated to the closed form of

$$\sum_{p=0}^{K-1} e^{-jpq} = \frac{1-e^{-jqK}}{1-e^{-jq}} . \tag{3-39}$$

We can now simplify Eq. (3-39)—here's the clever step. If we multiply and divide the numerator and denominator of Eq. (3-39)'s right-hand side by the appropriate half-angled exponentials, we break the exponentials into two parts and get

$$\sum_{p=0}^{K-1} e^{-jpq} = \frac{e^{-jqK/2}(e^{jqK/2} - e^{-jqK/2})}{e^{-jq/2}(e^{jq/2} - e^{-jq/2})}$$

$$= e^{-jq(K-1)/2} \cdot \frac{(e^{jqK/2} - e^{-jqK/2})}{(e^{jq/2} - e^{-jq/2})} . \tag{3-40}$$

Let's pause for a moment here to remind ourselves where we're going. We're trying to get Eq. (3-40) into a usable form because it's part of Eq. (3-38) that we're using to evaluate $X(m)$ in Eq. (3-36) in our quest for an understandable expression for the DFT of a rectangular function.

Equation (3-40) looks even more complicated than Eq. (3-39), but things can be simplified inside the parentheses. From Euler's equation: $\sin(\emptyset) = (e^{j\emptyset} - e^{-j\emptyset})/2j$, Eq. (3-40) becomes

$$\sum_{p=0}^{K-1} e^{-jpq} = e^{-jq(K-1)/2} \cdot \frac{2j \sin(qK/2)}{2j \sin(q/2)}$$

$$= e^{-jq(K-1)/2} \cdot \frac{\sin(qK/2)}{\sin(q/2)}. \tag{3-41}$$

Substituting Eq. (3-41) for the summation in Eq. (3-38), our expression for $X(q)$ becomes

$$X(q) = e^{jq(n_o)} \cdot e^{-jq(K-1)/2} \cdot \frac{\sin(qK/2)}{\sin(q/2)}$$

$$= e^{jq(n_o-(K-1)/2)} \cdot \frac{\sin(qK/2)}{\sin(q/2)}. \tag{3-42}$$

Returning our dummy variable q to its original value of $2\pi m/N$,

$$X(m) = e^{j(2\pi m/N)(n_o-(K-1)/2)} \cdot \frac{\sin(2\pi mK/2N)}{\sin(2\pi m/2N)}, \text{ or}$$

General form of the
Dirichlet kernel: → $\quad X(m) = e^{j(2\pi m/N)(n_o-(K-1)/2)} \cdot \frac{\sin(\pi mK/N)}{\sin(\pi m/N)}. \tag{3-43}$

So there it is (Whew!). Equation (3-43) is the general expression for the DFT of the rectangular function as shown in Figure 3-24. Our $X(m)$ is a complex expression (pun intended) where a ratio of sine terms is the amplitude of $X(m)$ and the exponential term is the phase angle of $X(m)$.[†] The ratio of sines factor in Eq. (3-43) lies on the periodic curve shown in Figure 3-25(a), and like all N-point DFT representations, the periodicity of $X(m)$ is N. This curve is known as the *Dirichlet kernel* (or the *aliased sinc function*) and has been thoroughly described in the literature[10,13,14]. (It's named after the nineteenth-century German mathematician Peter Dirichlet (pronounced dee-ree-'klay), who studied the convergence of trigonometric series used to represent arbitrary functions.)

We can zoom in on the curve at the $m = 0$ point and see more detail in Figure 3-25(b). The dots are shown in Figure 3-25(b) to remind us that the DFT of our rectangular function results in discrete amplitude values that lie on the curve. So when we perform DFTs, our discrete results are sampled values of the continuous sinc function's curve in Figure 3-25(a). As we'll show later, we're primarily interested in the absolute value, or magnitude, of the Dirichlet kernel in Eq. (3-43). That magnitude $|X(m)|$ is shown in Figure 3-25(c). Although we first saw the sinc function's curve in Figure 3-9 in Section 3.8, where we introduced the topic of DFT leakage, we'll encounter this curve often in our study of digital signal processing.

For now, there are just a few things we need to keep in mind concerning the Dirichlet kernel. First, the DFT of a rectangular function has a main lobe, centered about the $m = 0$ point. The peak amplitude of the main lobe is K. This peak value makes sense, right? The $m = 0$ sample of a DFT $X(0)$ is the sum of the original samples, and the sum of K unity-valued samples is K. We can show this in a more substantial way by evaluating Eq. (3-43) for $m = 0$. A difficulty arises when we plug $m = 0$ into Eq. (3-43) because we end up with $\sin(0)/\sin(0)$, which is the indeterminate ratio $0/0$. Well, hardcore mathematics to the rescue here. We can use L'Hospital's rule to take the derivative of the numerator and the denominator of Eq. (3-43), and *then* set $m = 0$ to determine the peak value of the magnitude of the Dirichlet kernel.[††] We proceed as

---

[†] N was an even number in Figure 3-24 depicting the $x(n)$. Had N been an odd number, the limits on the summation in Eq. (3-35) would have been $-(N-1)/2 \le n \le (N-1)/2$. Using these alternate limits would have led us to exactly the same $X(m)$ as in Eq. (3-43).

[††] L'Hospital is pronounced 'lō-pē-tòl, like baby doll.

$$|X(m)|_{m\to 0} = \frac{d}{dm}X(m) = \frac{d[\sin(\pi mK/N)]/dm}{d[\sin(\pi m/N)]/dm}$$

$$= \frac{\cos(\pi mK/N)}{\cos(\pi m/N)} \cdot \frac{d(\pi mK/N)/dm}{d(\pi m/N)/dm}$$

$$= \frac{\cos(0)}{\cos(0)} \cdot \frac{\pi K/N}{\pi/N} = 1 \cdot K = K \ , \qquad (3\text{-}44)$$

which is what we set out to show. (We could have been clever and evaluated Eq. (3-35) with $m = 0$ to get the result of Eq. (3-44). Try it, and keep in mind that $e^{j0} = 1$.) Had the amplitudes of the nonzero samples of $x(n)$ been

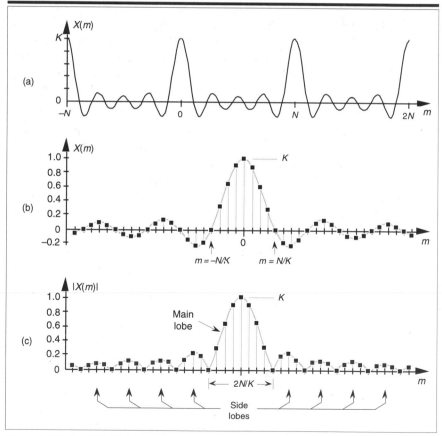

Figure 3-25 The Dirichlet kernel of $X(m)$: (a) periodic continuous curve on which the $X(m)$ samples lie; (b) $X(m)$ amplitudes about the $m = 0$ sample; (c) $|X(m)|$ magnitudes about the $m = 0$ sample.

different than unity, say some amplitude A_o, then, of course, the peak value of the Dirichlet kernel would be A_oK instead of just K. The next important thing to notice about the Dirichlet kernel is the main lobe's width. The first zero crossing of Eq. (3-43) occurs when the numerator's argument is equal to π. That is, when $\pi mK/N = \pi$. So the value of m at the first zero crossing is given by

$$m_{\text{first zero crossing}} = \frac{\pi N}{\pi K} = \frac{N}{K} \qquad (3\text{-}45)$$

as shown in Figure 3-25(b). Thus the main lobe width $2N/K$, as shown in Figure 3-25(c), is inversely proportional to K.[†]

Notice that the main lobe in Figure 3-25(a) is surrounded by a series of oscillations, called *sidelobes,* as in Figure 3-25(c). These sidelobe magnitudes decrease the farther they're separated from the main lobe. However, no matter how far we look away from the main lobe, these sidelobes never reach zero magnitude—and they cause a great deal of heartache for practitioners in digital signal processing. These sidelobes cause high-amplitude signals to overwhelm and hide neighboring low-amplitude signals in spectral analysis, and they complicate the design of digital filters. As we'll see in Chapter 5, the unwanted ripple in the passband and the poor stopband attenuation in simple digital filters are caused by the rectangular function's DFT sidelobes. (The development, analysis, and application of window functions came about to minimize the ill effects of those sidelobes in Figure 3-25.)

Let's demonstrate the relationship in Eq. (3-45) by way of a simple but concrete example. Assume that we're taking a 64-point DFT of the 64-sample rectangular function, with eleven unity values, shown in Figure 3-26(a). In this example, $N = 64$ and $K = 11$. Taking the 64-point DFT of the sequence in Figure 3-26(a) results in an $X(m)$ whose real and imaginary parts, $X_{\text{real}}(m)$ and $X_{\text{imag}}(m)$, are shown in Figure 3-26(b) and Figure 3-26(c) respectively. Figure 3-26(b) is a good illustration of how the real part of the DFT of a real input sequence has even symmetry, and Figure 3-26(c) confirms that the imaginary part of the DFT of a real input sequence has odd symmetry. (These symmetry properties were introduced in Section 3.2 and are discussed further in Appendix C, Section C.3.)

---

† This is a fundamental characteristic of Fourier transforms. The narrower the function in one domain, the wider its transform will be in the other domain.

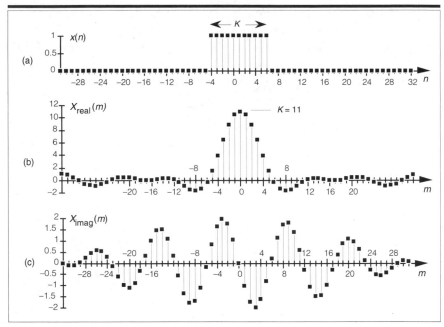

Figure 3-26 DFT of a rectangular function: (a) original function $x(n)$; (b) real part of the DFT of $x(n)$, $X_{real}(m)$; (c) imaginary part of the DFT of $x(n)$, $X_{imag}(m)$.

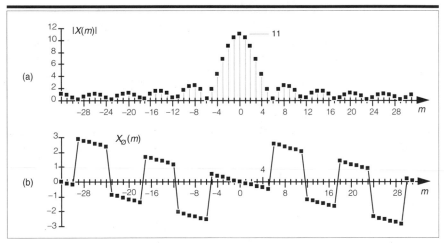

Figure 3-27 DFT of a generalized rectangular function: (a) magnitude $|X(m)|$; (b) phase angle in radians.

Although $X_{real}(m)$ and $X_{imag}(m)$ tell us everything there is to know about the DFT of $x(n)$, it's a bit easier to comprehend the true spectral nature of $X(m)$ by viewing its absolute magnitude. This magnitude, from Eq. (3-7), is provided in Figure 3-27(a) where the main and sidelobes are clearly evident now. As we expected, the peak value of the main lobe is 11 because we had $K = 11$ samples in $x(n)$. The width of the main lobe from Eq. (3-45) is 64/11, or 5.82. Thus, the first positive frequency zero-crossing location lies just below the $m = 6$ sample of our discrete $|X(m)|$ represented by the squares in Figure 3-27(a). The phase angles associated with $|X(m)|$, first introduced in Equations (3-6) and (3-8), are shown in Figure 3-27(b).

To understand the nature of the DFT of rectangular functions more fully, let's discuss a few more examples using less general rectangular functions that are more common in digital signal processing than the $x(n)$ in Figure 3-24.

3.13.2 DFT of a Symmetrical Rectangular Function

Equation (3-43) is a bit complicated because our original function $x(n)$ was so general. In practice, special cases of rectangular functions lead to simpler versions of Eq. (3-43). Consider the symmetrical $x(n)$ rectangular function in Figure 3-28. As shown in Figure 3-28, we often need to determine the DFT of a rectangular function that's centered about the $n = 0$ index point. In this case, the K unity-valued samples begin at $n = -n_o = -(K-1)/2$. So substituting $(K-1)/2$ for n_o in Eq. (3-43), yields

$$X(m) = e^{j(2\pi m/N)((K-1)/2-(K-1)/2)} \cdot \frac{\sin(\pi mK/N)}{\sin(\pi m/N)}$$

$$= e^{j(2\pi m/N)(0)} \cdot \frac{\sin(\pi mK/N)}{\sin(\pi m/N)} . \tag{3-46}$$

Because $e^{j0} = 1$, Eq. (3-46) becomes

Symmetrical form of the Dirichlet kernel: $\rightarrow$

$$X(m) = \frac{\sin(\pi mK/N)}{\sin(\pi m/N)} . \tag{3-47}$$

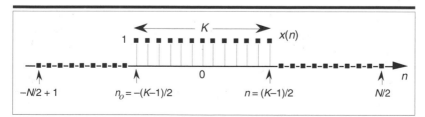

Figure 3-28 Rectangular $x(n)$ with K samples centered about $n = 0$.

Equation (3-47) indicates that the DFT of the symmetrical rectangular function in Figure 3-28 is itself a real function; that is, there's no complex exponential in Eq. (3-47), so this particular DFT contains no imaginary part or phase term. As we stated in Section 3.2, if $x(n)$ is real and even, $x(n) = x(-n)$, then $X_{real}(m)$ is nonzero and $X_{imag}(m)$ is always zero. We demonstrate this by taking the 64-point DFT of the sequence in Figure 3-29(a). Our $x(n)$ is 11 unity-valued samples centered about the $n = 0$ index. Here the DFT results in an $X(m)$ whose real and imaginary parts are shown in Figure 3-29(b) and Figure 3-29(c), respectively. As Eq. (3-47) predicted,

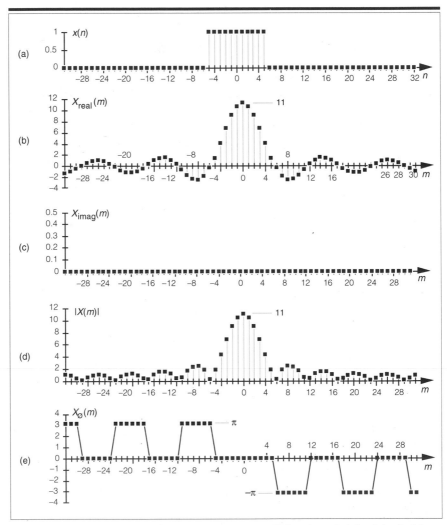

Figure 3-29 DFT of a rectangular function centered about $n = 0$: (a) original $x(n)$; (b) $X_{real}(m)$; (c) $X_{imag}(m)$; (d) magnitude of $X(m)$; (e) phase angle of $X(m)$ in radians.

$X_{real}(m)$ is nonzero and $X_{imag}(m)$ is zero. The magnitude and phase of $X(m)$ are depicted in Figure 3-29(d) and Figure 3-29(e).

Notice that the magnitudes in Figure 3-27(a) and Figure 3-29(d) are identical. This verifies the very important shifting theorem of the DFT; that is, the magnitude $|X(m)|$ depends only on the number of nonzero samples in $x(n)$, K, and *not* on their position relative to the $n = 0$ index value. Shifting the K unity-valued samples to center them about the $n = 0$ index merely affects the phase angle of $X(m)$, not its magnitude.

Speaking of phase angles, it's interesting to realize here that even though $X_{imag}(m)$ is zero in Figure 3-29(c), the phase angle of $X(m)$ is not always zero. In this case, $X(m)$'s individual phase angles in Figure 3-29(e) are either $+\pi$, zero, or $-\pi$ radians. With $+\pi$ and $-\pi$ radians both being equal to -1, we could easily reconstruct $X_{real}(m)$ from $|X(m)|$ and the phase angle $X_{\phi}(m)$ if we must. $X_{real}(m)$ is equal to $|X(m)|$ with the signs of $|X(m)|$'s alternate sidelobes reversed.[†] To gain some further appreciation of how the DFT of a rectangular function is a sampled version of the Dirichlet kernel, let's increase the number of our nonzero $x(n)$ samples. Figure 3-30(a) shows a 64-point $x(n)$ where 31 unity-valued samples are centered about the $n = 0$ index location. The magnitude of $X(m)$ is provided in Figure 3-30(b). By broadening the $x(n)$ function, i.e., increasing K, we've narrowed the Dirichlet kernel of $X(m)$. This follows from Eq. (3-45), right? The kernel's first zero crossing is inversely proportional to K, so, as we extend the width of K, we squeeze $|X(m)|$ in toward $m = 0$. In this

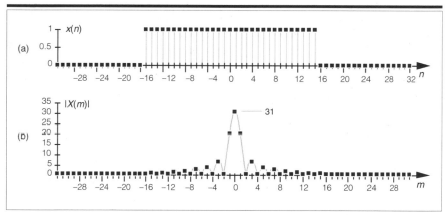

Figure 3-30 DFT of a symmetrical rectangular function with 31 unity values: (a) original $x(n)$; (b) magnitude of $X(m)$.

[†] The particular pattern of $+\pi$ and $-\pi$ values in Figure 3-29(e) is an artifact of the software used to generate that figure. A different software package may show a different pattern, but as long as the nonzero phase samples are either $+\pi$ or $-\pi$, the phase results will be correct.

example, $N = 64$ and $K = 31$. From Eq. (3-45) the first positive zero crossing of $X(m)$ occurs at $64/31$, or just slightly to the right of the $m = 2$ sample in Figure 3-30(b). Also notice that the peak value of $|X(m)| = K = 31$, as mandated by Eq. (3-44).

3.13.3 DFT of an All Ones Rectangular Function

The DFT of a special form of $x(n)$ is routinely called for, leading to yet another simplified form of Eq. (3-43). In the literature, we often encounter a rectangular function where $K = N$; that is, all N samples of $x(n)$ are nonzero, as shown in Figure 3-31. In this case, the N unity-valued samples begin at $n = -n_o = -(N-1)/2$. We obtain the expression for the DFT of the function in Figure 3-31 by substituting $K = N$ and $n_o = (N-1)/2$ in Eq. (3-43) to get

$$X(m) = e^{j(2\pi m / N)[(N-1)/2 - (N-1)/2]} \cdot \frac{\sin(\pi m N / N)}{\sin(\pi m / N)}$$

$$= e^{j(2\pi m / N)(0)} \cdot \frac{\sin(\pi m)}{\sin(\pi m / N)}, \text{ or}$$

All Ones form of the Dirichlet kernel (Type 1): → $X(m) = \dfrac{\sin(\pi m)}{\sin(\pi m / N)}.$ (3-48)

Equation (3-48) takes the first form of Eq. (3-34) that we alluded to at the beginning of this Section 3.13.[†] Figure 3-32 demonstrates the meaning of Eq. (3-48). The DFT magnitude of the all ones function, $x(n)$ in

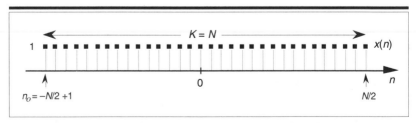

Figure 3-31 Rectangular function with N unity-valued samples.

[†] By the way, there's nothing *official* about calling Eq. (3-48) a Type 1 Dirichlet kernel. We're using the phrase *Type* 1 merely to distinguish Eq. (3-48) from other mathematical expressions for the Dirichlet kernel that we're about to encounter.

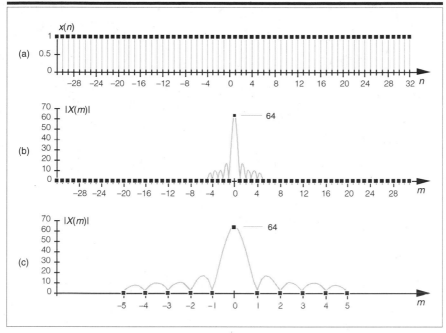

Figure 3-32 All ones function: (a) rectangular function with $N = 64$ unity-valued samples; (b) DFT magnitude of the all ones time function; (c) close-up view of the DFT magnitude of an all ones time function.

Figure 3-32(a), is shown in Figure 3-32(b) and Figure 3-32(c). Take note that if m was continuous Eq. (3-48) describes the shaded curves in Figure 3-32(b) and Figure 3-32(c). If m is restricted to be integers, then Eq. (3-48) represents the dots in those figures.

The Dirichlet kernel of $X(m)$ in Figure 3-32(b) is now as narrow as it can get. The main lobe's first positive zero crossing occurs at the $m = 64/64 = 1$ sample in Figure 3-32(b) and the peak value of $|X(m)| = N = 64$. With $x(n)$ being all ones, $|X(m)|$ is zero for all $m \neq 0$. The sinc function in Eq. (3-48) is of utmost importance—as we'll see at the end of this chapter, it defines the overall DFT frequency response to an input sinusoidal sequence, and it's also the amplitude response of a single DFT bin.

The form of Eq. (3-48) allows us to go one step further to identify the most common expression for the DFT of an all ones rectangular function found in the literature. To do this, we have to use an approximation principle found in the mathematics of trigonometry that you may have heard before. It states that when α is small, then $\sin(\alpha)$ is approximately equal to α, i.e., $\sin(\alpha) \approx \alpha$. This idea comes about when we consider a pie-shaped

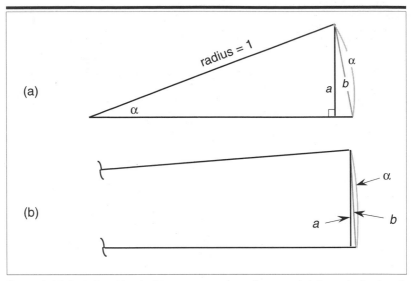

Figure 3-33 Relationships between an angle a, line *a* = sin(a), and a's chord *b*: (a) large angle a; (b) small angle a.

section of a circle whose radius is 1 as shown in Figure 3-33(a). That section is defined by the length of the arc α measured in radians and α's chord b. If we draw a right triangle inside the section, we can say that $a = \sin(\alpha)$. As α gets smaller the long sides of our triangle become almost parallel, the length of chord b approaches the length of arc α, and the length of line a approaches the length of b. So, as depicted in Figure 3-33(b), when α is small, $\alpha \approx b \approx a = \sin(\alpha)$. We use this $\sin(\alpha) \approx \alpha$ approximation when we look at the denominator of Eq. (3-48). When $\pi m/N$ is small, then $\sin(\pi m/N)$ is approximately equal to $\pi m/N$. So we can, when N is large, state

All Ones form of the
Dirichlet kernel (Type 2): →
$$X(m) \approx \frac{\sin(\pi m)}{\pi m / N} = N \cdot \frac{\sin(\pi m)}{\pi m}. \qquad (3\text{-}49)$$

It has been shown that when N is larger than, say, 10 in Eq. (3-48), Eq. (3-49) accurately describes the DFT's output.[†] Equation (3-49) is often normalized by dividing it by N, so we can express the normalized DFT of an all ones rectangular function as

---

[†] We can be comfortable with this result because, if we let $K = N$, we'll see that the peak value of $X(m)$ in Eq. (3-49), for $m = 0$, is equal to N, which agrees with Eq. (3-44).

All Ones form of the
Dirichlet kernel (Type 3): →

$$X(m) \approx \frac{\sin(\pi m)}{\pi m}. \tag{3-50}$$

Equation (3-50), taking the second form of Eq. (3-34) that is so often seen in the literature, also has the DFT magnitude shown in Figure 3-32(b) and Figure 3-32(c).

3.13.4 Time and Frequency Axes Associated with Rectangular Functions

Let's establish the physical dimensions associated with the n and m index values. So far in our discussion, the n index was merely an integer enabling us to keep track of individual $x(n)$ sample values. If the n index represents instants in time, we can identify the time period separating adjacent $x(n)$ samples to establish the time scale for the $x(n)$ axis and the frequency scale for the $X(m)$ axis. Consider the time-domain rectangular function given in Figure 3-34(a). That function comprises N time samples obtained t_s seconds apart, and the full sample interval is Nt_s seconds. Each

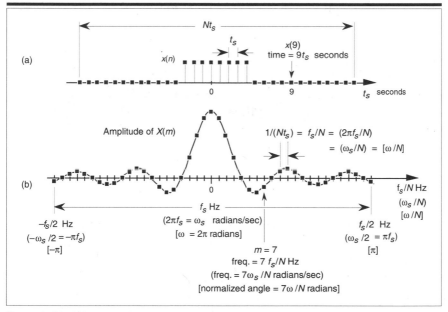

Figure 3-34 DFT time and frequency axis dimensions: (a) time-domain axis uses time index n; (b) various representations of the DFT's frequency axis.

Table 3-1 Characteristics of Various DFT Frequency Axis Representations

DFT Frequency Axis Representation	X(m) Frequency Variable	Resolution of X(m)	Repetition Interval of X(m)	Frequency Axis Range
Frequency in Hz	mf_s/N	f_s/N	f_s	$-f_s/2$ to $f_s/2$
Frequency in radians	$m\omega_s/N$ or $2\pi m f_s/N$	ω_s/N or $2\pi f_s/N$	ω_s or $2\pi f_s$	$-\omega_s/2$ to $\omega_s/2$ or $-\pi f_s$ to πf_s
Normalized angle in radians	$2\pi m/N$	$2\pi/N$	2π	$-\pi$ to π

$x(n)$ sample occurs at nt_s seconds for some value of n. For example, the $n = 9$ sample value, $x(9) = 0$, occurs at $9t_s$ seconds.

The frequency axis of $X(m)$ can be represented in a number of different ways. Three popular types of DFT frequency axis labeling are shown in Figure 3-34(b) and listed in Table 3-1. Let's consider each representation individually.

3.13.4.1 DFT Frequency Axis in Hz

If we decide to relate $X(m)$ to the time sample period t_s, or the sample rate $f_s = 1/t_s$, then the frequency axis variable is $m/Nt_s = mf_s/N$. So each $X(m)$ sample is associated with a cyclic frequency of mf_s/N Hz. In this case, the resolution of $X(m)$ is f_s/N. The DFT repetition period, or periodicity, is f_s Hz, as shown in Figure 3-34(b). If we substitute the cyclic frequency variable mf_s/N for the generic variable of m/N in Eq. (3-47), we obtain an expression for the DFT of a symmetrical rectangular function, where $K < N$, in terms of the sample rate f_s in Hz. That expression is

$$X(mf_s) = \frac{\sin(\pi m f_s K/N)}{\sin(\pi m f_s/N)}. \tag{3-51}$$

For an all ones rectangular function where $K = N$, the amplitude normalized $\sin(x)/x$ approximation in Eq. (3-50) can be rewritten in terms of sample rate f_s in Hz as

$$X(mf_s) \approx \frac{\sin(\pi m f_s)}{\pi m f_s}. \tag{3-52}$$

3.13.4.2 DFT Frequency Axis in Radians/Second

We can measure $X(m)$'s frequency in radians/s by defining the time-domain sample rate in radians/s as $\omega_s = 2\pi f_s$. So each $X(m)$ sample is associated with a radian frequency of $m\omega_s/N = 2\pi m f_s/N$ radians/s. In this case, $X(m)$'s resolution is $\omega_s/N = 2\pi f_s/N$ radians/s, and the DFT repetition period is $\omega_s = 2\pi f_s$ radians/s, as shown by the expressions in parenthesis in Figure 3-34(b). With $\omega_s = 2\pi f_s$, then $\pi f_s = \omega_s/2$. If we substitute $\omega_s/2$ for πf_s in Eq. (3-51), we obtain an expression for the DFT of a symmetrical rectangular function, where $K < N$, in terms of the sample rate ω_s in radians/s:

$$X(m\omega_s) = \frac{\sin(m\omega_s K/2N)}{\sin(m\omega_s/2N)}. \tag{3-53}$$

For an all ones rectangular function where $K = N$, the amplitude normalized $\sin(x)/x$ approximation in Eq. (3-50) can be stated in terms of a sample rate ω_s in radians/s as

$$X(m\omega_s) \approx \frac{2\sin(m\omega_s/2)}{m\omega_s}. \tag{3-54}$$

3.13.4.3 DFT Frequency Axis Using a Normalized Angle Variable

Many authors simplify the notation of their equations by using a normalized variable for the radian frequency $\omega_s = 2\pi f_s$. By normalized, we mean that the sample rate f_s is assumed to be equal to 1, and this establishes a normalized radian frequency ω_s equal to 2π. Thus, the frequency axis of $X(m)$ now becomes a normalized angle ω, and each $X(m)$ sample is associated with an angle of $m\omega/N$ radians. Using this convention, $X(m)$'s resolution is ω/N radians, and the DFT repetition period is $\omega = 2\pi$ radians, as shown by the expressions in brackets in Figure 3-34(b).

Unfortunately the usage of these three different representations of the DFT's frequency axis is sometimes confusing for the beginner. When reviewing the literature, the reader can learn to convert between these frequency axis notations by consulting Figure 3-34 and Table 3-1.

3.13.5 Alternate Form of the DFT of an All Ones Rectangular Function

Using the normalized radian angle notation for the DFT axis from the bottom row of Table 3-1 leads to another prevalent form of the DFT of the all ones rectangular function in Figure 3-31. Letting our normalized discrete frequency axis variable be $\omega_m = 2\pi m/N$, then $\pi m = N\omega_m/2$. Substituting the term $N\omega_m/2$ for πm in Eq. (3-48), we obtain

All Ones form of the Dirichlet kernel (Type 4): →

$$X(\omega) = \frac{\sin(N\omega_m/2)}{\sin(\omega_m/2)}.$$ (3-55)

Table 3-2 DFT of Various Rectangular Functions

Description	Expression	
DFT of a general rectangular function, where $K < N$, in terms of the integral frequency m variable	$X(m) = \dfrac{\sin(\pi mK/N)}{\sin(\pi m/N)}$ $\cdot e^{j(2\pi m/N)(n_o - (K-1)/2)}$	(3-46)
DFT of a symmetrical rectangular function, where $K < N$, in terms of the integral frequency variable m	$X(m) = \dfrac{\sin(\pi mK/N)}{\sin(\pi m/N)}.$	(3-47)
DFT of an all ones rectangular function in terms of the integer frequency variable m (Dirichlet kernel Type 1)	$X(m) = \dfrac{\sin(\pi m)}{\sin(\pi m/N)}.$	(3-48)
DFT of an all ones rectangular function in terms of the integer frequency variable m (Dirichlet kernel Type 2)	$X(m) \approx \dfrac{N\sin(\pi m)}{\pi m}.$	(3-49)
Amplitude normalized DFT of an all ones rectangular function in terms of the integral frequency variable m (Dirichlet kernel Type 3)	$X(m) \approx \dfrac{\sin(\pi m)}{\pi m}.$	(3-50)
DFT of a symmetrical rectangular function, where $K < N$, in terms of the sample rate f_s in Hz	$X(mf_s) = \dfrac{\sin(\pi mf_s K/N)}{\sin(\pi mf_s/N)}.$	(3-51)
Amplitude normalized DFT of an all ones rectangular function in terms of the sample rate f_s in Hz	$X(mf_s) \approx \dfrac{\sin(\pi mf_s)}{\pi mf_s}.$	(3-52)
DFT of a symmetrical rectangular function, where $K < N$, in terms of the sample rate ω_s in radians/s	$X(m\omega_s) = \dfrac{\sin(m\omega_s K/2N)}{\sin(m\omega_s/2N)}.$	(3-53)
Amplitude normalized DFT of an all ones rectangular function in terms of the sample rate ω_s in radians/s	$X(m\omega_s) \approx \dfrac{2\sin(m\omega_s/2)}{m\omega_s}.$	(3-54)
DFT of an all ones rectangular function in terms of the normalized discrete frequency variable ω_m (Dirichlet kernel Type 4).	$X(\omega) = \dfrac{\sin(N\omega_m/2)}{\sin(\omega_m/2)}.$	(3-55)

Equation (3-55), taking the third form of Eq. (3-34) sometimes seen in the literature, also has the DFT magnitude shown in Figure 3-32(b) and Figure 3-32(c).

We've covered so many different expressions for the DFT of various rectangular functions that it's reasonable to compile their various forms in Table 3-2.

3.13.6 Inverse DFT of a General Rectangular Function

Let's think now about computing the inverse DFT of a rectangular frequency-domain function; that is, given a rectangular $X(m)$ function, find a time-domain function $x(n)$. We can define the general form of a rectangular frequency-domain function, as we did for Figure 3-24, to be that shown in Figure 3-35. The inverse DFT of the rectangular function $X(m)$ in Figure 3-35 is

$$x(n)=\frac{1}{N}\sum_{m=-(N/2)+1}^{N/2}X(m)e^{j2\pi mn/N}. \qquad (3\text{-}56)$$

The same algebraic acrobatics we used to arrive at Eq. (3-43) can be applied to Eq. (3-56), guiding us to

$$x(n) = e^{-j(2\pi n/N)(m_o-(K-1)/2)}\cdot\frac{1}{N}\cdot\frac{\sin(\pi nK/N)}{\sin(\pi n/N)}, \qquad (3\text{-}57)$$

for the inverse DFT of the rectangular function in Figure 3-35. The descriptions we gave for Eq. (3-43) apply equally well to Eq. (3-57) with the exception of a $1/N$ scale factor term and a change in the sign of Eq. (3-43)'s phase angle. Let's use Eq. (3-57) to take a 64-point inverse DFT of the 64-sample rectangular function shown in Figure 3-36(a). The

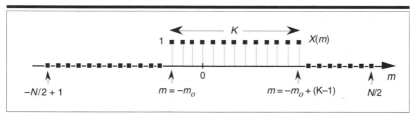

Figure 3-35 General rectangular frequency-domain function of width K samples defined over N samples where $K < N$.

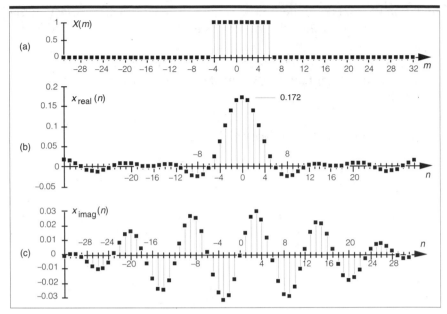

Figure 3-36 Inverse DFT of a general rectangular function: (a) original function $X(m)$; (b) real part, $x_{real}(n)$; (c) imaginary part, $x_{imag}(n)$.

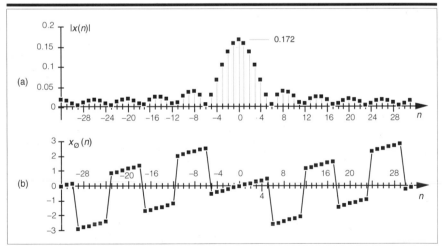

Figure 3-37 Inverse DFT of a generalized rectangular function: (a) magnitude $|x(n)|$; (b) phase angle of $x(n)$ in radians.

inverse DFT of the sequence in Figure 3-36(a) results in an $x(n)$ whose real and imaginary parts, $x_{real}(n)$ and $x_{imag}(n)$, are shown in Figure 3-36(b) and Figure 3-36(c), respectively. With $N = 64$ and $K = 11$ in this example, we've made the inverse DFT functions in Figure 3-36 easy to compare with those *forward* DFT functions in Figure 3-26. See the similarity between the real parts, $X_{real}(m)$ and $x_{real}(n)$, in Figure 3-26(b) and Figure 3-36(b)? Also notice the sign change between the imaginary parts in Figure 3-26(c) and Figure 3-36(c).

The magnitude and phase angle of $x(n)$ are shown in Figure 3-37(a) and Figure 3-37(b). Note the differences in main lobe peak magnitude between Figure 3-27(a) and Figure 3-37(a). The peak magnitude value in Figure 3-37(a) is $K/N = 11/64$, or 0.172. Also notice the sign change between the phase angles in Figure 3-27(b) and Figure 3-37(b). The illustrations in Figures 3-26, 3-27, 3-36, and 3-37 are good examples of the fundamental duality relationships between the forward and inverse DFT.

3.13.7 Inverse DFT of a Symmetrical Rectangular Function

The inverse DFT of the general rectangular function in Figure 3-36 is not very common in digital signal processing. However, in discussions concerning digital filters, we will encounter the inverse DFT of symmetrical rectangular functions. This inverse DFT process is found in the window design method of what are called low-pass finite impulse response (FIR) digital filters. That method begins with the definition of a symmetrical frequency function, $H(m)$, such as that in Figure 3-38. The inverse DFT of the frequency samples in Figure 3-38 is then taken to determine the time-domain coefficients for use in a low-pass FIR filter. (Time-Domain FIR filter coefficients are typically denoted $h(n)$ instead of $x(n)$, so we'll use $h(n)$ throughout the remainder of this inverse DFT discussion.)

In the case of the frequency domain $H(m)$ in Figure 3-38, the K unity-valued samples begin at $m = -m_o = -(K-1)/2$. So plugging $(K-1)/2$ in for m_o in Eq. (3-57), gives

$$h(n) = e^{j(2\pi n/N)((K-1)/2-(K-1)/2)} \cdot \frac{1}{N} \cdot \frac{\sin(\pi nK/N)}{\sin(\pi n/N)}$$

$$= e^{j(2\pi n/N)(0)} \cdot \frac{1}{N} \cdot \frac{\sin(\pi nK/N)}{\sin(\pi n/N)} . \tag{3-58}$$

Again, because $e^{j0} = 1$, Eq. (3-58) becomes

$$h(n) = \frac{1}{N} \cdot \frac{\sin(\pi nK/N)}{\sin(\pi n/N)} . \tag{3-59}$$

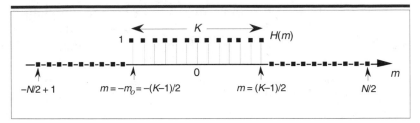

Figure 3-38 Frequency-domain rectangular function of width K samples defined over N samples.

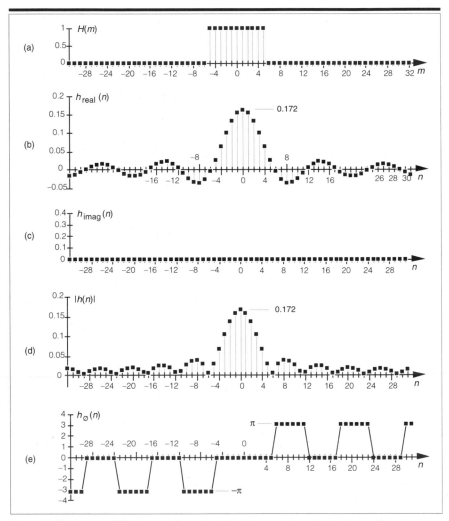

Figure 3-39 Inverse DFT of a rectangular function centered about $m = 0$: (a) original $H(m)$; (b) $h_{real}(n)$; (c) $h_{imag}(n)$; (d) magnitude of $h(n)$; (e) phase of $h(n)$ in radians.

Equation (3-59) tells us that the inverse DFT of the symmetrical rectangular function in Figure 3-38 is itself a real function that we can illustrate by example. We'll perform a 64-point inverse DFT of the sequence in Figure 3-39(a). Our $H(m)$ is 11 unity-valued samples centered about the $m = 0$ index. Here the inverse DFT results in an $h(n)$ whose real and imaginary parts are shown in Figure 3-39(b) and Figure 3-39(c), respectively. As Eq. (3-47) predicted, $h_{real}(n)$ is nonzero and $h_{imag}(n)$ is zero. The magnitude and phase of $h(n)$ are depicted in Figure 3-39(d) and Figure 3-39(e). (Again, we've made the functions in Figure 3-39 easy to compare with the forward DFT function in Figure 3-29.) It is $h(n)$'s real part that we're really after here. Those values of $h_{real}(n)$ are used for the time-domain filter coefficients in the design of FIR low-pass filters that we'll discuss in Section 5.3.

3.14 The DFT Frequency Response to a Complex Input

In this section, we'll determine the frequency response to an N-point DFT when its input is a discrete sequence representing a complex sinusoid expressed as $x_c(n)$. By frequency response we mean the DFT output samples when a complex sinusoidal sequence is applied to the DFT. We begin by depicting an $x_c(n)$ input sequence in Figure 3-40. This time sequence is of the form

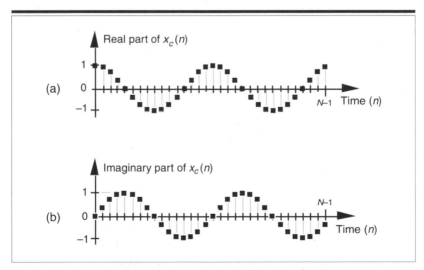

Figure 3-40 Complex time-domain sequence $x_c(n) = e^{j2\pi nk/N}$ having two complete cycles ($k = 2$) over N samples: (a) real part of $x_c(n)$; (b) imaginary part of $x_c(n)$.

$$x_c(n) = e^{j2\pi nk/N} ,$$ (3-60)

where k is the number of complete cycles occurring in the N samples. Figure 3-40 shows $x_c(n)$ if we happen to let $k = 2$. If we denote our DFT output sequence as $X_c(m)$, and apply our $x_c(n)$ input to the DFT expression in Eq. (3-2) we have

$$X_c(m) = \sum_{n=0}^{N-1} x_c(n)e^{-j2\pi nm/N} = \sum_{n=0}^{N-1} e^{j2\pi nk/N} \cdot e^{-j2\pi nm/N}$$

$$= \sum_{n=0}^{N-1} e^{j2\pi n(k-m)/N} .$$ (3-61)

If we let $N = K$, $n = p$, and $q = -2\pi(k-m)/N$, Eq. (3-61) becomes

$$X_c(m) = \sum_{p=0}^{K-1} e^{-jpq} .$$ (3-62)

Why did we make the substitutions in Eq. (3-61) to get Eq. (3-62)? Because, happily, we've already solved Eq. (3-62) when it was Eq. (3-39). That closed form solution was Eq. (3-41) that we repeat here as

$$X_c(m) = \sum_{p=0}^{K-1} e^{-jpq} = e^{-jq(K-1)/2} \cdot \frac{\sin(qK/2)}{\sin(q/2)} .$$ (3-63)

Replacing our original variables from Eq. (3-61), we have our answer:

DFT of a complex sinusoid: →
$$X_c(m) = e^{j[\pi(k-m)-\pi(k-m)/N]} \cdot \frac{\sin[\pi(k-m)]}{\sin[\pi(k-m)/N]}$$ (3-64)

Like the Dirichlet kernel in Eq. (3-43), the $X_c(m)$ in Eq. (3-64) is a complex expression where a ratio of sine terms is the amplitude of $X_c(m)$ and the exponential term is the phase angle of $X_c(m)$. At this point, we're interested only in the ratio of sines factor in Eq. (3-64). Its magnitude is shown in Figure 3-41. Notice that, because $x_c(n)$ is complex, there are no negative frequency components in $X_c(m)$. Let's think about the shaded curve in

Figure 3-41 for a moment. That curve is the continuous Fourier transform of the complex $x_c(n)$ and can be thought of as the continuous spectrum of the $x_c(n)$ sequence.[†] By continuous spectrum we mean a spectrum that's defined at all frequencies, not just at the periodic f_s/N analysis frequencies of an N-point DFT. The shape of this spectrum with its main lobe and sidelobes is a direct and unavoidable effect of analyzing any finite-length time sequence, such as our $x_c(n)$ in Figure 3-40.

We can conceive of obtaining this continuous spectrum analytically by taking the continuous Fourier transform of our discrete $x_c(n)$ sequence, a process some authors call the discrete-time Fourier transform (DTFT), but we can't actually calculate the continuous spectrum on a computer. That's because the DTFT is defined only for infinitely long time sequences, and the DTFT's frequency variable is continuous with infinitely fine-grained resolution. What we can do, however, is use the DFT to calculate an approximation of $x_c(n)$'s continuous spectrum. The DFT outputs represented by the dots in Figure 3-41 are a discrete sampled version of $x_c(n)$'s continuous spectrum. We could have sampled that continuous spectrum more often, i.e., approximated it more closely, with a larger DFT by appending additional zeros to the original $x_c(n)$ sequence. We actually did that in Figure 3-21.

Figure 3-41 shows us why, when an input sequence's frequency is exactly at the $m = k$ bin center, the DFT output is zero for all bins except where $m = k$. If our input sequence frequency was $k+0.25$ cycles in the sample interval, the DFT will sample the continuous spectrum shown in Figure 3-42 where all of the DFT output bins would be nonzero. This effect is a demonstration of DFT leakage described in Section 3.8.

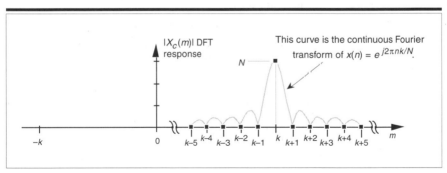

Figure 3-41 N-point DFT frequency magnitude response to a complex sinusoid having integral k cycles in the N-point time sequence $x_c(n) = e^{j2\pi nk/N}$.

[†] Just as we used L'Hospital's rule to find the peak value of the Dirichlet kernel in Eq. (3-44), we could also evaluate Eq. (3-64) to show that the peak of $X_c(m)$ is N when $m = k$.

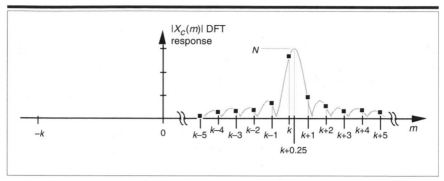

Figure 3-42 *N*-point DFT frequency magnitude response showing spectral leakage of a complex sinusoid having *k*+0.25 cycles in the *N*-point time sequence $x_c(n)$.

Again, just as there are several different expressions for the DFT of a rectangular function that we listed in Table 3-2, we can express the amplitude response of the DFT to a complex input sinusoid in different ways to arrive at Table 3-3.

Table 3-3 Various Forms of the Amplitude Response of the DFT to a Complex Input Sinusoid Having *k* Cycles in the Sample Interval

Description	*Expression*	
Complex input DFT amplitude response in terms of the integral frequency variable *m* [From Eq. (3-64)]	$X_c(m) = \dfrac{\sin[\pi(k-m)]}{\sin[\pi(k-m)/N]}$	(3-65)
Alternate form of the complex input DFT amplitude response in terms of the integral frequency variable *m* [based on Eq. (3-49)]	$X_c(m) \approx \dfrac{N\sin[\pi(k-m)]}{\pi(k-m)}$	(3-66)
Amplitude normalized complex input DFT response in terms of the integral frequency variable *m*	$X_c(m) \approx \dfrac{\sin[\pi(k-m)]}{\pi(k-m)}$	(3-67)
Complex input DFT response in terms of the sample rate f_s in Hz	$X_c(mf_s) \approx \dfrac{N\sin[\pi(k-m)f_s]}{\pi(k-m)f_s}$	(3-68)
Amplitude normalized complex input DFT response in terms of the sample rate f_s in Hz	$X_c(mf_s) \approx \dfrac{\sin[\pi(k-m)f_s]}{\pi(k-m)f_s}$	(3-69)
Amplitude normalized complex input DFT response in terms of the sample rate ω_s	$X_c(m\omega_s) \approx \dfrac{2\sin[(k-m)\omega_s/2]}{(k-m)\omega_s}$	(3-70)

At this point, the thoughtful reader may notice that the DFT's response to a complex input of k cycles per sample interval in Figure 3-41 looks suspiciously like the DFT's response to an all ones rectangular function in Figure 3-32(c). The reason the shapes of those two response curves look the same is because their shapes *are* the same. If our DFT input sequence was a complex sinusoid of $k = 0$ cycles, i.e., a sequence of identical constant values, the ratio of sines term in Eq. (3-64) becomes

$$\frac{\sin[\pi(0-m)]}{\sin[\pi(0-m)/N]} = \frac{-\sin(\pi m)}{-\sin(\pi m/N)} = \frac{\sin(\pi m)}{\sin(\pi m/N)} ,$$

which is identical to the all ones form of the Dirichlet kernel in Eq. (3-48). The shape of our $X_c(m)$ DFT response *is* the sinc function of the Dirichlet kernel.

3.15 The DFT Frequency Response to a Real Cosine Input

Now that we know what the DFT frequency response is to a complex sinusoidal input, it's easy to determine the DFT frequency response to a real input sequence. Say we want the DFT's response to a real cosine sequence, like that shown in Figure 3-40(a), expressed as

$$x_r(n) = \cos(2\pi nk/N) , \tag{3-71}$$

where k is the integral number of complete cycles occurring in the N samples. Remembering Euler's relationship $\cos(\varnothing) = (e^{j\varnothing} + e^{-j\varnothing})/2$, we can show the desired DFT as $X_r(m)$ where

$$X_r(m) = \sum_{n=0}^{N-1} x_r(n)e^{j2\pi nm/N} = \sum_{n=0}^{N-1} \cos(2\pi nk/N) \cdot e^{-j2\pi nm/N}$$

$$= \sum_{n=0}^{N-1} (e^{j2\pi nk/N} + e^{-j2\pi nk/N})/2 \cdot e^{-j2\pi nm/N}$$

$$= \frac{1}{2}\sum_{n=0}^{N-1} e^{j2\pi n(k-m)/N} + \frac{1}{2}\sum_{n=0}^{N-1} e^{-j2\pi n(k+m)/N} . \tag{3-72}$$

Fortunately, in the previous section we just finished determining the closed form of a summation expression like those in Eq. (3-72), so we can write the closed form for $X_r(m)$ as

DFT of a real cosine: →

$$X_r(m) = e^{j[\pi(k-m)-\pi(k-m)/N]} \cdot \frac{1}{2} \frac{\sin[\pi(k-m)]}{\sin[\pi(k-m)/N]}$$

$$+ e^{j[\pi(k+m)-\pi(k+m)/N]} \cdot \frac{1}{2} \frac{\sin[\pi(k+m)]}{\sin[\pi(k+m)/N]} \cdot \qquad (3\text{-}73)$$

We show the magnitude of those two ratio of sines terms as the sinc functions in Figure 3-43. Here again, the DFT is sampling the input cosine sequence's continuous spectrum and, because $k = m$, only one DFT bin is nonzero. Because the DFT's input sequence is real, $X_r(m)$ has both positive and negative frequency components. The positive frequency portion of Figure 3-43 corresponds to the first ratio of sines term in Eq.

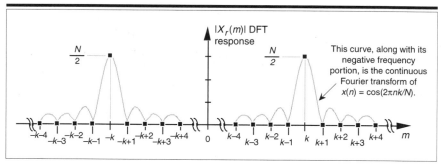

Figure 3-43 *N*-point DFT frequency magnitude response to a real cosine having integral *k* cycles in the *N*-point time sequence $x_r(n) = \cos(2\pi nk/N)$.

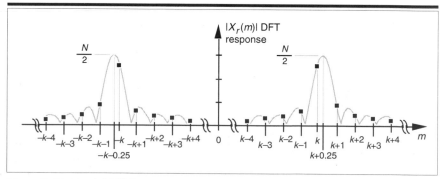

Figure 3-44 *N*-point DFT frequency magnitude response showing spectral leakage of a real cosine having *k*+0.25 cycles in the *N*-point time sequence $x_r(n)$.

Table 3-4 Various Forms of the *Positive Frequency* Amplitude Response of the DFT to a Real Cosine Input Having k Cycles in the Sample Interval

Description	Expression	
Real input DFT amplitude response in terms of the integral frequency variable m [From Eq. (3-73)]	$X_r(m) = \dfrac{\sin[\pi(k-m)]}{2\sin[\pi(k-m)/N]}$	(3-74)
Alternate form of the real input DFT amplitude response in terms of the integral frequency variable m [based on Eq. (3-49)]	$X_r(m) \approx \dfrac{N\sin[\pi(k-m)]}{2\pi(k-m)}$	(3-75)
Amplitude normalized real input DFT response in terms of the integral frequency variable m	$X_r(m) \approx \dfrac{\sin[\pi(k-m)]}{2\pi(k-m)}$	(3-76)
Real input DFT response in terms of the sample rate f_s in Hz	$X_r(mf_s) \approx \dfrac{N\sin[\pi(k-m)f_s]}{2\pi(k-m)f_s}$	(3-77)
Amplitude normalized real input DFT response in terms of the sample rate f_s in Hz	$X_r(mf_s) \approx \dfrac{\sin[\pi(k-m)f_s]}{2\pi(k-m)f_s}$	(3-78)
Amplitude normalized real input DFT response in terms of the sample rate ω_s in radians/s	$X_r(m\omega_s) \approx \dfrac{\sin[(k-m)\omega_s/2]}{(k-m)\omega_s}$	(3-79)

(3-73) and the second ratio of sines term in Eq. (3-73) produces the negative frequency components of $X_r(m)$.

DFT leakage is again demonstrated if our input sequence frequency were shifted from the center of the kth bin to $k+0.25$ as shown in Figure 3-44. (We used this concept of real input DFT amplitude response to introduce the effects of DFT leakage in Section 3.8.)

In Table 3-4, the various mathematical expressions for the (positive frequency) amplitude response of the DFT to a real cosine input sequence are simply those expressions in Table 3-3 reduced in amplitude by a factor of 2.

3.16 The DFT Single-Bin Frequency Response to a Real Cosine Input

Now that we understand the DFT's overall N-point (or N-bin) frequency response to a real cosine of k cycles per sample interval, we conclude this chapter by determining the frequency response of a *single* DFT bin. We can think of a single DFT bin as a kind of bandpass filter, and this useful notion is used, for example, to describe DFT scalloping loss (Section 3.10), employed in the design of frequency-domain filter banks, and applied in a telephone frequency multiplexing technique known as transmultiplexing[15]. To determine a DFT single-bin's frequency response, consider applying a real $x_r(n)$ cosine sequence to a DFT and monitoring the output magnitude of just the $m = k$ bin. Let's say the initial frequency of the input cosine sequence starts at a frequency of $k<m$ cycles and increases up to a frequency of $k>m$ cycles in the sample interval. If we measure the DFT's

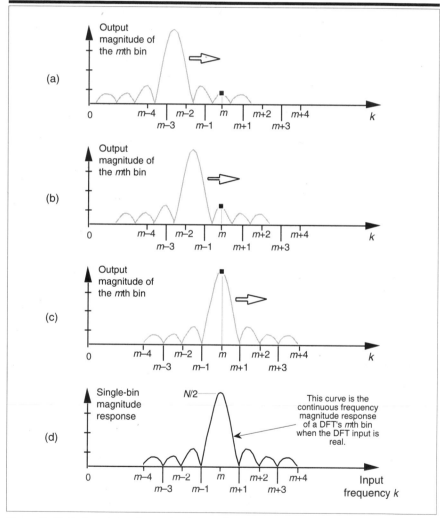

Figure 3-45 Determining the output magnitude of the *m*th bin of an *N*-point DFT:
(a) when the real $x_r(n)$ has $k = m-2.5$ cycles in the time sequence;
(b) when the real $x_r(n)$ has $k = m-1.5$ cycles in the time sequence;
(c) when the real $x_r(n)$ has $k = m$ cycles in the time sequence;
(d) the DFT single-bin frequency magnitude response of the $m = k$ bin.

$m = k$ bin during that frequency sweep, we'll see that its output magnitude must track the input cosine sequence's continuous spectrum, shown as the shaded curves in Figure 3-45.

Figure 3-45(a) shows the $m = k$ bin's output when the input $x_r(n)$'s frequency is $k = m$–2.5 cycles per sample interval. Increasing $x_r(n)$'s frequency to $k = m$–1.5 cycles per sample interval results in the $m = k$ bin's output, shown in Figure 3-45(b). Continuing to sweep $x_r(n)$'s frequency higher Figure 3-45(c) shows the $m = k$ bin's output when the input frequency is $k = m$. Throughout our input frequency sweeping exercise, we can see that the $m = k$ bin's output magnitude must trace out the cosine sequence's continuous spectrum, shown by the solid curve in Figure 3-45(d). This means that a DFT's single-bin frequency magnitude response, to a real input sinusoid, is that solid sinc function curve defined by Eqs. (3-74) through (3-79).

References

[1] Bracewell, R. "The Fourier Transform," *Scientific American*, June 1989.

[2] Struik, D. *A Concise History of Mathematics*, Dover Publications Inc., New York, 1967, p. 142.

[3] Williams, C. S. *Designing Digital Filters*. Section 8.6, Prentice-Hall, Englewood Cliffs, New Jersey, 1986, p. 122.

[4] Press, W., et al. *Numerical Recipes—The Art of Scientific Computing*. Cambridge University Press, 1989, p. 426.

[5] Geckinli, N. C., and Yavuz, D. "Some Novel Windows and a Concise Tutorial Comparison of Window Families," *IEEE Trans. on Acoust. Speech, and Signal Proc.*, Vol. ASSP-26, No. 6, December 1978. (By the way, on page 505 of this paper, the phrase "such that W(f) ≥ 0 ∀ f" indicates that W(f) is never negative. The symbol ∀ means "for all.")

[6] O'Donnell, J. "Looking Through the Right Window Improves Spectral Analysis," *EDN*, November 1984.

[7] Kaiser, J. F. "Digital Filters," in *System Analysis by Digital Computer*. Ed. by F. F. Kuo and J. F. Kaiser, John Wiley and Sons, New York, 1966, pp. 218–277.

[8] Rabiner, L. R., and Gold, B. *The Theory and Application of Digital Signal Processing*. Prentice-Hall, Englewood Cliffs, New Jersey, 1975, p. 88.

[9] Schoenwald, J. "The Surface Acoustic Wave Filter: Window Functions," *RF Design*, March 1986.

[10] Harris, F. "On the Use of Windows for Harmonic Analysis with the Discrete Fourier Transform," *Proceedings of the IEEE*, Vol. 66, No. 1, January 1978.

[11] Nuttall, A. H. "Some Windows with Very Good Sidelobe Behavior," *IEEE Trans. on Acoust. Speech, and Signal Proc.*, Vol. ASSP-29, No. 1, February 1981.

[12] Yanagimoto, Y. "Receiver Design for a Combined RF Network and Spectrum Analyzer," *Hewlett-Packard Journal*, October, 1993.

[13] Gullemin, E. A. *The Mathematics of Circuit Analysis*. John Wiley and Sons, New York, 1949, p. 511.

[14] Lanczos, C. *Discourse on Fourier Series*, Chapter 1, Hafner Publishing Co., New York, 1966, p. 7–47.

[15] Freeny, S. "TDM/FDM Translation As an Application of Digital Signal Processing," *IEEE Communications Magazine*, January 1980.

The Fast Fourier Transform

Although the DFT is the most straightforward mathematical procedure for determining the frequency content of a time-domain sequence, it's terribly inefficient. As the number of points in the DFT is increased to hundreds, or thousands, the amount of necessary number crunching becomes excessive. In 1965 a paper was published by Cooley and Tukey describing a very efficient algorithm to implement the DFT[1]. That algorithm is now known as the fast Fourier transform (FFT).[†] Before the advent of the FFT, thousand-point DFTs took so long to perform that their use was restricted to the larger research and university computer centers. Thanks to Cooley, Tukey, and the semiconductor industry, 1024-point DFTs can now be performed in a few seconds on home computers.

Volumes have been written about the FFT, and, like no other innovation, the development of this algorithm transformed the discipline of digital signal processing by making the power of Fourier analysis affordable. In this chapter, we'll show why the most popular FFT algorithm (called the *radix-2* FFT) is superior to the classical DFT algorithm, present a series of recommendations to enhance our use of the FFT in practice, and provide a list of sources for FFT routines in various software languages. We conclude this chapter, for those readers wanting to know the internal details, with a derivation of the radix-2 FFT and introduce several different ways in which this FFT is implemented.

[†] Actually, the FFT has an interesting history. While analyzing X-ray scattering data, a couple of physicists in the 1940s were taking advantage of the symmetries of sines and cosines using a mathematical method based on a technique published in the early 1900s. Remarkably, over 20 years passed before the FFT was (re)discovered. Reference [2] tells the full story.

4.1 Relationship of the FFT to the DFT

Although many different FFT algorithms have been developed, in this section we'll see why the radix-2 FFT algorithm is so popular and learn how it's related to the classical DFT algorithm. The radix-2 FFT algorithm is a very efficient process for performing DFTs under the constraint that the DFT size be an integral power of two. (That is, the number of points in the transform is $N = 2^k$, where k is some positive integer.) Let's see just why the radix-2 FFT is the favorite spectral analysis technique used by signal-processing practitioners.

Recall that our DFT Example 1 in Section 3.1 illustrated the number of redundant arithmetic operations necessary for a simple 8-point DFT. (For example, we ended up calculating the product of $1.0607 \cdot 0.707$ four separate times.) On the other hand, the radix-2 FFT eliminates these redundancies and greatly reduces the number of necessary arithmetic operations. To appreciate the FFT's efficiency, let's consider the number of complex multiplications necessary for our old friend, the expression for an N-point DFT,

$$X(m) = \sum_{n=0}^{N-1} x(n)e^{-j2\pi nm/N}. \qquad (4\text{-}1)$$

For an 8-point DFT, Eq. (4-1) tells us that we'd have to perform N^2 or 64 complex multiplications. (That's because, for each of the eight $X(m)$s, we have to sum eight complex products as n goes from 0 to 7.) As we'll verify in later sections of this chapter, the number of complex multiplications, for an N-point FFT, is approximately

$$\frac{N}{2} \cdot \log_2 N. \qquad (4\text{-}2)$$

(We say approximately because some multiplications turn out to be multiplications by $+1$ or -1, which amount to mere sign changes.) Well, this $(N/2)\log_2 N$ value is a significant reduction from the N^2 complex multiplications required by Eq. (4-1), particularly for large N. To show just how significant, Figure 4-1 compares the number of complex multiplications required by DFTs and radix-2 FFTs as a function of the number of input data points N. When $N = 512$, for example, the DFT requires 200 times more complex multiplications than those needed by the FFT. When $N = 8192$, the DFT must calculate 1000 complex multiplications for *each* complex multiplication in the FFT!

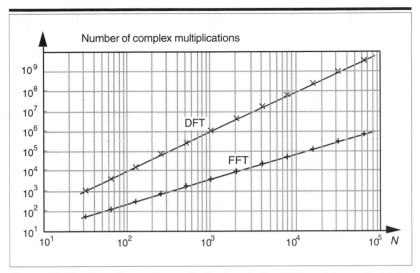

Figure 4-1 Number of complex multiplications in the DFT and the radix-2 FFT as a function of N.

It's appropriate now to make clear that the FFT is not an approximation to the DFT. It's exactly equal to the DFT; it *is* the DFT. Moreover, all of the performance characteristics of the DFT described in the previous chapter, output symmetry, linearity, output magnitudes, leakage, scalloping loss, etc., also describe the behavior of the FFT.

4.2 Hints on Using FFTs in Practice

Based on how useful FFTs are, here's a list of practical pointers, or tips, on acquiring input data samples and using the radix-2 FFT to analyze real-world signals or data.

4.2.1 Sample Fast Enough and Long Enough

When digitizing continuous signals with an A/D converter, for example, we know, from Chapter 2, that our sampling rate must be greater than twice the bandwidth of the continuous A/D input signal to prevent frequency-domain aliasing. Depending on the application, practitioners typically sample at 2.5 to four times the signal bandwidth. If we know that the bandwidth of the continuous signal is not too large relative to the maximum sample rate of our A/D converter, it's easy to avoid aliasing. If we don't know the continuous A/D input signal's bandwidth, how do we tell if we're having aliasing problems? Well, we should mistrust any FFT results that have significant spectral components at frequencies near half the sample rate. Ideally, we'd like to work with signals whose spectral amplitudes decrease with increasing frequency. Be very suspicious of aliasing if there are any spectral components whose frequencies appear to depend on the sample rate. If we suspect that aliasing is occurring or that the continuous signal contains broadband noise, we'll have to use an analog low-pass filter prior to A/D conversion. The cutoff frequency of the low-pass filter must, of course, be greater than the frequency band of interest but less than half the sample rate.

Although we know that an N-point radix-2 FFT requires $N = 2^k$ input samples, just how many samples must we collect before we perform our FFT? The answer is that the data collection time interval must be long enough to satisfy our desired FFT frequency resolution for the given sample f_s rate. The data collection time interval is the reciprocal of the desired FFT frequency resolution, and the longer we sample at a fixed f_s sample rate, the finer our frequency resolution will be; that is, the total data collection time interval is N/f_s seconds, and our N-point FFT bin-to-bin frequency resolution is f_s/N Hz. So, for example, if we need a spectral resolution of 5 Hz, then, $f_s/N = 5$ Hz, and

$$N = \frac{f_s}{\text{desired resolution}} = \frac{f_s}{5} = 0.2f_s. \tag{4-3}$$

In this case, if f_s is, say, 10 kHz, then N must be at least 2,000, and we'd choose N equal to 2048 because this number is a power of 2.

4.2.2 Manipulating the Time Data Prior to Transformation

When using the radix-2 FFT, if we don't have control over the length of our time-domain data sequence, and that sequence length is not an integral power of two, we have two options. We could discard enough data samples so that the remaining FFT input sequence length is some integral power of two. This scheme is not recommended because ignoring data samples degrades our resultant frequency-domain resolution. (The larger N, the better our frequency resolution, right?) A better approach is to append enough zero-valued samples to the end of the time data sequence to match the number of points of the next largest radix-2 FFT. For example, if we have 1000 time samples to transform, rather than analyzing only 512 of them with a 512-point FFT, we should add 24 trailing zero-valued samples to the original sequence and use a

1024-point FFT. (This *zero-stuffing* technique is discussed in more detail in Section 3.11.)

FFTs suffer the same ill effects of spectral leakage that we discussed for the DFT in Section 3.8. We can multiply the time data by a window function to alleviate this leakage problem. Be prepared, though, for the frequency resolution degradation inherent when windows are used. By the way, if appending zeros is necessary to extend a time sequence, we have to make sure that we append the zeros *after* multiplying the original time data sequence by a window function. Applying a window function to the appended zeros will distort the resultant window and worsen our FFT leakage problems.

Although windowing will reduce leakage problems, it will not eliminate them altogether. Even when windowing is employed, high-level spectral components can obscure nearby low-level spectral components. This is especially evident when the original time data has a nonzero average, i.e., it's riding on a DC bias. When the FFT is performed in this case, a large-amplitude DC spectral component at 0 Hz will overshadow its spectral neighbors. We can eliminate this problem by calculating the average of the windowed time sequence and subtract that average value from each sample in the original windowed sequence. (Alternatively, the averaging and subtraction process could just as well be performed before windowing.) This technique makes the new time sequence's average (mean) value equal to zero and eliminates any high-level, 0-Hz component in the FFT results.

4.2.3 Enhancing FFT Results

If we're using the FFT to detect signal energy in the presence of noise and enough time-domain data is available, we can improve the sensitivity of our processing by averaging multiple FFTs. This technique, discussed in Section 10.7, can be implemented very efficiently to detect signal energy that's actually below the average noise level; that is, given enough time-domain data, we can detect signal components that have negative signal-to-noise ratios.

If our original time-domain data is real-valued only, we can take advantage of the 2N-Point Real FFT technique in Section 10.5 to speed up our processing; that is, a 2N-point real sequence can be transformed with a single N-point complex radix-2 FFT. Thus we can get the frequency resolution of a 2N-point FFT for just about the computational price of performing a standard N-point FFT. Another FFT speed enhancement is the possible use of the frequency-domain windowing technique discussed in

Section 10.3. If we need the FFT of unwindowed time-domain data and, at the same time, we also want the FFT of that same time data with a window function applied, we don't have to perform two separate FFTs. We can perform the FFT of the unwindowed data, and then we can perform frequency-domain windowing to reduce spectral leakage on any, or all, of the FFT bin outputs.

4.2.4 Interpreting FFT Results

The first step in interpreting FFT results is to compute the absolute frequency of the individual FFT bin centers. Like the DFT, the FFT bin spacing is the ratio of the sampling rate (f_s) over the number of points in the FFT, or f_s/N. With our FFT output designated by $X(m)$, where $m = 0, 1, 2, 3, \ldots, N–1$, the absolute frequency of the mth bin center is mf_s/N. If the FFT's input time samples are real, only the $X(m)$ outputs from $m = 0$ to $m = N/2$ are independent. So, in this case, we need determine only the absolute FFT bin frequencies for m over the range of $0 \leq m \leq N/2$. If the FFT input samples are complex, all N of the FFT outputs are independent, and we should compute the absolute FFT bin frequencies for m over the full range of $0 \leq m \leq N–1$.

If necessary, we can determine the true amplitude of time-domain signals from their FFT spectral results. To do so, we have to keep in mind that radix-2 FFT outputs are complex and of the form

$$X(m) = X_{real}(m) + jX_{imag}(m) . \tag{4-4}$$

Also, the FFT output magnitude samples,

$$X_{mag}(m) = |X(m)| = \sqrt{X_{real}(m)^2 + X_{imag}(m)^2} , \tag{4-5}$$

are all inherently multiplied by the factor $N/2$, as described in Section 3.4, when the input samples are real. If the FFT input samples are complex, the scaling factor is N. So to determine the correct amplitudes of the time-domain sinusoidal components, we'd have to divide the FFT magnitudes by the appropriate scale factor, $N/2$ for real inputs and N for complex inputs.

If a window function was used on the original time-domain data, some of the FFT input samples will be attenuated. This reduces the resultant FFT output magnitudes from their true unwindowed values. To calculate the correct amplitudes of various time-domain sinusoidal components, then, we'd have to further divide the FFT magnitudes by the appropriate processing loss factor associated with the window function used. Processing loss factors for the most popular window functions are listed in Reference [3].

Should we want to determine the power spectrum $X_{PS}(m)$ of an FFT result, we'd calculate the magnitude-squared values using

$$X_{PS}(m) = |X(m)|^2 = X_{real}(m)^2 + X_{imag}(m)^2 . \qquad (4\text{-}6)$$

Doing so would allow us to compute the power spectrum in decibels with

$$X_{dB}(m) = 10 \cdot \log_{10}(|X(m)|^2) \ \ dB . \qquad (4\text{-}7)$$

The normalized power spectrum in decibels can be calculated using

$$\text{normalized } X_{dB}(m) = 10 \cdot \log_{10}\left(\frac{|X(m)|^2}{(|X(m)|_{max})^2}\right) , \qquad (4\text{-}8)$$

or

$$\text{normalized } X_{dB}(m) = 20 \cdot \log_{10}\left(\frac{|X(m)|}{|X(m)|_{max}}\right) . \qquad (4\text{-}9)$$

In Eqs. (4-8) and (4-9), the term $|X(m)|_{max}$ is the largest FFT output magnitude sample. In practice, we find that plotting $X_{dB}(m)$ is very informative because of the enhanced low-magnitude resolution afforded by the logarithmic decibel scale, as described in Appendix E. If either Eq. (4-8) or Eq. (4-9) is used, no compensation need be performed for the above-mentioned N or $N/2$ FFT scale or window processing loss factors. Normalization through division by $(|X(m)|_{max})^2$ or $|X(m)|_{max}$ eliminates the effect of any absolute FFT or window scale factors.

Knowing that the phase angles $X_\varnothing(m)$ of the individual FFT outputs are given by

$$X_\varnothing(m) = \tan^{-1}\left(\frac{X_{imag}(m)}{X_{real}(m)}\right) , \qquad (4\text{-}10)$$

it's important to watch out for $X_{real}(m)$ values that are equal to zero. That would invalidate our phase angle calculations in Eq. (4-10) due to division by a zero condition. In practice, we want to make sure that our calculations (or software compiler) detect occurrences of $X_{real}(m) = 0$ and set the corresponding $X_{\varnothing}(m)$ to 90° if $X_{imag}(m)$ is positive, set $X_{\varnothing}(m)$ to 0° if $X_{imag}(m)$ is zero, and set $X_{\varnothing}(m)$ to –90° if $X_{imag}(m)$ is negative. While we're on the subject of FFT output phase angles, be aware that FFT outputs containing significant noise components can cause large fluctuations in the computed $X_{\varnothing}(m)$ phase angles. This means that the $X_{\varnothing}(m)$ samples are only meaningful when the corresponding $|X(m)|$ is well above the average FFT output noise level.

4.3 FFT Software Programs

For readers seeking actual FFT software routines without having to buy those high-priced signal processing software packages, public domain radix-2 FFT routines are readily available. References [4–7] provide standard FFT program listings using the FORTRAN language. Reference [8] presents an efficient FFT program written in FORTRAN for real-only input data sequences. Reference [9] provides a standard FFT program written in HP BASIC™, and reference [10] presents an FFT routine written in Applesoft BASIC™. Readers interested in the Ada language can find FFT-related subroutines in reference [11].

4.4 Derivation of the Radix-2 FFT Algorithm

This section and those that follow provide a detailed description of the internal data structures and operations of the radix-2 FFT for those readers interested in developing software FFT routines or designing FFT hardware. To see just exactly how the FFT evolved from the DFT, we return to the equation for an N-point DFT,

$$X(m) = \sum_{n=0}^{N-1} x(n)e^{-j2\pi nm/N} . \qquad (4\text{-}11)$$

A straightforward derivation of the FFT proceeds with the separation of the input data sequence $x(n)$ into two parts. When $x(n)$ is segmented into its even and odd indexed elements, we can, then, break Eq. (4-11) into two parts as

$$X(m) = \sum_{n=0}^{(N/2)-1} x(2n)e^{-j2\pi(2n)m/N} + \sum_{n=0}^{(N/2)-1} x(2n+1)e^{-j2\pi(2n+1)m/N} . \qquad (4\text{-}12)$$

Pulling the constant phase angle outside the second summation,

$$X(m) = \sum_{n=0}^{(N/2)-1} x(2n)e^{-j2\pi(2n)m/N} + e^{-j2\pi m/N}\sum_{n=0}^{(N/2)-1} x(2n+1)e^{-j2\pi(2n)m/N}. \quad (4\text{-}13)$$

Well, here the equations get so long and drawn out that we'll use the standard notation to simplify things. We'll define $W_N = e^{-j2\pi/N}$ to represent the complex phase angle factor that is constant with N. So Eq. (4-13) becomes

$$X(m) = \sum_{n=0}^{(N/2)-1} x(2n)W_N^{2nm} + W_N^m \sum_{n=0}^{(N/2)-1} x(2n+1)W_N^{2nm}. \quad (4\text{-}14)$$

Because $W_N^2 = e^{-j2\pi 2/(N)} = e^{-j2\pi/(N/2)}$, we can substitute $W_{N/2}$ for W_N^2 in Eq. (4-14), as

$$X(m) = \sum_{n=0}^{(N/2)-1} x(2n)W_{N/2}^{nm} + W_N^m \sum_{n=0}^{(N/2)-1} x(2n+1)W_{N/2}^{nm}. \quad (4\text{-}15)$$

So we now have two $N/2$ summations whose results can be combined to give us the N-point DFT. We've reduced some of the necessary number crunching in Eq. (4-15) relative to Eq. (4-11) because the W terms in the two summations of Eq. (4-15) are identical. There's a further benefit in breaking the N-point DFT into two parts because the upper half of the DFT outputs are easy to calculate. Consider the $X(m+N/2)$ output. If we plug $m+N/2$ in for m in Eq. (4-15), then

$$X(m+N/2) = \sum_{n=0}^{(N/2)-1} x(2n)W_{N/2}^{n(m+N/2)}$$

$$+ W_N^{(m+N/2)}\sum_{n=0}^{(N/2)-1} x(2n+1)W_{N/2}^{n(m+N/2)}. \quad (4\text{-}16)$$

It looks like we're complicating things, right? Well, just hang in there for a moment. We can now simplify the phase angle terms inside the summations because

$$W_{N/2}^{n(m+N/2)} = W_{N/2}^{nm} W_{N/2}^{nN/2} = W_{N/2}^{nm} (e^{-j2\pi n2N/2N})$$

$$= W_{N/2}^{nm}(1) = W_{N/2}^{nm}, \tag{4-17}$$

for any integer n. Looking at the so-called *twiddle factor* in front of the second summation in Eq. (4-16), we can simplify it as

$$W_N^{(m+N/2)} = W_N^m W_N^{N/2} = W_N^m (e^{-j2\pi N/2N}) = W_N^m(-1) = -W_N^m. \tag{4-18}$$

OK, using Eqs. (4-17) and (4-18), we represent Eq. (4-16)'s $X(m+N/2)$ as

$$X(m + N/2) = \sum_{n=0}^{(N/2)-1} x(2n)W_{N/2}^{nm} - W_N^m \sum_{n=0}^{(N/2)-1} x(2n+1)W_{N/2}^{nm}. \tag{4-19}$$

Now, let's repeat Eqs. (4-15) and (4-19) to see the similarity;

$$X(m) = \sum_{n=0}^{(N/2)-1} x(2n)W_{N/2}^{nm} + W_N^m \sum_{n=0}^{(N/2)-1} x(2n+1)W_{N/2}^{nm}, \tag{4-20}$$

and

$$X(m + N/2) = \sum_{n=0}^{(N/2)-1} x(2n)W_{N/2}^{nm} - W_N^m \sum_{n=0}^{(N/2)-1} x(2n+1)W_{N/2}^{nm}. \tag{4-20'}$$

So here we are. We need not perform any sine or cosine multiplications to get $X(m+N/2)$. We just change the sign of the twiddle factor W_N^m and use the results of the two summations from $X(m)$ to get $X(m+N/2)$. Of course, m goes from 0 to $(N/2)-1$ in Eq. (4-20) which means, for an N-point DFT, we perform an $N/2$-point DFT to get the first $N/2$ outputs and use those to get the last $N/2$ outputs. For $N = 8$, Eqs. (4-20) and (4-20') are implemented as shown in Figure 4-2.

If we simplify Eqs. (4-20) and (4-20') to the form

$$X(m) = A(m) + W_N^m B(m), \tag{4-21}$$

and

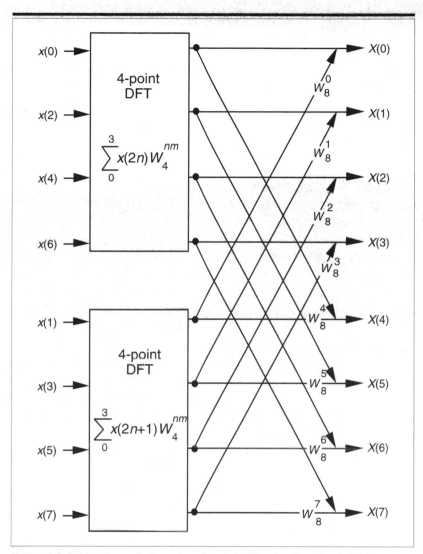

Figure 4-2 FFT Implementation of an 8-point DFT using two 4-point DFTs.

$$X(m+N/2) = A(m) - W_N^m \, B(m) \,, \tag{4-21'}$$

we can go further and think about breaking the two 4-point DFTs into four 2-point DFTs. Let's see how we can subdivide the upper 4-point DFT in Figure 4-2 whose four outputs are $A(m)$ in Eqs. (4-21) and (4-21'). We segment the inputs to the upper 4-point DFT into their odd and even components

$$A(m) = \sum_{n=0}^{(N/2)-1} x(2n)W_{N/2}^{nm} = \sum_{2n=0}^{(N/4)-1} x(4n)W_{N/2}^{4nm} + \sum_{2n=0}^{(N/4)-1} x(4n+1)W_{N/2}^{(4n+1)m}.$$

or

$$A(m) = \sum_{2n=0}^{(N/4)-1} x(4n)W_{N/4}^{2nm} + W_{N/2}^{m} \sum_{2n=0}^{(N/4)-1} x(4n+1)W_{N/4}^{2nm}. \qquad (4\text{-}22)$$

If we let p = 2n, the index of the summations in Eq. (4-22) can be simplified to give $A(m)$ the more familiar form of

$$A(m) = \sum_{p=0}^{(N/4)-1} x(2p)W_{N/4}^{pm} + W_{N/2}^{m} \sum_{p=0}^{(N/4)-1} x(2p+1)W_{N/4}^{pm}. \qquad (4\text{-}23)$$

Notice the similarity between Eq. (4-23) and Eq. (4-20). This capability to subdivide an $N/2$-point DFT into two $N/4$-point DFTs gives the FFT its capacity to greatly reduce the number of necessary multiplications to implement DFTs. (We're going to prove this shortly.) Following the same steps that we used to obtain $A(m)$, we can show that Eq. (4-21)'s $B(m)$ is given by

$$B(m) = \sum_{p=0}^{(N/4)-1} x(2p)W_{N/4}^{pm} - W_{N/2}^{m} \sum_{p=0}^{(N/4)-1} x(2p+1)W_{N/4}^{pm}. \qquad (4\text{-}24)$$

For our $N = 8$ example, Eqs. (4-23) and (4-24) are implemented as shown in Figure 4-3. The FFT's well-known *butterfly* pattern of signal flows is certainly evident, and we see the further shuffling of the input data in Figure 4-3. The twiddle factor $W_{N/2}^{m}$ in Eqs. (4-23) and (4-24), for our $N = 8$ example, ranges from W_4^0 to W_4^3 because the m index, for $A(m)$ and $B(m)$, goes from 0 to 3. For any N-point DFT, we can break each of the $N/2$-point DFTs into two $N/4$-point DFTs to further reduce the number of sine and cosine multiplications. Eventually, we would arrive at an array of 2-point DFTs where no further computational savings could be realized. This is why the number of points in our FFTs are constrained to be some power of 2 and why this FFT algorithm is referred to as the radix-2 FFT.

Moving right along, let's go one step further, and then we'll be finished with our $N = 8$ point FFT derivation. The 2-point DFT functions in

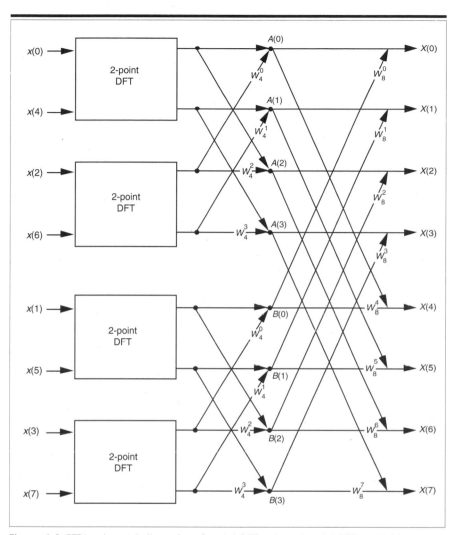

Figure 4-3 FFT implementation of an 8-point DFT as two 4-point DFTs and four 2-point DFTs.

Figure 4-3 cannot be partitioned into smaller parts—we've reached the end of our DFT reduction process arriving at the butterfly of a single 2-point DFT as shown in Figure 4-4. From the definition of W_N, $W_N^0 = e^{-j2\pi 0/N} = 1$ and $W_N^{N/2} = e^{-j2\pi N/2N} = e^{-j\pi} = -1$. So the 2-point DFT blocks in Figure 4-3 can be replaced by the butterfly in Figure 4-4 to give us a full 8-point FFT implementation of the DFT as shown in Figure 4-5.

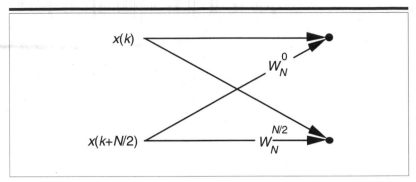

Figure 4-4 Single 2-point DFT butterfly.

OK, we've gone through a fair amount of algebraic foot shuffling here. To verify that the derivation of the FFT is valid, we can apply the 8-point data sequence of Chapter 3's DFT Example 1 to the 8-point FFT represented by Figure 4-5. The data sequence representing $x(n) = \sin(2\pi1000nt_s) + 0.5\sin(2\pi2000nt_s+3\pi/4)$ is

$$x(0) = 0.3535, \qquad x(1) = 0.3535,$$
$$x(2) = 0.6464, \qquad x(3) = 1.0607,$$
$$x(4) = 0.3535, \qquad x(5) = -1.0607,$$
$$x(6) = -1.3535, \qquad x(7) = -0.3535. \qquad (4\text{-}25)$$

We begin grinding through this example by applying the input values from Eq. (4-25) to Figure 4-5, giving the data values shown on left side of Figure 4-6. The outputs of the second stage of the FFT are

$A(0) = 0.707 + W_4^0\,(-0.707) = 0.707 + (1 + j0)(-0.707) = 0 + j0,$

$A(1) = 0.0 + W_4^1\,(1.999) = 0.0 + (0 - j1)(1.999) = 0 - j1.999,$

$A(2) = 0.707 + W_4^2\,(-0.707) = 0.707 + (-1 + j0)(-0.707) = 1.414 + j0,$

$A(3) = 0.0 + W_4^3\,(1.999) = 0.0 + (0 + j1)(1.999) = 0 + j1.999,$

$B(0) = -0.707 + W_4^0\,(0.707) = -0.707 + (1 + j0)(0.707) = 0 + j0,$

$B(1) = 1.414 + W_4^1\,(1.414) = 1.414 + (0 - j1)(1.414) = 1.414 - j1.414,$

$B(2) = -0.707 + W_4^2\,(0.707) = -0.707 + (-1 + j0)(0.707) = -1.414 + j0,$ and

$B(3) = 1.414 + W_4^3\,(1.414) = 1.414 + (0 + j1)(1.414) = 1.414 + j1.414 .$

Calculating the outputs of the third stage of the FFT to arrive at our final answer

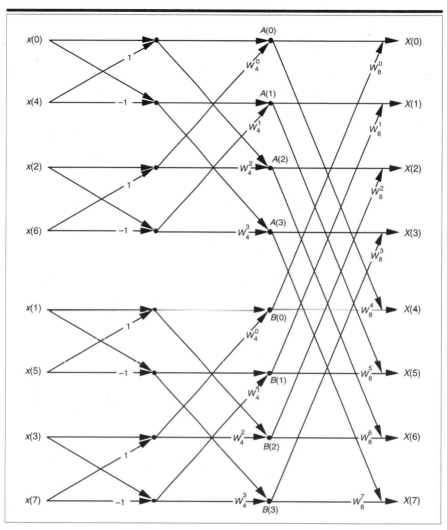

Figure 4-5 Full decimation-in-time FFT implementation of an 8-point DFT.

$$X(0) = A(0) + W_8^0 B(0) = 0 + j0 + (1 + j0)(0 + j0) = 0 + j0 + 0 + j0 = 0 \angle 0°,$$

$$X(1) = A(1) + W_8^1 B(1) = 0 - j1.999 + (0.707 - j0.707)(1.414 - j1.414)$$
$$= 0 - j1.999 + 0 - j1.999 = 0 - j4 = 4 \angle{-90°}$$

$$X(2) = A(2) + W_8^2 B(2) = 1.414 + j0 + (0 - j1)(-1.414 + j0)$$
$$= 1.414 + j0 + 0 + j1.4242 = 1.414 + j1.414 = 2 \angle 45°,$$

$$X(3) = A(3) + W_8^3 B(3) = 0 + j1.999 + (-0.707 - j0.707)(1.414 + j1.414)$$
$$= 0 + j1.999 + 0 - j1.999 = 0 \angle 0°,$$

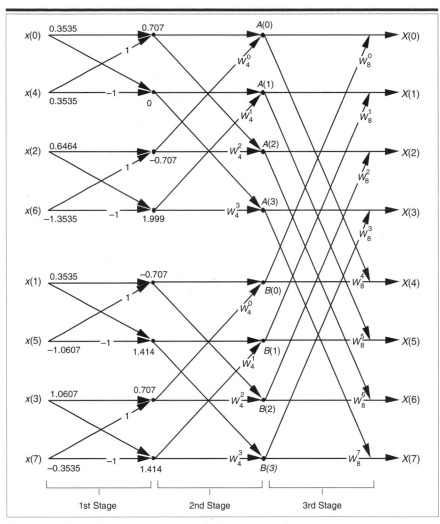

Figure 4-6 8-point FFT of Example 1 from Section 3.1.

$$X(4) = A(0) + W_8^4 B(0) = 0 + j0 + (-1 + j0)(0 + j0)$$
$$= 0 + j0 + 0 + j0 = 0 \angle 0°,$$

$$X(5) = A(1) + W_8^5 B(1) = 0 - j1.999 + (-0.707 + j0.707)(1.414 - j1.414)$$
$$= 0 - j1.999 + 0 + j1.999 = 0 \angle 0°,$$

$$X(6) = A(2) + W_8^6 B(2) = 1.414 + j0 + (0 + j1)(-1.414 + j0)$$
$$= 1.414 + j0 + 0 - j1.414 = 1.414 - j1.414 = 2 \angle -45°, \text{ and}$$

$$X(7) = A(3) + W_8^7 B(3) = 0 + j1.999 + (0.707 + j0.707)(1.414 + j1.414)$$
$$= 0 + j1.999 + 0 + j1.999 = 0 + j4 = 4 \angle 90°.$$

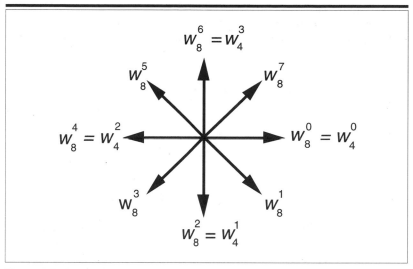

Figure 4-7 Cyclic redundancies in the twiddle factors of an 8-point FFT.

So, happily, the FFT gives us the correct results, and again we remind the reader that the FFT is not an approximation to a DFT; it is a DFT with a reduced number of necessary arithmetic operations. You've seen from the above example that the 8-point FFT example required less effort than the 8-point DFT Example 1 in Section 3.1. Some authors like to explain this arithmetic reduction by the redundancies inherent in the twiddle factors W_N^m. They illustrate this with the *starburst* pattern in Figure 4-7 showing the equivalencies of some of the twiddle factors in an 8-point DFT.

4.5 FFT Input/Output Data Index Bit Reversal

OK, let's look into some of the special properties of the FFT that are important to FFT software developers and FFT hardware designers. Notice that Figure 4-5 was titled "Full decimation-in-time FFT implementation of an 8-point DFT." The *decimation-in-time* phrase refers to how we broke the DFT input samples into odd and even parts in the derivation of Eqs. (4-20), (4-23), and (4-24). This time decimation leads to the scrambled order of the input data's index n in Figure 4-5. The pattern of this shuffled order can be understood with the help of Table 4-1. The shuffling of the input data is known as *bit reversal* because the scrambled order of the input data index can be obtained by reversing the bits of the binary representation of the normal input data index order. Sounds confusing, but it's really not—Table 4-1 illustrates the input index bit reversal for our 8-point FFT example. Notice the normal index order in the left

Table 4-1 Input Index Bit Reversal for an 8-point FFT

Normal order of index n	Binary bits of index n	Reversed bits of index n	Bit-reversed order of index n
0	000	000	0
1	001	100	4
2	010	010	2
3	011	110	6
4	100	001	1
5	101	101	5
6	110	011	3
7	111	111	7

column of Table 4-1 and the scrambled order in the right column that corresponds to the final decimated input index order in Figure 4-5. We've
transposed the original binary bits representing the normal index order
by reversing their positions. The most significant bit becomes the least
significant bit and the least significant bit becomes the most significant
bit, the next to the most significant bit becomes the next to the least significant bit, and the next to the least significant bit becomes the next to
the most significant bit, and so on.[†]

4.6 Radix-2 FFT Butterfly Structures

Let's explore the butterfly signal flows of the decimation-in-time FFT a bit
further. To simplify the signal flows, let's replace the twiddle factors in
Figure 4-5 with their equivalent values referenced to W_N^m, where $N = 8$. We
can show just the exponents m of W_N^m, to get the FFT structure shown in
Figure 4-8. That is, W_4^1 from Figure 4-5 is equal to W_8^2 and is shown as a 2
in Figure 4-8, W_4^2 from Figure 4-5 is equal to W_8^4 and is shown as a 4 in
Figure 4-8, etc. The 1s and –1s in the first stage of Figure 4-5 are replaced
in Figure 4-8 by 0s and 4s, respectively. Other than the twiddle factor notation, Figure 4-8 is identical to Figure 4-5. We can shift around the signal
nodes in Figure 4-5 and arrive at an 8-point decimation-in-time FFT as
shown in Figure 4-9. Notice that the input data in Figure 4-9 is in its normal order and the output data indices are bit-reversed. In this case, a bit-
reversal operation needs to be performed at the output of the FFT to
unscramble the frequency-domain results.

[†] Many that are first shall be last; and the last first. [Mark 10:31]

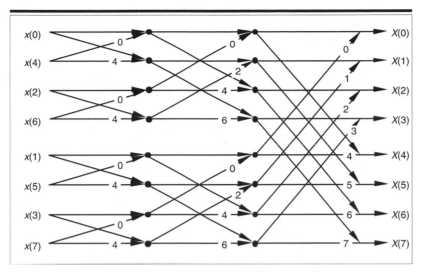

Figure 4-8 8-point decimation-in-time FFT with bit-reversed inputs.

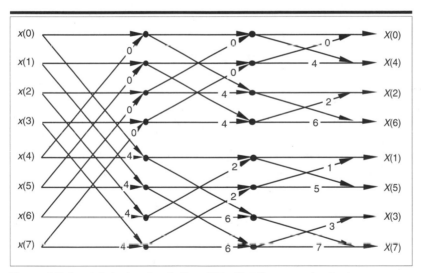

Figure 4-9 8-point decimation-in-time FFT with bit-reversed outputs.

Figure 4-10 shows an FFT signal-flow structure that avoids the bit-reversal problem altogether, and the graceful weave of the traditional FFT butterflies is replaced with a tangled, but effective, configuration.

Not too long ago, hardware implementations of the FFT spent most of their time (clock cycles) performing multiplications, and the bit-reversal process necessary to access data in memory wasn't a significant portion of

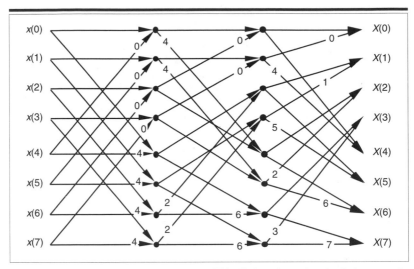

Figure 4-10 8-point decimation-in-time FFT with inputs and outputs in normal order.

the overall FFT computational problem. Now that high-speed multiplier/accumulator integrated circuits can multiply two numbers in a single clock cycle, FFT data multiplexing and memory addressing have become much more important. This has led to the development of efficient algorithms to perform bit reversal[12–15].

There's another derivation for the FFT that leads to butterfly structures looking like those we've already covered, but the twiddle factors in the butterflies are different. This alternate FFT technique is known as the decimation-in-frequency algorithm. Where the decimation-in-time FFT algorithm is based on subdividing the input data into its odd and even components, the decimation-in-frequency FFT algorithm is founded upon calculating the odd and even output frequency samples separately. The derivation of the decimation-in-frequency algorithm is straightforward and included in many tutorial papers and textbooks, so we won't go through the derivation here[4,5,15,16]. We will, however, illustrate decimation-in-frequency butterfly structures (analogous to the structures in Figures 4-8 through 4-10) in Figures 4-11 though 4-13.

So an equivalent decimation-in-frequency FFT structure exists for each decimation-in-time FFT structure. It's important to note that the number of necessary multiplications to implement the decimation-in-frequency FFT algorithms is the same as the number necessary for the decimation-in-time FFT algorithms. There are so many different FFT butterfly structures described in the literature, that it's easy to become confused about

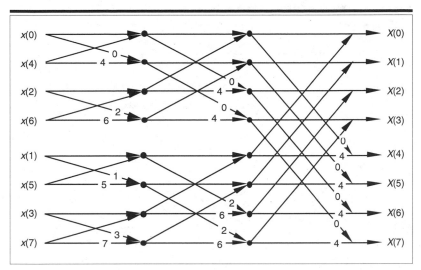

Figure 4-11 8-point decimation-in-frequency FFT with bit-reversed inputs

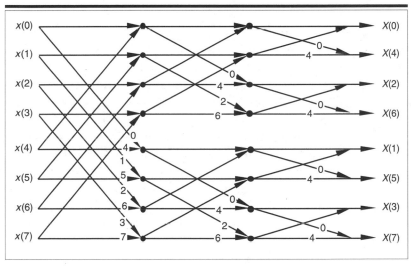

Figure 4-12 8-point decimation-in-frequency FFT with bit-reversed outputs.

which structures are decimation-in-time and which are decimation-in-frequency. Depending on how the material is presented, it's easy for a beginner to fall into the trap of believing that decimation-in-time FFTs always have their inputs bit-reversed and decimation-in-frequency FFTs always have their outputs bit-reversed. This is not true, as the above figures show. Decimation-in-time or -frequency is determined by whether

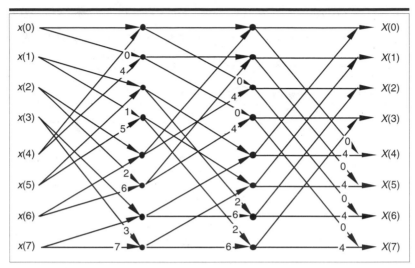

Figure 4-13 8-point decimation-in-frequency FFT with inputs and outputs in normal order.

the DFT inputs or outputs are partitioned when deriving a particular FFT butterfly structure from the DFT equations.

Let's take one more look at a single butterfly. The FFT butterfly structures in Figures 4-8, 4-9, 4-11, and 4-12 are the direct result of the derivations of the decimation-in-time and decimation-in-frequency algorithms. Although it's not very obvious at first, the twiddle factor exponents shown in these structures do have a consistent pattern. Notice how they always take the general forms shown in Figure 4-14(a).[†] To implement the decimation-in-time butterfly of Figure 4-14(a), we'd have to perform two complex multiplications and two complex additions. Well, there's a better way. Consider the decimation-in-time butterfly in Figure 4-14(a). If the top input is x and the bottom input is y, the top butterfly output would be

$$x' = x + W_N^k y \, , \tag{4-26}$$

and the bottom butterfly output would be

$$y' = x + W_N^{k+N/2} y \, . \tag{4-27}$$

[†] Remember, for simplicity the butterfly structures in Figures 4-8 through 4-13 show only the twiddle factor exponents, k and $k+N/2$, and not the entire complex twiddle factors.

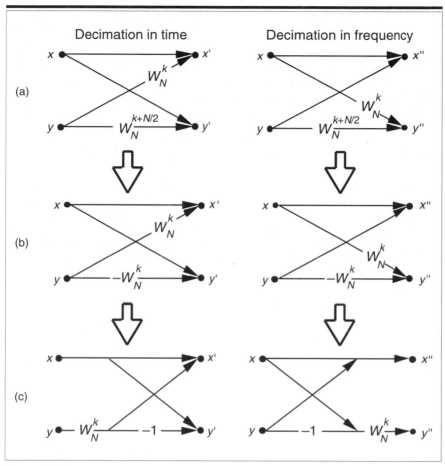

Figure 4-14 Decimation-in-time and decimation-in-frequency butterfly structures: (a) original form; (b) simplified form; (c) optimized form.

Fortunately, the operations in Eqs. (4-26) and (4-27) can be simplified because the two twiddle factors are related by

$$W_N^{k+N/2} = W_N^k W_N^{N/2} = W_N^k (e^{-j2\pi N/2N}) = W_N^k(-1) = -W_N^k. \qquad (4-28)$$

So we can replace the $W_N^{k+N/2}$ twiddle factors in Figure 4-14(a) with $-W_N^k$ to give us the simplified butterflies shown in Figure 4-14(b). Realizing that the twiddle factors in Figure 4-14(b) differ only by their signs, the optimized butterflies in Figure 4-14(c) can be used. Notice that these *optimized* butterflies require two complex additions but only one complex multiplication, thus reducing our computational work-load.[†]

We'll often see the optimized butterfly structures of Figure 4-14(c) in the literature instead of those in Figure 4-14(a). These optimized butterflies give us an easy way to recognize decimation-in-time and decimation-in-frequency algorithms. When we do come across the optimized butterflies from Figure 4-14(c), we'll know that the algorithm is decimation-in-time if the twiddle factor precedes the –1, or else the algorithm is decimation-in-frequency if the twiddle factor follows the –1.

Sometimes we'll encounter FFT structures in the literature that use the notation shown in Figure 4-15 [5, 17]. These wingless butterflies are equivalent to those shown in Figure 4-14(c). The signal-flow convention in Figure 4-15 is such that the plus output of a circle is the sum of the two samples that enter the circle from the left, and the minus output of a circle is the difference of the samples that enter the circle. So the outputs of the decimation-in-time butterflies in Figure 4-14(c) and Figure 4-15(a) are given by

$$x' = x + W_N^k\, y,\ \text{and}\ y' = x - W_N^k\, y\ . \tag{4-29}$$

The outputs of the decimation-in-frequency butterflies in Figure 4-14(c) and Figure 4-15(b) are

$$x'' = x + y,\ \text{and}\ y'' = W_N^k(x - y) = W_N^k x - W_N^k y. \tag{4-30}$$

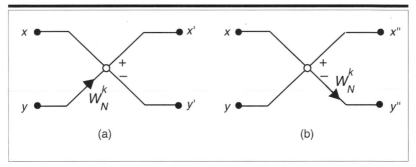

Figure 4-15 Alternate FFT butterfly notation: (a) decimation-in-time; (b) decimation-in-frequency.

[†] It's because there are $(N/2)\log_2 N$ butterflies in an N-point FFT that we said the number of complex multiplications performed by an FFT is $(N/2)\log_2 N$ in Eq. (4-2).

So which FFT structure is the best one to use? It depends on the application, the hardware implementation, and convenience. If we're using a software routine to perform FFTs on a general-purpose computer, we usually don't have a lot of choices. Most folks just use whatever existing FFT routines happen to be included in their commercial software package. Their code may be optimized for speed, but you never know. Examination of the software code may be necessary to see just how the FFT is implemented. If we *feel the need for speed,* we should check to see if the software calculates the sines and cosines each time it needs a twiddle factor. Trigonometric calculations normally take many machine cycles. It may be possible to speed up the algorithm by calculating the twiddle factors ahead of time and storing them in a table. That way, they can be *looked up,* instead of being calculated each time they're needed in a butterfly. If we're writing our own software routine, checking for butterfly output data overflow and careful magnitude scaling may allow our FFT to be performed using integer arithmetic that can be faster on some machines.[†] Care must be taken, however, when using integer arithmetic; some Reduced Instruction Set Computer (RISC) processors actually take longer to perform integer calculations because they're specifically designed to operate on floating-point numbers.

If we're using commercial array processor hardware for our calculations, the code in these processors is *always* optimized because their purpose in life is high speed. Array processor manufacturers typically publicize their products by specifying the speed at which their machines perform a 1024-point FFT. Let's look at some of our options in selecting a particular FFT structure in case we're designing special-purpose hardware to implement an FFT.

The FFT butterfly structures previously discussed typically fall into one of two categories: in-place FFT algorithms and double-memory FFT algorithms. An in-place algorithm is depicted in Figure 4-5. The output of a butterfly operation can be stored in the same hardware memory locations that previously held the butterfly's input data. No intermediate storage is necessary. This way, for an N-point FFT, only $2N$ memory locations are needed. (The 2 comes from the fact that each butterfly node represents a data value that has both a real and an imaginary part.) The rub with the in-place algorithms is that data routing and memory addressing is rather complicated. A double-memory FFT

[†] Overflow is what happens when the result of an arithmetic operation has too many bits, or digits, to be represented in the hardware registers designed to contain that result. FFT data overflow is described in Section 9.3.

structure is that depicted in Figure 4-10. With this structure, intermediate storage is necessary because we no longer have the standard butterflies, and $4N$ memory locations are needed. However, data routing and memory address control is much simpler in double-memory FFT structures than the in-place technique. The use of high-speed, floating-point integrated circuits to implement pipelined FFT architectures takes better advantage of their pipelined structure when the double-memory algorithm is used[18].

There's another class of FFT structures, known as constant-geometry algorithms, that make the addressing of memory both simple and constant for each stage of the FFT. These structures are of interest to those folks who build special-purpose FFT hardware devices[4,19]. From the standpoint of general hardware the decimation-in-time algorithms are optimum for real input data sequences, and decimation-in-frequency is appropriate when the input is complex[8]. When the FFT input data is symmetrical in time, special FFT structures exist to eliminate unnecessary calculations. These special butterfly structures based on input data symmetry are described in the literature[20].

For two-dimensional FFT applications, such as processing photographic images, the decimation-in-frequency algorithms appear to be the optimum choice[21]. Your application may be such that FFT input and output bit reversal is not an important factor. Some FFT applications allow manipulating a bit-reversed FFT output sequence in the frequency domain without having to unscramble the FFT's output data. Then an inverse transform that's expecting bit-reversed inputs will give a time-domain output whose data sequence is correct. This situation avoids the need to perform any bit reversals at all. Multiplying two FFT outputs to implement convolution or correlation are examples of this possibility.[†] As we can see, finding the optimum FFT algorithm and hardware architecture for an FFT is a fairly complex problem to solve, but the literature provides guidance[4,22,23].

References

[1] Cooley, J. and Tukey, J. "An Algorithm for the Machine Calculation of Complex Fourier Series," *Math. Comput.*, Vol. 19, No. 90, Apr. 1965, pp. 297–301.

[2] Cooley, J., Lewis, P., and Welch, P. "Historical Notes on the Fast Fourier Transform," *IEEE Trans. on Audio and Electroacoustics*, Vol. AU-15, No. 2, June 1967.

[†] See Section 10.10 for an example of using the FFT to perform convolution.

[3] Harris, F. J. "On the Use of Windows for Harmonic Analysis with the Discrete Fourier Transform," *Proceedings of the IEEE,* Vol. 66, No. 1, pp. 54, January 1978.

[4] Oppenheim , A. V., and Schafer, R. W. *Discrete-Time Signal Processing,* Prentice-Hall, Englewood Cliffs, New Jersey, 1989, pp. 608.

[5] Rabiner, L. R. and Gold, B. *Theory and Application of Digital Signal Processing,* Prentice-Hall, Englewood Cliffs, New Jersey, 1975, pp. 367.

[6] Stearns, S. *Digital Signal Analysis,* Hayden Book Co., Rochelle Park, New Jersey, 1975, pp. 265.

[7] *Programs for Digital Signal Processing,* Chapter 1, IEEE Press, New York, 1979.

[8] Sorenson, H. V., Jones, D. L., Heideman, M. T., and Burrus, C. S. "Real-Valued Fast Fourier Transform Algorithms," *IEEE Trans. on Acoust. Speech, and Signal Proc.,* Vol. ASSP-35, No. 6, June 1987.

[9] Bracewell, R. *The Fourier Transform and It's Applications,* 2nd Edition, Revised, McGraw-Hill, New York, 1986, pp. 405.

[10] Cobb, F. "Use Fast Fourier Transform Programs to Simplify, Enhance Filter Analysis," *EDN,* 8 March 1984.

[11] Carlin, F. "Ada and Generic FFT Generate Routines Tailored to Your Needs," *EDN,* 23 April 1992.

[12] Evans, D. "An Improved Digit-Reversal Permutation Algorithm for the Fast Fourier and Hartley Transforms," *IEEE Trans. on Acoust. Speech, and Signal Proc.,* Vol. ASSP-35, No. 8, August 1987.

[13] Burris, C. S. "Unscrambling for Fast DFT Algorithms," *IEEE Trans. on Acoust. Speech, and Signal Proc.,* Vol. 36, No. 7, July 1988.

[14] Rodriguez, J. J. "An Improved FFT Digit-Reversal Algorithm," *IEEE Trans. on Acoust. Speech, and Signal Proc.,* Vol. ASSP-37, No. 8, August 1989.

[15] Land, A. "Bit Reverser Scrambles Data for FFT," *EDN,* March 2, 1995.

[16] JG-AE Subcommittee on Measurement Concepts, "What Is the Fast Fourier Transform?," *IEEE Trans. on Audio and Electroacoustics,* Vol. AU-15, No. 2, June 1967.

[17] Cohen, R., and Perlman, R. "500 kHz Single-Board FFT System Incorporates DSP-Optimized Chips," *EDN,* 31 October 1984.

[18] Eldon, J., and Winter, G. E. "Floating-point Chips Carve Out FFT Systems," *Electronic Design,* 4 August 1983.

[19] Lamb, K. "CMOS Building Blocks Shrink and Speed Up FFT Systems," *Electronic Design,* 6 August 1987.

[20] Markel, J. D. "FFT Pruning," *IEEE Trans. on Audio and Electroacoustics,* Vol. AU-19, No. 4, December 1971.

[21] Wu, H. R., and Paoloni, F. J. "The Structure of Vector Radix Fast Fourier Transforms," *IEEE Trans. on Acoust. Speech, and Signal Proc.,* Vol. ASSP-37, No. 8, August 1989.

[22] Ali, Z. M. "High Speed FFT Processor," *IEEE Trans. on Communications,* Vol. COM-26, No. 5, May 1978.

[23] Bergland, G. "Fast Fourier Transform Hardware Implementations—An Overview," *IEEE Trans. on Audio and Electroacoustics,* Vol. AU-17, June 1969.

CHAPTER FIVE

Finite Impulse Response Filters

The filtering of digitized data, if not the most fundamental, is certainly the oldest discipline in the field of digital signal processing. Digital filtering's origins go back forty years. The growing availability of digital computers in the early 1950s led to efforts in the smoothing of discrete sampled data and the analysis of discrete data control systems. However it wasn't until the early to mid-1960s, around the time the Beatles came to America, that the analysis and development of digital equivalents of analog filters began in earnest. That's when digital signal processing experts realized that computers could go beyond the mere analysis of digitized signals into the domain of actually changing signal characteristics through filtering. Today, digital filtering is so widespread that the quantity of literature pertaining to it exceeds that of any other topic in digital signal processing. In this chapter, we introduce the fundamental attributes of digital filters, learn how to quantify their performance, and review the principles associated with the design of finite impulse response digital filters.

So let's get started by illustrating the concept of filtering a time-domain signal as shown in Figure 5-1.

In general, filtering is the processing of a time-domain signal resulting in some change in that signal's original spectral content. The change is usually the reduction, or filtering out, of some unwanted input spectral components; that is, filters allow certain frequencies to pass while attenuating other frequencies. Figure 5-1 shows both analog and digital versions of a filtering process. Where an analog filter operates on a continuous signal, a digital filter processes a sequence of discrete sample values. The digital filter in Figure 5-1(b), of course, can be a software program in a computer, a programmable hardware processor, or a dedicated integrated circuit. Traditional linear digital filters typically come in two flavors: finite

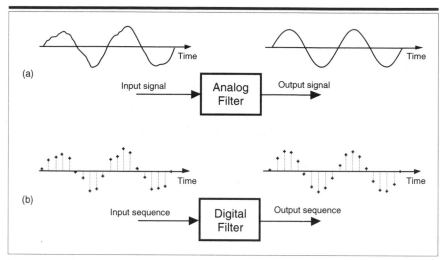

Figure 5-1 Filters: (a) an analog filter with a noisy tone input and a reduced-noise tone output; (b) the digital equivalent of the analog filter.

impulse response (FIR) filters and infinite impulse response (IIR) filters. Because FIR filters are the simplest type of digital filter to analyze, we'll examine them in this chapter and cover IIR filters in Chapter 6.

5.1 An Introduction to Finite Impulse Response FIR Filters

First of all, FIR digital filters use only current and past input samples, and none of the filter's previous output samples, to obtain a current output sample value. (That's also why FIR filters are sometimes called *nonrecursive* filters.) Given a finite duration of nonzero input values, the effect is that an FIR filter will always have a finite duration of nonzero output values and that's how FIR filters got their name. So, if the FIR filter's input suddenly becomes a sequence of all zeros, the filter's output will eventually be all zeros. While not sounding all that unusual, this characteristic is however, very important, and we'll soon find out why, as we learn more about digital filters.

FIR filters use addition to calculate their outputs in a manner much the same as the process of averaging uses addition. In fact, averaging is a kind of FIR filter that we can illustrate with an example. Let's say we're counting the number of cars that pass over a bridge every minute, and we need to know the average number of cars per minute over five-minute intervals; that is, every minute we'll calculate the average num-

Table 5-1 Values for the Averaging Example

Minute index	Number of cars/minute over the last minute	Number of cars/minute averaged over the last five minutes
1	10	–
2	22	–
3	24	–
4	42	–
5	37	27
6	77	40.4
7	89	53.8
8	22	53.4
9	63	57.6
10	9	52

ber of cars/minute over the last five minutes. If the results of our car counting for the first ten minutes are those values shown in the center column of Table 5-1, then the average number of cars/minute over the previous five one-minute intervals is listed in the right column of the table. We've added the number of cars for the first five one-minute intervals and divided by five to get our first five-minute average output value, (10+22+24+42+37)/5 = 27. Next we've averaged the number of cars/minute for the second to the sixth one-minute intervals to get our second five-minute average output of 40.4. Continuing, we average the number of cars/minute for the third to the seventh one-minute intervals to get our third average output of 53.8, and so on. With the number of cars/minute for the one-minute intervals represented by the dashed line in Figure 5-2, we show our five-minute average output as the solid line. (Figure 5-2 shows cars/minute input values beyond the first ten minutes listed in Table 5-1 to illustrate a couple of important ideas to be discussed shortly.)

There's much to learn from this simple averaging example. In Figure 5-2, notice that the sudden changes in our input sequence of cars/minute are flattened out by our averager. The averager output sequence is considerably smoother than the input sequence. Knowing that sudden transitions in a time sequence represent high frequency components, we can say that our averager is behaving like a low-pass filter and smoothing sudden changes in the input. Is our averager an FIR filter? It sure is—no previous averager output value is used to determine a current output

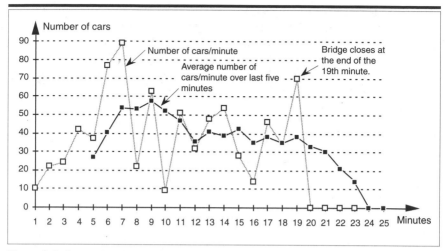

Figure 5-2 Averaging the number of cars/minute. The dashed line shows the individual cars/minute, and the solid line is the number of cars/minute averaged over the last five minutes.

value; only input values are used to calculate output values. In addition, we see that, if the bridge were suddenly closed at the end of the 19th minute, the dashed line immediately goes to zero cars/minute at the end of the 20th minute, and the averager's output in Figure 5-2 approaches and settles to a value of zero by the end of the 24th minute.

Figure 5-2 shows the first averager output sample occurring at the end of the 5th minute because that's when we first have five input samples to calculate a valid average. The 5th output of our averager can be denoted as $y_{ave}(5)$ where

$$y_{ave}(5) = \frac{1}{5}[x(1) + x(2) + x(3) + x(4) + x(5)] \ . \tag{5-1}$$

In the general case, if the kth input sample is $x(k)$, then the nth output is

$$y_{ave}(n) = \frac{1}{5}[x(n-4) + x(n-3) + x(n-2) + x(n-1) + x(n)] = \frac{1}{5}\sum_{k=n-4}^{n} x(k) \ . \tag{5-2}$$

Look at Eq. (5-2) carefully now. It states that the nth output is the average of the nth input sample and the four previous input samples.

We can formalize the digital filter nature of our averager by creating the block diagram in Figure 5-3 showing how the averager calculates its output samples.

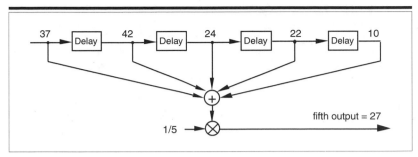

Figure 5-3 Averaging filter block diagram when the fifth input sample value, 37, is applied.

This block diagram, referred to as the filter *structure*, is a physical depiction of how we might calculate our averaging filter outputs with the input sequence of values shifted, in order, from left to right along the top of the filter as new output calculations are performed. This structure, implementing Eqs. (5-1) and (5-2), shows those values used when the first five input sample values are available. The delay elements in Figure 5-3, called *unit delays*, merely indicate a shift register arrangement where input sample values are temporarily stored during an output calculation.

In averaging, we add five numbers and divide the sum by five to get our answer. In a conventional FIR filter implementation, we can just as well multiply each of the five input samples by the coefficient 1/5 and then perform the summation as shown in Figure 5-4(a). Of course, the two methods in Figures 5-3 and 5-4(a) are equivalent because Eq. (5-2) describing the structure shown in Figure 5-3 is equivalent to

$$y_{ave}(n) = \frac{1}{5}x(n-4) + \frac{1}{5}x(n-3) + \frac{1}{5}x(n-2) + \frac{1}{5}x(n-1) + \frac{1}{5}x(n)$$

$$= \sum_{k=n-4}^{n} \frac{1}{5}x(k) \tag{5-3}$$

that describes the structure in Figure 5-4(a).[†]

Let's make sure we understand what's happening in Figure 5-4(a). Each of the first five input values are multiplied by 1/5, and the five products are summed to give the 5th filter output value. The left to right

---

[†] We've used the venerable distributive law for multiplication and addition of scalars, $a(b+c+d) = ab+ac+ad$, in moving Eq. (5-2)'s factor of 1/5 inside the summation in Eq. (5-3).

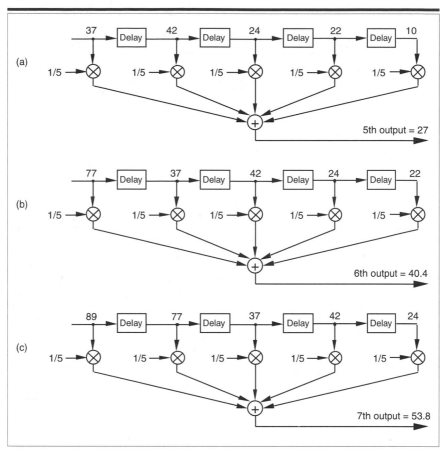

Figure 5-4 Alternate averaging filter structure: (a) input values used for the 5th output value; (b) input values used for the 6th output value; (c) input values used for the 7th output value.

sample shifting is illustrated in Figures 5-4(b) and 5-4(c). To calculate the filter's 6th output value, the input sequence is right shifted discarding the 1st input value of 10, and the 6th input value 77 is accepted on the left. Likewise, to calculate the filter's 7th output value, the input sequence is right shifted discarding the 2nd value of 22, and the 7th input value 89 arrives on the left. So, when a new input sample value is applied, the filter discards the oldest sample value, multiplies the samples by the coefficients of 1/5, and sums the products to get a single new output value. The filter's structure using this bucket brigade shifting process is often called a transversal filter due to the cross-directional flow of the input samples. Because we *tap off* five separate input sample values to calculate an output value, the structure in Figure 5-4 is called a 5-tap FIR filter in digital filter vernacular.

One important and, perhaps, most interesting aspect of understanding FIR filters is learning how to predict their behavior when sinusoidal samples of various frequencies are applied to the input, i.e., how to estimate their frequency-domain response. Two factors affect an FIR filter's frequency response: the number of taps and the specific values used for the multiplication coefficients. We'll explore these two factors using our averaging example and, then, see how we can use them to design FIR filters. This brings us to the point where we have to introduce the C word: convolution. (Actually, we already slipped a convolution equation in on the reader without saying so. It was Eq. (5-3), and we'll examine it in more detail later.)

5.2 Convolution in FIR Filters

OK, here's where we get serious about understanding the mathematics behind FIR filters. We can graphically depict Eq. (5-3)'s and Figure 5-4's calculations as shown in Figure 5-5. Also, let's be formal and use the standard notation of digital filters for indexing the input samples and the filter coefficients by starting with an initial index value of zero; that is, we'll call the initial input value the 0th sample $x(0)$. The next input sample is represented by the term $x(1)$, the following input sample is called $x(2)$, and so on. Likewise, our five coefficient values will be indexed from zero to four, $h(0)$ through $h(4)$. (This indexing scheme makes the equations describing our example consistent with conventional filter notation found in the literature.)

In Eq. (5-3) we used the factor of 1/5 as the filter coefficients multiplied by our averaging filter's input samples. The left side of Figure 5-5 shows the alignment of those coefficients, black squares, with the filter input sample values represented by the white squares. Notice in Figure 5-5(a) through 5-5(e) that we're marching the input samples to the right, and, at each step, we calculate the filter output sample value using Eq. (5-3). The output samples on the right side of Figure 5-5 match the first five values represented by the black squares in Figure 5-2. The input samples in Figure 5-5 are those values represented by the white squares in Figure 5-2. Notice the time order of the inputs in Figure 5-5 has been reversed from the input sequence order in Figure 5-2! That is, the input sequence has been flipped in the time domain in Figure 5-5. This time order reversal is what happens to the input data using the filter structure in Figure 5-4.

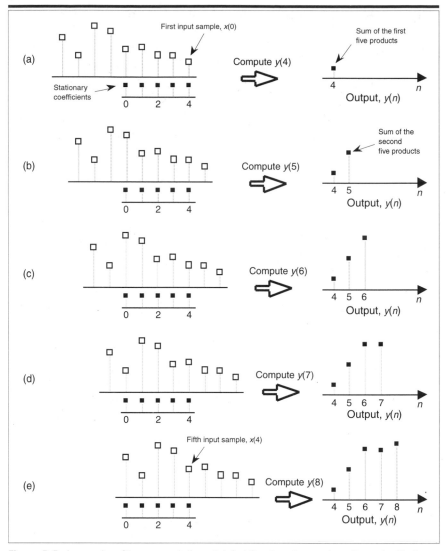

Figure 5-5 Averaging filter convolution: (a) first five input samples aligned with the stationary filter coefficients, index $n = 4$; (b) input samples shift to the right and index $n = 5$; (c) index $n = 6$; (d) index $n = 7$; (e) index $n = 8$.

Repeating the first part of Eq. (5-3) and omitting the subscript on the output term, our original FIR filter's $y(n)$th output is given by

$$y(n) = \frac{1}{5}x(n-4) + \frac{1}{5}x(n-3) + \frac{1}{5}x(n-2) + \frac{1}{5}x(n-1) + \frac{1}{5}x(n) \, . \qquad (5\text{-}4)$$

Because we'll explore filters whose coefficients are not all the same value, we need to represent the individual filter coefficients by a variable, such as the term $h(k)$, for example. Thus we can rewrite the averaging filter's output from Eq. (5-4) in a more general way as

$$y(n) = h(4)x(n-4) + h(3)x(n-3) + h(2)x(n-2) + h(1)x(n-1) + h(0)x(n)$$

$$= \sum_{k=0}^{4} h(k)x(n-k) \ , \qquad (5\text{-}5)$$

where $h(0)$ through $h(4)$ all equal 1/5. Equation (5-5) is a concise way of describing the filter structure in Figure 5-4 and the process illustrated in Figure 5-5.

Let's take Eq. (5-5) one step further and say, for a general M-tap FIR filter, the nth output is

$$y(n) = \sum_{k=0}^{M-1} h(k)x(n-k) \ . \qquad (5\text{-}6)$$

Well, there it is. Eq. (5-6) is the infamous convolution equation as it applies to digital FIR filters. Beginners in the field of digital signal processing often have trouble understanding the concept of convolution. It need not be that way. Eq. (5-6) is merely a series of multiplications followed by the addition of the products. The process is actually rather simple. We just flip the time order of an input sample sequence and start stepping the flipped sequence across the filter's coefficients as shown in Figure 5-5. For each new filter input sample, we sum a series of products to compute a single filter output value.

Let's pause for a moment and introduce a new term that's important to keep in mind, the *impulse response*. The impulse response of a filter is exactly what its name implies—it's the filter's output time-domain sequence when the input is a single unity-valued sample (impulse) preceded and followed by zero-valued samples. Figure 5-6 illustrates this idea in the same way we determined the filter's output sequence in Figure 5-5. The left side of Figure 5-6 shows the alignment of the filter coefficients, black squares, with the filter input impulse sample values represented by the white squares. Again, in Figure 5-6(a) through 5-6(e) we're shifting the input samples to the right, and, at each step, we calculate the filter output sample value using Eq. (5-4). The output samples on the right side of

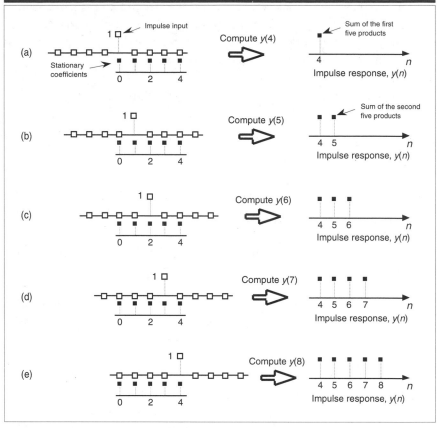

Figure 5-6 Convolution of filter coefficients and an input impulse to obtain the filter's output impulse response: (a) impulse sample aligned with the first filter coefficient, index $n = 4$; (b) impulse sample shifts to the right and index $n = 5$; (c) index $n = 6$; (d) index $n = 7$; (e) index $n = 8$.

Figure 5-6 are the filter's impulse response. Notice the key point here: the FIR filter's impulse response is identical to the five filter coefficient values. For this reason, the terms *FIR filter coefficients* and *impulse response* are synonymous. Thus, when someone refers to the impulse response of an FIR filter, they're also talking about the coefficients.

Returning to our averaging filter, recall that coefficients (or impulse response) $h(0)$ through $h(4)$ were all equal to $1/5$. As it turns out, our filter's performance can be improved by using coefficients whose values are not all the same. By performance we mean how well the filter passes desired signals and attenuates unwanted signals. We judge that performance by determining the shape of the filter's frequency-domain response that we obtain by the convolution property of linear systems. To describe this concept, let's repeat Eq. (5-6) using the abbreviated notation of

$$y(n) = h(k) * x(n) \tag{5-7}$$

where the * symbol means convolution. (Equation 5-7 is read as "y of n equals the convolution of h of k and x of n.") The process of convolution, as it applies to FIR filters is as follows: The discrete Fourier transform (DFT) of the convolution of a filter's impulse response (coefficients), and an input sequence is equal to the product of the spectrum of the input sequence and the DFT of the impulse response. The idea we're trying to convey here is that if two time-domain sequences $h(k)$ and $x(n)$ have DFTs of $H(m)$ and $X(m)$, respectively, then the DFT of $y(n) = h(k) * x(n)$ is $H(m) \cdot X(m)$. Making this point in a more compact way, we state this relationship with the expression

$$y(n) = h(k) * x(n) \xrightleftharpoons[\text{IDFT}]{\text{DFT}} H(m) \cdot X(m). \tag{5-8}$$

With IDFT indicating the inverse DFT, Eq. (5-8) indicates that two sequences resulting from $h(k) * x(n)$ and $H(m) \cdot X(m)$ are Fourier transform pairs. So taking the DFT of $h(k) * x(n)$ gives us the product $H(m) \cdot X(m)$ that is the spectrum of our filter output $Y(m)$. Likewise, we can determine $h(k) * x(n)$ by taking the inverse DFT of $H(m) \cdot X(m)$. The very important conclusion to learn from Eq. (5-8) is that convolution in the time-domain is equivalent to multiplication in the frequency-domain. To help us appreciate this principle, Figure 5-7 sketches the relationship between convolution in the time domain and multiplication in the frequency domain. The process of convolution with regard to linear systems is discussed in more detail in Section 5.9. The beginner is encouraged to review that material to get a general idea of why and when the convolution process can be used to analyze digital filters.

Equation (5-8) and the relationships in Figure 5-7 tell us what we need to do to determine the frequency response of an FIR filter. The product $X(m) \cdot H(m)$ is the DFT of the filter output. Because $X(m)$ is the DFT of the filter's input sequence, the frequency response of the filter is then defined as $H(m)$, the DFT of filter's impulse response $h(k)$.[†] Getting back to our original problem, we can determine our averaging filter's frequency-domain response by taking the DFT of the individual filter coefficients (impulse response) in Eq. (5-4). If we take the five $h(k)$ coefficient values of 1/5 and append 59 zeros, we have the sequence depicted in Figure 5-8(a). Performing a 64-point DFT on that sequence, and normalizing the

---

[†] We use the *term impulse response* here, instead of *coefficients*, because this concept also applies to IIR filters. IIR filter frequency responses are also equal to the DFT of their impulse responses.

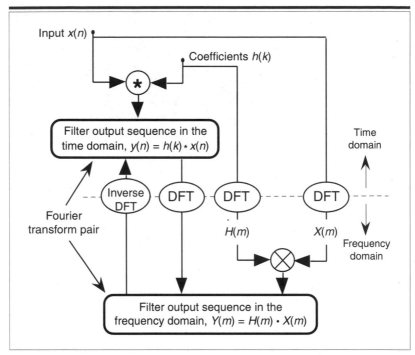

Figure 5-7 Relationships of convolution as applied to FIR digital filters.

DFT magnitudes, gives us the filter's frequency magnitude response $|H(m)|$ in Figure 5-8(b) and phase response shown in Figure 5-8(c).[†] $H(m)$ is our old friend, the $\sin(x)/x$ function from Section 3.13.

Let's relate the discrete frequency response samples in Figures 5-8(b) and 5-8(c) to the physical dimension of the sample frequency f_s. We know, from Section 3.5 and our experience with the DFT, that the $m = N/2$ discrete frequency sample, $m = 32$ in this case, is equal to the folding frequency, or half the sample rate, $f_s/2$. Keeping this in mind, we can convert the discrete frequency axis in Figure 5-8 to that shown in Figure 5-9. In Figure 5-9(a), notice that the filter's magnitude response is, of course, periodic in the frequency domain with a period of the equivalent sample rate f_s. Because we're primarily interested in the filter's response between 0 and half the sample rate, Figure 5-9(c) shows that frequency band in greater

---

[†] There's nothing sacred about using a 64-point DFT here. We could just as well have appended only enough zeros to take a 16- or 32-point FFT. We chose 64 points to get a frequency resolution that would make the shape of the response in Figure 5-8(b) reasonably smooth. Remember, the more points in the FFT, the finer the frequency resolution—right?

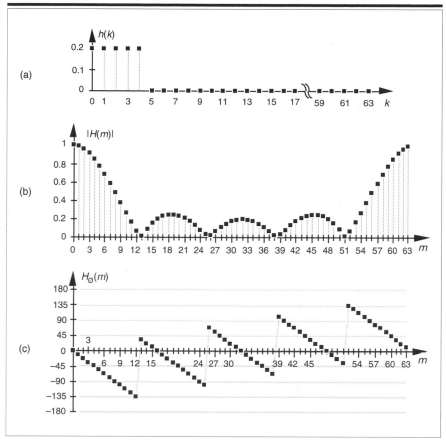

Figure 5-8 Averaging FIR filter: (a) filter coefficient sequence h(k) with
appended zeros; (b) normalized discrete frequency magnitude
response | H(m) | of the h(k) filter coefficients; (c) phase-angle
response of H(m) in degrees.

detail affirming the notion that averaging behaves like a low-pass filter. It's
a relatively poor low-pass filter compared to an arbitrary, *ideal* low-pass
filter indicated by the dashed lines in Figure 5-9(c), but our averaging filter will attenuate higher frequency inputs relative to its response to low-
frequency input signals.

We can demonstrate this by way of example. Suppose we applied a
low-frequency sinewave to the averaging FIR filter, as shown in Figure
5-10(a), and that the input sinewave's frequency is $f_s/32$ and its peak
amplitude is unity. The filter output in this case would be a sinewave of
frequency $f_s/32$, but it's peak amplitude would be reduced to a value of
0.96, and the output sinewave's first sample value is delayed by a phase

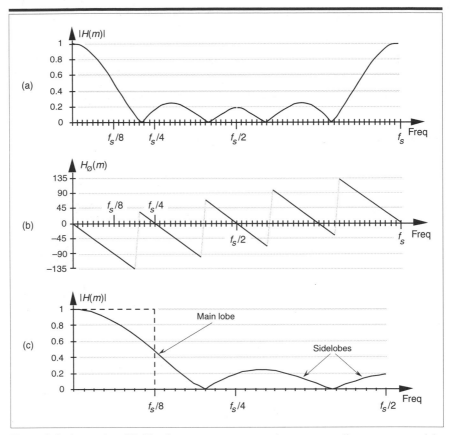

Figure 5-9 Averaging FIR filter frequency response shown as continuous curves: (a) normalized frequency magnitude response, $|H(m)|$; (b) phase angle response of $H(m)$ in degrees; (c) the filter's magnitude response between zero Hz and half the sample rate, $f_s/2$ Hz.

angle of 22.5°. Next, if we applied a higher frequency sinewave of $3f_s/32$ to the FIR filter as shown in Figure 5-10(b), the filter output would, then, be a sinewave of frequency $3f_s/32$, but it's peak amplitude is even further reduced to a value of 0.69. In addition, the Figure 5-10(b) output sinewave's first sample value has an even larger phase angle delay of 67.5°. Although the output amplitudes and phase delays in Figure 5-10 were measured values from actually performing a 5-tap FIR filter process on the input sinewave samples, we could have obtained those amplitude and phase delay values directly from Figures 5-8(a) and 5-8(b). The emphasis here is that we don't have to implement an FIR filter and apply various sinewave inputs to discover what its frequency

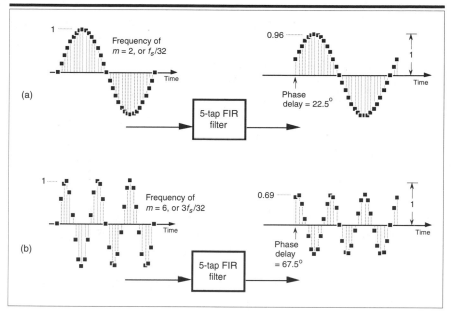

Figure 5-10 Averaging FIR filter input and output responses: (a) with an input sinewave of frequency $f_s/32$; (b) with an input sinewave of frequency $3f_s/32$.

response will be. We need merely take the DFT of the FIR filter's coefficients (impulse response) to determine the filter's frequency response as we did for Figure 5-8.

Figure 5-11 is another depiction of how well our 5-tap averaging FIR filter performs, where the dashed line is the filter's magnitude response $|H(m)|$, and the shaded line is the $|X(m)|$ magnitude spectrum of the filter's input values (the white squares in Figure 5-2). The solid line is the magnitude spectrum of the filter's output sequence which is shown by the black squares in Figure 5-2. So in Figure 5-11, the solid output spectrum is the product of the dashed filter response curve and the shaded input spectrum, or $|X(m) \cdot H(m)|$. Again, we see that our averager does indeed attenuate the higher frequency portion of the input spectrum.

Let's pause for a moment to let all of this soak in a little. So far we've gone through the averaging filter example to establish that

- FIR filters perform time-domain convolution by summing the products of the shifted input samples and a sequence of filter coefficients,

- an FIR filter's output sequence is equal to the convolution of the input sequence and a filter's impulse response (coefficients),

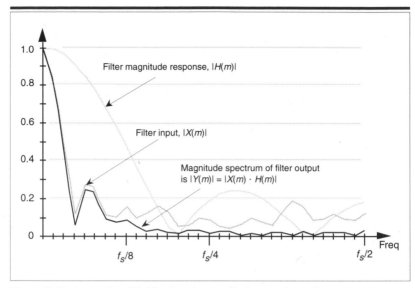

Figure 5-11 Averaging FIR filter input magnitude spectrum, frequency magnitude response, and output magnitude spectrum.

- an FIR filter's frequency response is the DFT of the filter's impulse response,[†]

- an FIR filter's output spectrum is the product of the input spectrum and the filter's frequency response, and

- convolution in the time domain and multiplication in the frequency domain are Fourier transform pairs.

OK, here's where FIR filters start to get really interesting. Let's change the values of the five filter coefficients to modify the frequency response of our 5-tap, low-pass filter. In fact, Figure 5-12(a) shows our original five filter coefficients and two other arbitrary sets of 5-tap coefficients. Figure 5-12(b) compares the frequency magnitude responses of those three sets of coefficients. Again, the frequency responses are obtained by taking the DFT of the three individual sets of coefficients and plotting the magnitude of the transforms, as we did for Figure 5-9(c). So we see three important characteristics in Figure 5-12. First, as we expected, different sets of coefficients give us different frequency magnitude responses. Second, a

---

[†] In Section 6.3, while treating an FIR filter as a special case of an IIR filter, we'll arrive at a mathematical expression for an FIR filter's frequency response in terms of its coefficient values.

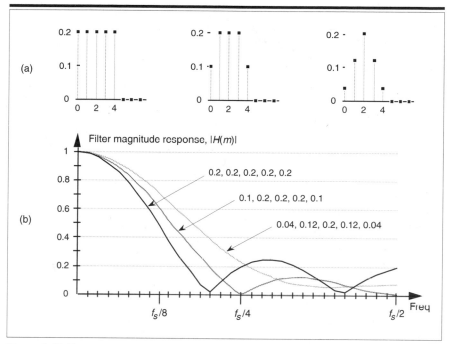

Figure 5-12 Three sets of 5-tap low-pass filter coefficients: (a) sets of coefficients: 0.2, 0.2, 0.2, 0.2, 0.2; 0.1, 0.2, 0.2, 0.2, 0.1; and 0.04, 0.12, 0.2, 0.12, 0.04; (b) frequency magnitude response of three low-pass FIR filters using those sets of coefficients.

sudden change in the values of the coefficient sequence, such as the 0.2 to 0 transition in the first coefficient set, causes ripples, or sidelobes, in the frequency response. Third, if we minimize the suddenness of the changes in the coefficient values, such as the third set of coefficients in Figure 5-12(a), we reduce the sidelobe ripples in the frequency response. However, reducing the sidelobes results in increasing the main lobe width of our low-pass filter. (As we'll see, this is exactly the same effect encountered in the discussion of window functions used with the DFT in Section 3.9.)

To reiterate the function of the filter coefficients, Figure 5-13 shows the 5-tap FIR filter structure using the third set of coefficients from Figure 5-12. The implementation of constant-coefficient transversal FIR filters does not get any more complicated than that shown in Figure 5-13. It's that simple. We can have a filter with more than five taps, but the input signal sample shifting, the multiplications by the constant coefficients, and the summation are all there is to it. (By constant coefficients, we don't mean coefficients whose values are all the same; we mean coefficients whose values remain unchanged, or time-invariant. There is a class of

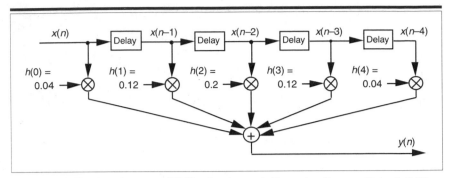

Figure 5-13 Five-tap, low-pass FIR filter implementation using the coefficients 0.04, 0.12, 0.2, 0.12, and 0.04.

digital filters, called adaptive filters, whose coefficient values are periodically changed to adapt to changing input signal parameters. While we won't discuss these adaptive filters in this introductory text, their descriptions are available in the literature[1–5].)

So far, our description of an FIR filter implementation has been presented from a hardware perspective. In Figure 5-13, to calculate a single filter output sample, five multiplications and five additions must take place before the arrival of the next input sample value. In a software implementation of a 5-tap FIR filter, however, all of the input data samples would be previously stored in memory. The software filter routine's job, then, is to access different five-sample segments of the $x(n)$ input data space, perform the calculations shown in Figure 5-13, and store the resulting filter $y(n)$ output sequence in an array of memory locations.[†]

Now that we have a basic understanding of what a digital FIR filter is, let's see what effect is had by using more than five filter taps by learning to design FIR filters.

5.3 Low-Pass FIR Filter Design

OK, instead of just accepting a given set of FIR filter coefficients and analyzing their frequency response, let's reverse the process and design our

---

[†] In reviewing the literature of FIR filters, the reader will often find the term z^{-1} replacing the delay function in Figure 5-13. This equivalence is explained in the next chapter when we study IIR filters.

own low-pass FIR filter. The design procedure starts with the determination of a *desired* frequency response followed by calculating the filter coefficients that will give us that response. There are two predominant techniques used to design FIR filters, the window method, and the so-called optimum method. Let's discuss them in that order.

5.3.1 Window Design Method

The window method of FIR filter design (also called the Fourier series method) begins with our deciding what frequency response we want for our low-pass filter. We can start by considering a continuous low-pass filter, and simulating that filter with a digital filter. We'll define the continuous frequency response $H(f)$ to be ideal, i.e., a low-pass filter with unity gain at low frequencies and zero gain (infinite attenuation) beyond some *cutoff frequency*, as shown in Figure 5-14(a). Representing this $H(f)$ response by a discrete frequency response is straightforward enough because the idea of a discrete frequency response is essentially the same as a continuous frequency response—with one important difference. As described in Sections 2.2 and 3.13, discrete frequency-domain representations are always periodic with the period being the sample rate f_s. The discrete representation of our ideal, continuous low-pass filter $H(f)$ is the periodic response $H(m)$ depicted by the frequency-domain samples in Figure 5-14(b).

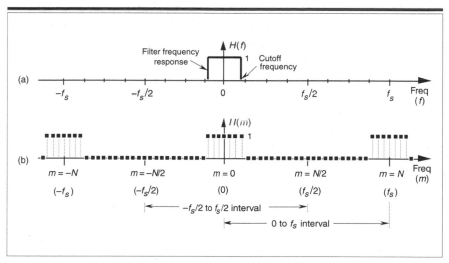

Figure 5-14 Low-pass filter frequency responses: (a) continuous frequency response $H(f)$; (b) periodic, discrete frequency response $H(m)$.

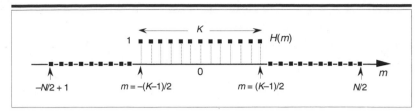

Figure 5-15 Arbitrary, discrete low-pass FIR filter frequency response defined over N frequency-domain samples covering the frequency range of f_s Hz.

We have two ways to determine our low-pass filter's time-domain coefficients. The first way is algebraic:

1. Develop an expression for the discrete frequency response $H(m)$.

2. Apply that expression to the inverse DFT equation to get the time domain $h(k)$.

3. Evaluate that $h(k)$ expression as a function of time index n.

The second method is to define the individual frequency-domain samples representing $H(m)$ and then have a software routine perform the inverse DFT of those samples, giving us the FIR filter coefficients. In either method, we need only define the periodic $H(m)$ over a single period of f_s Hz. As it turns out, defining $H(m)$ in Figure 5-14(b) over the frequency span $-f_s/2$ to $f_s/2$ is the easiest form to analyze algebraically, and defining $H(m)$ over the frequency span 0 to f_s is the best representation if we use the inverse DFT to obtain our filter's coefficients. Let's try both methods to determine the filter's time-domain coefficients.

In the algebraic method, we can define an arbitrary discrete frequency response $H(m)$ using N samples to cover the $-f_s/2$ to $f_s/2$ frequency range and establish K unity-valued samples for the passband of our low-pass filter as shown in Figure 5-15. To determine $h(k)$ algebraically we need to take the inverse DFT of $H(m)$ in the form of

$$h(k) = \frac{1}{N} \sum_{m=-(N/2)+1}^{N/2} H(m)e^{j2\pi mk/N} , \qquad (5\text{-}9)$$

where our time-domain index is k. The solution to Eq. (5-9), derived in Section 3.13 as Eq. (3-59), is repeated here as

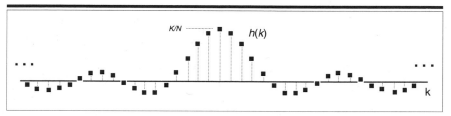

Figure 5-16 Time-domain *h(k)* coefficients obtained by evaluating Eq. (5-10).

$$h(k) = \frac{1}{N} \cdot \frac{\sin(\pi k K / N)}{\sin(\pi k / N)}.$$
(5-10)

If we evaluate Eq. (5-10) as a function of k, we get the sequence shown in Figure 5-16 taking the form of the classic $\sin(x)/x$ function. By reviewing the material in Section 3.13, it's easy to see the great deal of algebraic manipulation required to arrive at Eq. (5-10) from Eq. (5-9). So much algebra, in fact, with its many opportunities for making errors, that digital filter designers like to avoid evaluating Eq. (5-9) algebraically. They prefer to use software routines to perform inverse DFTs (in the form of an inverse FFT) to determine $h(k)$, and so will we.

We can demonstrate the software inverse DFT method of FIR filter design with an example. Let's say we need to design a low-pass FIR filter simulating the continuous frequency response shown in Figure 5-17(a). The discrete representation of the filter's frequency response $H(m)$ is shown in Figure 5-17(b), where we've used $N = 32$ points to represent the frequency-domain variable $H(f)$. Because it's equivalent to Figure 5-17(b) but avoids the negative values of the frequency index m, we represent the discrete frequency samples over the range 0 to f_s in Figure 5-17(c), as opposed to the $-f_s/2$ to $+f_s/2$ range in Figure 5-17(b). OK, we're almost there. Using a 32-point inverse FFT to implement a 32-point inverse DFT of the $H(m)$ sequence in Figure 5-17(c), we get the 32 $h(k)$ values depicted by the dots from $k = -15$ to $k = 16$ in Figure 5-18(a).[†] We have one more step to perform. Because we want our final 31-tap $h(k)$ filter coefficients to be symmetrical with their peak value in the center of the coefficient sample set, we drop the $k = 16$ sample and shift the k index to the left from Figure 5-18(a) giving us the desired

---

[†] If you want to use this FIR design method but only have a forward FFT software routine available, Section 10.6 shows a slick way to perform an inverse FFT with the forward FFT algorithm.

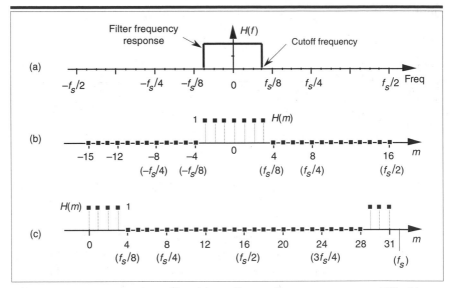

Figure 5-17 An ideal low-pass filter: (a) continuous frequency response $H(f)$; (b) discrete response $H(m)$ over the range of $-f_s/2$ to $f_s/2$ Hz; (c) discrete response $H(m)$ over the range 0 to f_s Hz.

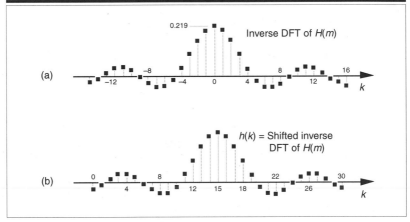

Figure 5-18 Inverse DFT of the discrete response in Figure 5-17(c): (a) normal inverse DFT indexing for k; (b) symmetrical coefficients used for a 31-tap low-pass FIR filter.

$\sin(x)/x$ form of $h(k)$ as shown in Figure 5-18(b). This shift of the index k will not change the frequency magnitude response of our FIR filter. (Remember from our discussion of the DFT Shifting Theorem in Section 3.6 that a shift in the time-domain manifests itself only as a linear phase shift in the frequency domain with no change in the frequency domain magnitude.) The sequence in Figure 5-18(b), then, is now the coefficients we use in the convolution process of Figure 5-5 to implement a low-pass FIR filter.

It's important to demonstrate that the more $h(k)$ terms we use as filter coefficients, the closer we'll approximate our ideal low-pass filter response. Let's be conservative, just use the center nine $h(k)$ coefficients, and see what our filter response looks like. Again, our filter's magnitude response in this case will be the DFT of those nine coefficients as shown on the right side of Figure 5-19(a). The ideal filter's frequency response is also shown for reference as the dashed curve. (To show the details of its shape, we've used a continuous curve for $|H(m)|$ in Figure 5-19(a), but we have to remember that $|H(m)|$ is really a sequence of discrete values.) Notice that using nine coefficients gives us a low-pass filter, but it's certainly far from ideal. Using more coefficients to improve our situation, Figure 5-19(b) shows 19 coefficients and their corresponding frequency magnitude response that is beginning to look more like our desired rectangular response. Notice that magnitude fluctuations, or ripples, are evident in the passband of our $H(m)$ filter response. Continuing, using all 31

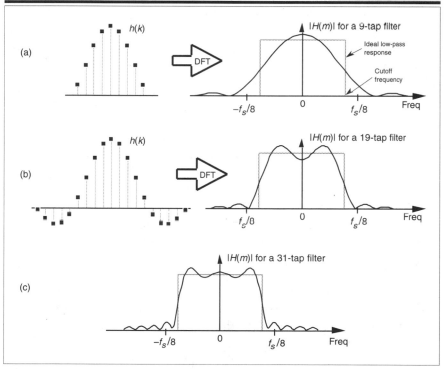

Figure 5-19 Coefficients and frequency responses of three low-pass filters: (a) 9-tap FIR filter; (b) a 19-tap FIR filter; (c) frequency response of the full 31-tap FIR filter.

of the $h(k)$ values for our filter coefficients results in the frequency response in Figure 5-19(c). Our filter's response is getting better (approaching the ideal), but those conspicuous passband magnitude ripples are still present.

It's important that we understand why those passband ripples are in the low-pass FIR filter response in Figure 5-19. Recall the above discussion of convolving the 5-tap averaging filter coefficients, or impulse response, with an input data sequence to obtain the averager's output. We established that convolution in the time domain is equivalent to multiplication in the frequency domain, that we symbolized with Eq. (5-8), and repeat it here as

$$h(k) * x(n) \xrightarrow[\text{IDFT}]{\text{DFT}} H(m) \cdot X(m) . \tag{5-11}$$

This association between convolution in the time-domain and multiplication in the frequency domain, sketched in Figure 5-7, indicates that, if two time-domain sequences $h(k)$ and $x(n)$ have DFTs of $H(m)$ and $X(m)$, respectively, then the DFT of $h(k) * x(n)$ is $H(m) \cdot X(m)$. No restrictions whatsoever need be placed on what the time-domain sequences $h(k)$ and $x(n)$ in Eq. (5-11) actually represent. As detailed later in Section 5.9, convolution in one domain is equivalent to multiplication in the other domain allowing us to state that multiplication in the time domain is equivalent to convolution in the frequency domain, or

$$h(k) \cdot x(n) \xrightarrow[\text{IDFT}]{\text{DFT}} H(m) * X(m) . \tag{5-12}$$

Now we're ready to understand why the magnitude ripples are present in Figure 5-19.

Rewriting Eq. (5-12) and replacing the $h(k)$ and $x(n)$ expressions with $h^\infty(k)$ and $w(k)$, respectively,

$$h^\infty(k) \cdot w(k) \xrightarrow[\text{IDFT}]{\text{DFT}} H^\infty(m) * W(m) . \tag{5-13}$$

Let's say that $h^\infty(k)$ represents an infinitely long $\sin(x)/x$ sequence of ideal low-pass FIR filter coefficients and that $w(k)$ represents a window sequence that we use to truncate the $\sin(x)/x$ terms as shown in Figure 5-20. Thus, the $w(k)$ sequence is a finite-length set of unity values and its DFT is $W(m)$. The length of $w(k)$ is merely the number of coefficients, or taps, we intend

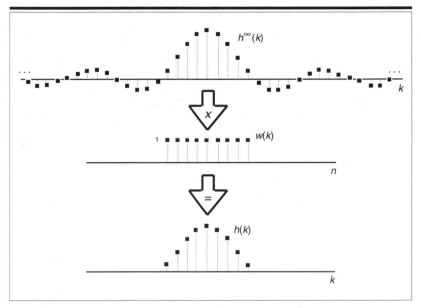

Figure 5-20 Infinite $h^\infty(k)$ sequence windowed by $w(k)$ to define the final filter coefficients $h(k)$.

to use to implement our low-pass FIR filter. With $h^\infty(k)$ defined as such, the product $h^\infty(k) \cdot w(k)$ represents the truncated set of filter coefficients $h(k)$ in Figures 5-19(a) and 5-19(b). So, from Eq. (5-13), the FIR filter's true frequency response $H(m)$ is the convolution

$$H(m) = H^\infty(m) * W(m) . \tag{5-14}$$

We depict this convolution in Figure 5-21 where, to keep the figure from being so busy, we show $H^\infty(m)$ (the DFT of the $h^\infty(k)$ coefficients) as the dashed rectangle. Keep in mind that it's really a sequence of constant-amplitude sample values.

Let's look at Figure 5-21(a) very carefully to see why all three $|H(m)|$s exhibit passband ripple in Figure 5-19. We can view a particular sample value of the $H(m) = H^\infty(m) * W(m)$ convolution as being the sum of the products of $H^\infty(m)$ and $W(m)$ for a particular frequency shift of $W(m)$. $H^\infty(m)$ and the unshifted $W(m)$ are shown in Figure 5-21(a.) With an assumed value of unity for all of $H^\infty(m)$, a particular $H(m)$ value is now merely the sum of the $W(m)$ samples that overlap the $H^\infty(m)$ rectangle. So, with a $W(m)$ frequency shift of 0 Hz, the sum of the $W(m)$ samples that overlap the $H^\infty(m)$ rectangle in Figure 5-21(a) is the value of $H(m)$ at 0 Hz.

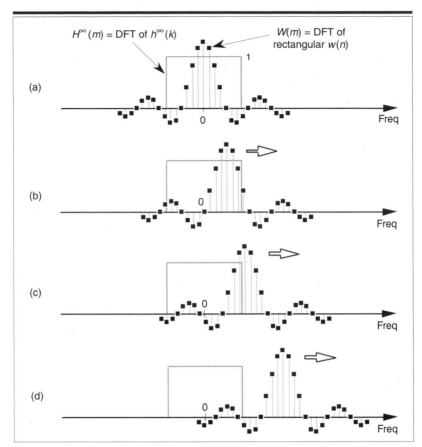

Figure 5-21 Convolution $W(m) * H^{\infty}(m)$: (a) unshifted $W(m)$ and $H^{\infty}(m)$; (b) shift of $W(m)$ leading to ripples within $H(m)$'s positive frequency passband; (c) shift of $W(m)$ causing response roll-off near $H(m)$'s positive cutoff frequency; (d) shift of $W(m)$ causing ripples beyond $H(m)$'s positive cutoff frequency.

As $W(m)$ is shifted to the right to give us additional positive frequency $H(m)$ values, we can see that the sum of the positive and negative values of $W(m)$ under the rectangle oscillate during the shifting of $W(m)$. As the convolution shift proceeds, Figure 5-21(b) shows why there are ripples in the passband of $H(m)$—again, the sum of the positive and negative $W(m)$ samples under the $H^{\infty}(m)$ rectangle continue to vary as the $W(m)$ function is shifted. The $W(m)$ frequency shift, indicated in Figure 5-21(c), where the peak of $W(m)$'s main lobe is now outside the $H^{\infty}(m)$ rectangle, corresponds to the frequency where $H(m)$'s passband begins to roll off. Figure 5-21(d) shows that, as the $W(m)$ shift continues, there will be ripples in $H(m)$

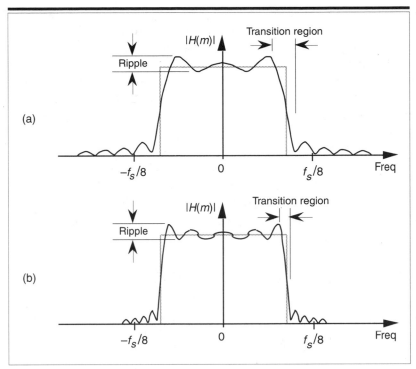

Figure 5-22 Passband ripple and transition regions: (a) for a 31-tap low-pass filter; (b) for a 63-tap low-pass filter.

beyond the positive cutoff frequency.[†] The point of all of this is that the ripples in $H(m)$ are caused by the sidelobes of $W(m)$.

Figure 5-22 helps us answer the question: How many $\sin(x)/x$ coefficients do we have to use (or how wide must $w(k)$ be) to get nice sharp falling edges and no ripples in our $H(m)$ passband? The answer is that we can't get there from here. It doesn't matter how many $\sin(x)/x$ coefficients (filter taps) we use, there will always be filter passband ripple. As long as $w(k)$ is a finite number of unity values (i.e., a rectangular window of finite width) there will be sidelobe ripples in $W(m)$, and this will induce passband ripples in the final $H(m)$ frequency response. To illustrate that increasing the number of $\sin(x)/x$ coefficients doesn't reduce passband ripple, we repeat the 31-tap, low-pass filter response in Figure 5-22(a). The frequency response, using 63 coefficients, is shown in

---

[†] In Figure 5-21(b), had we started to shift $W(m)$ to the left in order to determine the negative frequency portion of $H(m)$, we would have obtained the mirror image of the positive frequency portion of $H(m)$.

Figure 5-22(b), and the passband ripple remains. We can make the filter's transition region more narrow using additional $h(k)$ filter coefficients, but we cannot eliminate the passband ripple. That ripple, known as Gibbs' phenomenon, manifests itself anytime a function ($w(k)$ in this case) with a instantaneous discontinuity is represented by a Fourier series[6–8]. No finite set of sinusoids will be able to change fast enough to be exactly equal to an instantaneous discontinuity. Another way to state this Gibbs' dilemma is that, no matter how wide our $w(k)$ window is, its DFT of $W(m)$ will always have sidelobe ripples. As shown in Figure 5-22(b), we can use more coefficients by extending the width of the rectangular $w(k)$ to narrow the filter transition region, but a wider $w(k)$ does not eliminate the filter passband ripple nor does it even reduce their peak-to-peak ripple magnitudes, as long as $w(k)$ has sudden discontinuities.

5.3.2 Windows Used in FIR Filter Design

OK. The good news is that we can minimize FIR passband ripple with window functions the same way we minimized DFT leakage in Section 3.9. Here's how. Looking back at Figure 5-20, by truncating the infinitely long $h^\infty(k)$ sequence through multiplication by the rectangular $w(k)$, our final $h(k)$ exhibited ripples in the frequency-domain passband. Figure 5-21 shows us that the passband ripples were caused by $W(m)$'s sidelobes that, in turn, were caused by the sudden discontinuities from zero to one and one to zero in $w(k)$. If we think of $w(k)$ in Figure 5-20 as a rectangular window, then, it is $w(k)$'s abrupt amplitude changes that are the source of our filter passband ripple. The window FIR design method is the technique of reducing $w(k)$'s discontinuities by using window functions other than the rectangular window.

Consider Figure 5-23 to see how a nonrectangular window function can be used to design low-ripple FIR digital filters. Imagine if we replaced Figure 5-20's rectangular $w(k)$ with the Blackman window function whose discrete values are defined as[†]

$$\omega(k) = 0.42 - 0.5\cos\left(\frac{2\pi k}{N}\right) + 0.08\cos\left(\frac{4\pi k}{N}\right), \text{ for } k = 0, 1, 2, \ldots, N-1 \ . \quad (5\text{-}15)$$

[†] As we mentioned in Section 3.9, specific expressions for window functions depend on the range of the sample index k. Had we defined k to cover the range $-N/2, < k < N/2$, for example, the expression for the Blackman window would have a sign change and be $w(k) = 0.42 + 0.5\cos(2\pi k/N) + 0.08\cos(4\pi k/N)$.

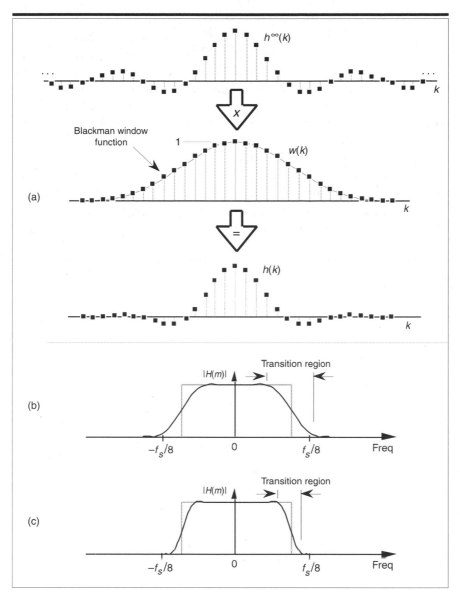

Figure 5-23 Coefficients and frequency response of a 31-tap Blackman-windowed FIR Filter: (a) defining the windowed filter coefficients $h(k)$; (b) low-ripple 31-tap frequency response; (c) low-ripple 63-tap frequency response.

This situation is depicted for $N = 31$ in Figure 5-23(a) where Eq. (5-15)'s $w(k)$ looks very much like the Hanning window function in Figure 3-17(a). This Blackman window function results in the 31 smoothly tapered $h(k)$ coefficients at the bottom of Figure 5-23(a). Notice two things about the resulting $H(m)$ in Figure 5-23(b). First, the good news. The passband ripples are greatly reduced from those evident in Figure 5-22(a)—so our Blackman window function did its job. Second, the price we paid for reduced passband ripple is a wider $H(m)$ transition region. We can get a steeper filter response roll-off by increasing the number of taps in our FIR filter. Figure 5-23(c) shows the improved frequency response had we used a 63-coefficient Blackman window function for a 63-tap FIR filter. So using a nonrectangular window function reduces passband ripple at the expense of slower passband to stopband roll-off.

A graphical comparison of the frequency responses for the rectangular and Blackman windows is provided in Figure 5-24. (The curves in Figure 5-24 were obtained for the window functions defined by 32 discrete samples, to which 480 zeros were appended, applied to a 512-point DFT.) The

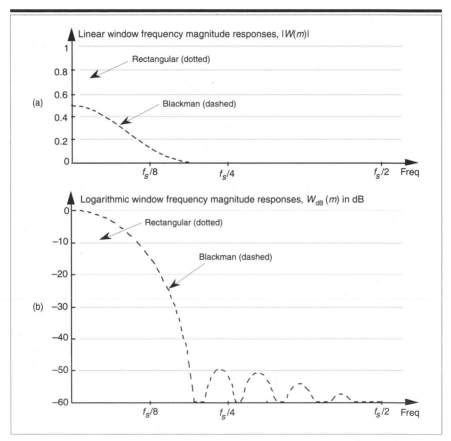

Figure 5-24 Rectangular vs. Blackman window frequency magnitude responses: (a) $|W(m)|$ on a linear scale; (b) normalized logarithmic scale of $W_{dB}(m)$.

sidelobe magnitudes of the Blackman window's $|W(m)|$ are too small to see on a linear scale. We can see those sidelobe details by plotting the two windows' frequency responses on a logarithmic scale and normalizing each plot so that their main lobe peak values are both zero dB. For a given window function, we can get the log magnitude response of $W_{dB}(m)$ by using the expression

$$W_{dB}(m) = 20 \cdot \log_{10}\left(\frac{|W(m)|}{|W(0)|}\right). \tag{5-16}$$

(The $|W(0)|$ term in Eq. (5-16) is the magnitude of $W(m)$ at the peak of the main lobe when $m = 0$.) Figure 5-24(b) shows us the greatly reduced sidelobe levels of the Blackman window and how that window's main lobe is almost three times as wide as the rectangular window's main lobe.

Of course, we could have used any of the other window functions, discussed in Section 3.9, for our low-pass FIR filter. That's why this FIR filter design technique is called the window design method. We pick a window function and multiply it by the $\sin(x)/x$ values from $H^{\infty}(m)$ in Figure 5-23(a) to get our final $h(k)$ filter coefficients. It's that simple. Before we leave the window method of FIR filter design, let's introduce two other interesting window functions.

Although the Blackman window and those windows discussed in Section 3.9 are useful in FIR filter design, we have little control over their frequency responses; that is, our only option is to select some window function and accept its corresponding frequency response. Wouldn't it be nice to have more flexibility in trading off, or striking a compromise between, a window's main lobe width and sidelobe levels? Fortunately, there are two popular window functions that give us this opportunity. Called the Chebyshev and the Kaiser window functions, they're defined by the following formidable expressions:

$$w(k) = \text{the } N\text{-point inverse DFT of}$$

Chebyshev window:→
(also called the Dolph–
Chebyshev and the
Tchebyschev window)

$$\frac{\cos\left[N \cdot \cos^{-1}\left[\alpha \cdot \cos\left(\pi\frac{m}{N}\right)\right]\right]}{\cosh[N \cdot \cosh^{-1}(\alpha)]},$$

$$\text{where } \alpha = \cosh\left(\frac{1}{N}\cosh^{-1}(10^{\gamma})\right) \text{ and } m = 0, 1, 2, ..., N-1, \tag{5-17}$$

Kaiser window:→
(also called the
Kaiser–Bessel window)

$$\omega(k) = \frac{I_o\left[\beta\sqrt{1-\left(\dfrac{k-p}{p}\right)^2}\right]}{I_o(\beta)},$$

for $k = 0, 1, 2, \ldots, N-1$, and $p = (N-1)/2$.   (5-18)

Two typical Chebyshev and Kaiser window functions and their frequency magnitude responses are shown in Figure 5-25. For comparison, the rectangular and Blackman window functions are also shown in that figure. (Again, the curves in Figure 5-25(b) were obtained for window

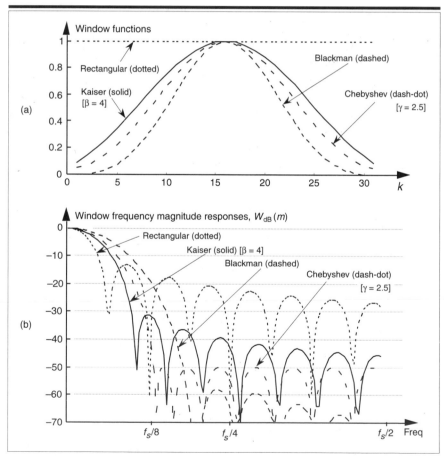

Figure 5-25 Typical window functions used with digital filters: (a) window coefficients in the time domain; (b) frequency-domain magnitude responses in dB.

functions defined by 32 discrete samples, with 480 zeros appended, applied to a 512-point DFT.)

Equation (5-17) was originally based on the analysis of antenna arrays using the mathematics of Chebyshev polynomials[9–11]. Equation (5-18) evolved from Kaiser's study of prolate spheroid functions using zeroth-order Bessel functions[12–13]. Don't be intimidated by the complexity of Eqs. (5-17) and (5-18)—at this point, we need not be concerned with the mathematical details of their development. We just need to realize that the *control* parameters γ and β, in Eqs. (5-17) and (5-18), give us control over the windows' main lobe widths and the sidelobe levels.

Let's see how this works for Chebyshev window functions, having four separate values of γ, and their frequency responses shown in Figure 5-26. FIR filter designers applying the window method typically use predefined software routines to obtain their Chebyshev window coefficients. Commercial digital signal processing software packages allow the user to specify three things: the window function (Chebyshev in this case), the desired number of coefficients (the number of taps in the FIR filter), and the value of γ. Selecting different values for γ enables us to adjust the sidelobe levels and see what effect those values have on main lobe width, a capability that we didn't have with the Blackman window or the window functions discussed in Section 3.9. The Chebyshev window function's stopband attenuation, in decibels, is equal to

$$\text{Atten}_{\text{Cheb}} = -20\gamma . \tag{5-19}$$

So, for example, if we needed our sidelobe levels to be no greater than –60 dB below the main lobe, we use Eq. (5-19) to establish a γ value of 3.0 and let the software generate the Chebyshev window coefficients.[†]

The same process applies to the Kaiser window, as shown in Figure 5-27. Commercial software packages allow us to specify β in Eq. (5-18) and provide us with the associated window coefficients. The curves in Figure 5-27(b), obtained for Kaiser window functions defined by 32 discrete samples, show that we can select the desired sidelobe levels and see what effect this has on the main lobe width.

Chebyshev or Kaiser, which is the best window to use? It depends on the application. Returning to Figure 5-25(b), notice that, unlike the constant sidelobe peak levels of the Chebyshev window, the Kaiser window's

---

[†] By the way, some digital signal processing software packages require that we specify $\text{Atten}_{\text{Cheb}}$ in decibels instead of γ. That way, we don't have to bother using Eq. (5-19) at all.

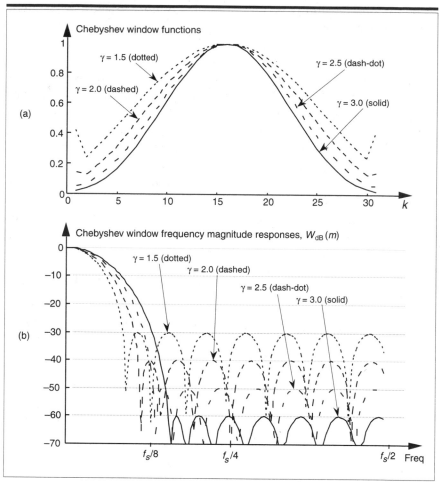

Figure 5-26 Chebyshev window functions for various γ values: (a) window coefficients in the time domain; (b) frequency-domain magnitude responses in dB.

sidelobes decrease with increased frequency. However, the Kaiser sidelobes are higher than the Chebyshev window's sidelobes near the main lobe. Our primary trade-off here is trying to reduce the sidelobe levels without broadening the main lobe too much. Digital filter designers typically experiment with various values of γ and β for the Chebyshev and Kaiser windows to get the optimum $W_{dB}(m)$ for a particular application. (For that matter, the Blackman window's very low sidelobe levels outweigh its wide main lobe in many applications.) Different window functions have their own individual advantages and disadvantages for FIR filter design. Regardless of the nonrectangular window function used, they

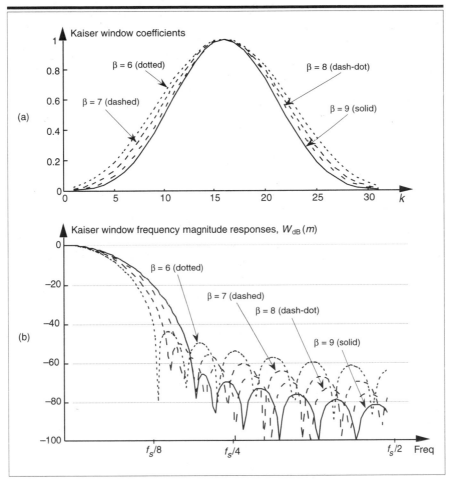

Figure 5-27 Kaiser window functions for various β values: (a) window coefficients in the time domain; (b) frequency-domain magnitude responses in dB.

always decrease the FIR filter passband ripple over that of the rectangular window. For the enthusiastic reader, a very thorough discussion of window functions can be found in reference [14].

5.4 Bandpass FIR Filter Design

The window method of low-pass FIR filter design can be used as the first step in designing a bandpass FIR filter. Let's say we want a 31-tap FIR filter with the frequency response shown in Figure 5-22(a), but instead of being centered about zero Hz, we want the filter's passband to be centered

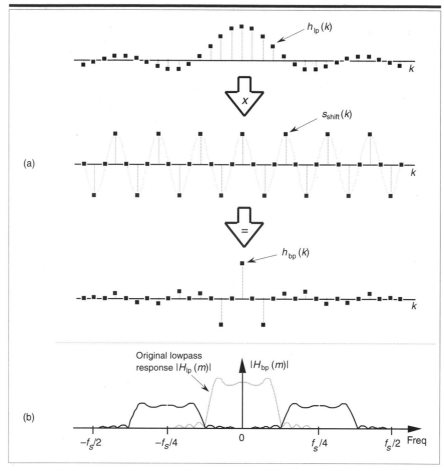

Figure 5-28 Bandpass filter with frequency response centered at $f_s/4$:
(a) generating 31-tap filter coefficients $h_{bp}(k)$; (b) frequency
magnitude response $|H_{bp}(m)|$.

about $f_s/4$ Hz. If we define a low-pass FIR filter's coefficients as $h_{lp}(k)$, our
problem is to find the $h_{bp}(k)$ coefficients of a bandpass FIR filter. As shown
in Figure 5-28, we can shift $H_{lp}(m)$'s frequency response by multiplying the
filter's $h_{lp}(k)$ low-pass coefficients by a sinusoid of $f_s/4$ Hz. That sinusoid
is represented by the $s_{shift}(k)$ sequence in Figure 5-28(a) whose values are
a sinewave sampled at a rate of four samples per cycle. Our final 31-tap
$h_{bp}(k)$ FIR bandpass filter coefficients are

$$h_{bp}(k) = h_{lp}(k) \cdot s_{shift}(k) , \qquad (5\text{-}20)$$

whose frequency magnitude response $|H_{bp}(m)|$ is shown as the solid curves in Figure 5-28(b). The actual magnitude of $|H_{bp}(m)|$ is half that of the original $|H_{lp}(m)|$ because half the values in $h_{bp}(k)$ are zero when $s_{shift}(k)$ corresponds exactly to $f_s/4$. This effect has an important practical implication. It means that, when we design an N-tap bandpass FIR filter centered at a frequency of $f_s/4$ Hz, we only need to perform approximately $N/2$ multiplications for each filter output sample. (There's no reason to multiply an input sample value, $x(n-k)$, by zero before we sum all the products from Eq. (5-6) and Figure 5-13, right? We just don't bother to perform the unnecessary multiplications at all.) Of course, when the bandpass FIR filter's center frequency is other than $f_s/4$, we're forced to perform the full number of N multiplications for each FIR filter output sample.

Notice, here, that the $h_{lp}(k)$ low-pass coefficients in Figure 5-28(a) have not been multiplied by any window function. In practice, we'd use an $h_{lp}(k)$ that has been windowed prior to implementing Eq. (5-20) to reduce the passband ripple. If we wanted to center the bandpass filter's response at some frequency other than $f_s/4$, we merely need to modify $s_{shift}(k)$ to represent sampled values of a sinusoid whose frequency is equal to the desired bandpass center frequency. That new $s_{shift}(k)$ sequence would then be used in Eq. (5-20) to get the new $h_{bp}(k)$.

5.5 Highpass FIR Filter Design

Going one step further, we can use the bandpass FIR filter design technique to design a highpass FIR filter. To obtain the coefficients for a highpass filter, we need only modify the shifting sequence $s_{shift}(k)$ to make it represent a sampled sinusoid whose frequency is $f_s/2$. This process is shown in Figure 5-29. Our final 31-tap highpass FIR filter's $h_{hp}(k)$ coefficients are

$$h_{hp}(k) = h_{lp}(k) \cdot s_{shift}(k)$$

$$= h_{lp}(k) \cdot (1, -1, 1, -1, 1, -1, \text{etc.}) , \qquad (5\text{-}21)$$

whose $|H_{hp}(m)|$ frequency response is the solid curve in Figure 5-29(b). Because $s_{shift}(k)$ in Figure 5-29(a) has alternating plus and minus ones, we can see that $h_{hp}(k)$ is merely $h_{lp}(k)$ with the sign changed for every other coefficient. Unlike $|H_{bp}(m)|$ in Figure 5-28(b), the $|H_{hp}(m)|$ response in Figure 5-29(b) has the same amplitude as the original $|H_{lp}(m)|$.

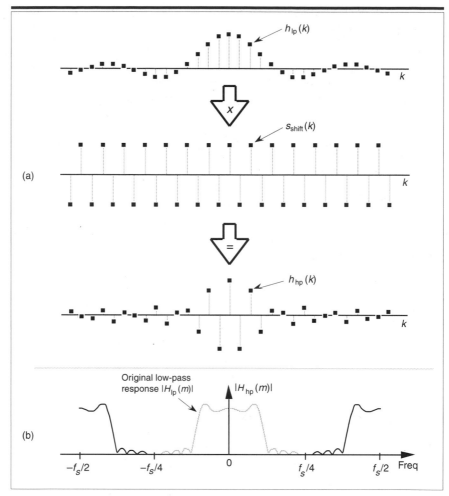

Figure 5-29 Highpass filter with frequency response centered at $f_s/2$:
(a) generating 31-tap filter coefficients $h_{hp}(k)$; (b) frequency
magnitude response $|H_{hp}(m)|$.

Again, notice that the $h_{lp}(k)$ low-pass coefficients in Figure 5-29(a) have
not been modified by any window function. In practice, we'd use a windowed $h_{lp}(k)$ to reduce the passband ripple before implementing Eq. (5-21).

5.6 Remez Exchange FIR Filter Design Method

Let's introduce one last FIR filter design technique that has found wide
acceptance in practice. The Remez Exchange FIR filter design method
(also called the Parks-McClellan, or Optimal method) is a popular tech-

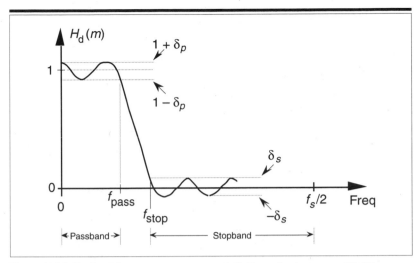

Figure 5-30 Desired frequency response definition of a low-pass FIR filter using the Remez Exchange design method.

nique used to design high-performance FIR filters.[†] To use this design method, we have to visualize a desired frequency response $H_d(m)$ like that shown in Figure 5-30.

We have to establish a desired passband cutoff frequency f_{pass} and the frequency where the attenuated stopband begins, f_{stop}. In addition, we must establish the variables δ_p and δ_s that define our desired passband and stopband ripple. Passband and stopband ripple, in decibels, are related to δ_p and δ_s by[15]

$$\text{Passband ripple} = 20 \cdot \log_{10}(1 + \delta_p) , \qquad (5\text{-}22)$$

and

$$\text{Stopband ripple} = -20 \cdot \log_{10}(\delta_s) \cdot \qquad (5\text{-}22')$$

(Some of the early journal papers describing the Remez design method used the equally valid expression $-20 \cdot \log_{10}(\delta_p)$ to define the passband ripple in decibels. However, Eq. (5-22) is the most common form used today.) Next, we apply these parameters to a computer software routine that generates the filter's N time-domain $h(k)$ coefficients where N is the minimum number of filter taps to achieve the desired filter response.

[†] *Remez* is pronounced re-'mã, like "away."

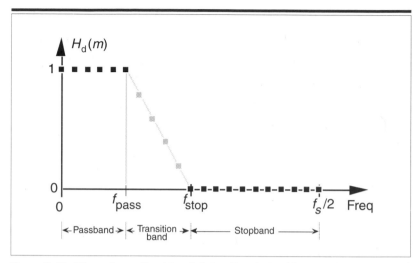

Figure 5-31 Alternate method for defining the desired frequency response of a low-pass FIR filter using the Remez Exchange technique.

On the other hand, some software Remez routines assume that we want δ_p and δ_s to be as small as possible and require us only to define the desired values of the $H_d(m)$ response as shown by the solid black dots in Figure 5-31. The software then adjusts the values of the undefined (shaded dots) values of $H_d(m)$ to minimize the error between our desired and actual frequency response while minimizing δ_p and δ_s. The filter designer has the option to define some of the $H_d(m)$ values in the transition band, and the software calculates the remaining undefined $H_d(m)$ transition band values. With this version of the Remez algorithm, the issue of most importance becomes how we define the transition region. We want to minimize its width while, at the same time, minimizing passband and stopband ripple. So exactly how we design an FIR filter using the Remez Exchange technique is specific to the available filter design software. Although the mathematics involved in the development of the Remez Exchange method is rather complicated, we don't have to worry about that here[16–20]. Just remember that the Remez Exchange design method gives us a Chebyshev-type filter whose actual frequency response is as close as possible to the desired $H_d(m)$ response for a given number of filter taps.

To illustrate the advantage of the Remez method, the solid curve in Figure 5-32 shows the frequency response of a 31-tap FIR designed using this technique. For comparison, Figure 5-32 also shows the frequency responses of two 31-tap FIR filters for the same passband width using the

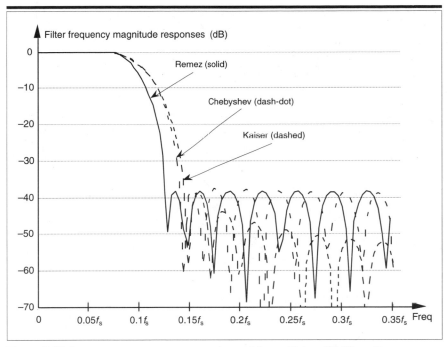

Figure 5-32 Frequency response comparison of three 31-tap FIR filters: Remez, Chebyshev windowed, and Kaiser windowed.

Chebyshev and Kaiser windowing techniques. Notice how the three filters have roughly the same stopband sidelobe levels, near the main lobe, but that the Remez filter has the more desirable (steeper) transition band roll-off.

5.7 Half-Band FIR Filters

There's a specialized FIR filter that's proved useful in decimation applications[21–25]. Called a *half-band FIR filter*, its frequency response is symmetrical about the $f_s/4$ point as shown in Figure 5-33(a). As such, the sum of f_{pass} and f_{stop} is $f_s/2$. This symmetry has the beautiful property that the time-domain FIR impulse response has every other filter coefficient being zero, except at the peak. This enables us to avoid approximately half the number of multiplications when implementing this kind of filter. By way of example, Figure 5-33(b) shows the coefficients for a 31-tap half-band filter where Δf was defined to be approximately $f_s/32$ using the Remez Exchange method. (To preserve further symmetry, the parameters δ_p and δ_s were specified to be equal to each other.)

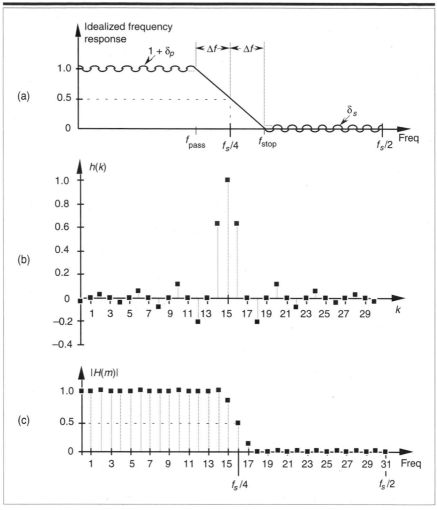

Figure 5-33 Half-Band FIR filter: (a) idealized continuous frequency response; (b) coefficients for a 31-tap half-band FIR filter; (c) discrete frequency magnitude response of a 31-tap half-band FIR filter.

Notice how the alternating $h(k)$ coefficients are zero, so we have to perform only 17 multiplications for each filter output sample instead of the expected 31 multiplications. For the general case of an S-tap half-band FIR filter, we'll need perform only $(S + 1)/2 + 1$ multiplications for each output sample.[†] Taking the DFT of those $h(k)$ coefficients gives us the half-band FIR frequency response shown in Figure 5-33(c).

[†] Section 10.8 shows a technique to further reduce the number of necessary multiplications for certain types of FIR filters including half-band filters.

5.8 Phase Response of FIR Filters

Although we illustrated a couple of output phase shift examples for our original averaging FIR filter in Figure 5-10, the subject of FIR phase response deserves additional attention. One of the dominant features of FIR filters is their linear phase response that we can demonstrate by way of example. Given the 25 $h(k)$ FIR filter coefficients in Figure 5-34(a), we can perform a DFT to determine the filter's $H(m)$ frequency response. The normalized real part, imaginary part, and magnitude of

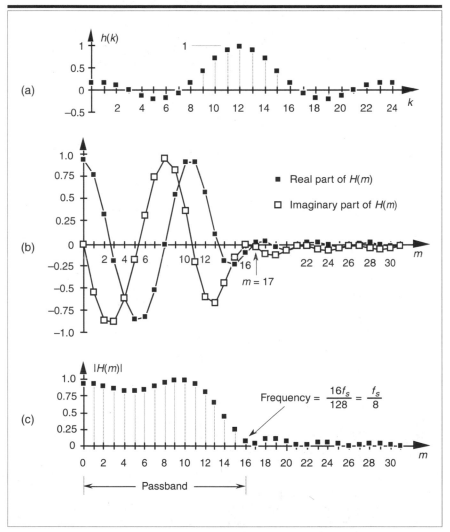

Figure 5-34 FIR filter frequency response $H(m)$: (a) $h(k)$ filter coefficients; (b) real and imaginary parts of $H(m)$; (c) magnitude of $H(m)$.

$H(m)$ are shown in Figures 5-34(b) and 5-34(c), respectively.[†] Being complex values, each $H(m)$ sample value can be described by its real and imaginary parts, or equivalently, by its magnitude $|H(m)|$ and its phase $H_\emptyset(m)$ shown in Figure 5-35(a).

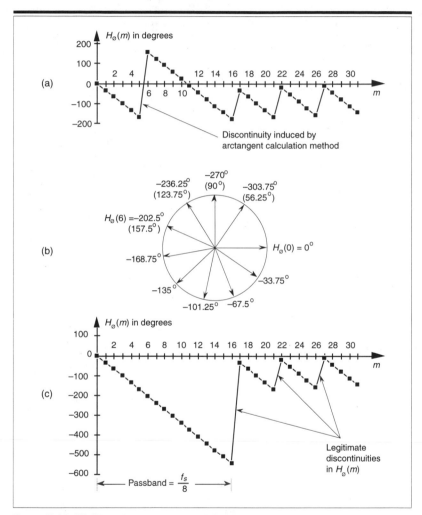

Figure 5-35 FIR filter phase response $H_\emptyset(m)$ in degrees: (a) calculated $H_\emptyset(m)$; (b) polar plot of $H_\emptyset(m)$'s first ten phase angles in degrees; (c) actual $H_\emptyset(m)$.

[†] Any DFT size greater than the $h(k)$ width of 25 is sufficient to obtain $H(m)$. The $h(k)$ sequence was padded with 103 zeros to take a 128-point DFT resulting in the $H(m)$ sample values in Figure 5-34.

The phase of a complex quantity is, of course, the arctangent of the imaginary part divided by the real part, or $\varnothing = \tan^{-1}(\text{imag}/\text{real})$. Thus the phase of $H_\varnothing(m)$ is determined from the samples in Figure 5-34(b).

The phase response in Figure 5-35(a) certainly looks linear over selected frequency ranges, but what do we make of those sudden jumps, or discontinuities, in this phase response? If we were to plot the angles of $H_\varnothing(m)$ starting with the $m = 0$ sample on a polar graph, using the nonzero real part of $H(0)$, and the zero-valued imaginary part of $H(0)$, we'd get the zero-angled $H_\varnothing(0)$ phasor shown on the right side of Figure 5-35(b). Continuing to use the real and imaginary parts of $H(m)$ to plot additional phase angles results in the phasors going clockwise around the circle in increments of $-33.75°$. It's at the $H_\varnothing(6)$ that we discover the cause of the first discontinuity in Figure 5-35(a). Taking the real and imaginary parts of $H(6)$, we'd plot our phasor oriented at an angle of $-202.5°$. But Figure 5-35(a) shows that $H_\varnothing(6)$ is equal to $157.5°$. The problem lies in the software routine used to generate the arctangent values plotted in Figure 5-35(a). The software adds $360°$ to any negative angles in the range of $-180° > \varnothing \geq -360°$, i.e., angles in the upper half of the circle. This makes $\varnothing$ a positive angle in the range of $0° < \varnothing \leq 180°$ and that's what gets plotted. (This apparent discontinuity between $H_\varnothing(5)$ and $H_\varnothing(6)$ is called *phase wrapping*.) So the true $H_\varnothing(6)$ of $-202.5°$ is converted to a $+157.5°$ as shown in parentheses in Figure 5-35(b). If we continue our polar plot for additional $H_\varnothing(m)$ values, we'll see that their phase angles continue to decrease with an angle increment of $-33.75°$. If we compensate for the software's behavior and plot phase angles more negative than $-180°$, by *unwrapping* the phase, we get the true $H_\varnothing(m)$ shown in Figure 5-35(c).[†] Notice that $H_\varnothing(m)$ is, indeed, linear over the passband of $H(m)$. It's at $H_\varnothing(17)$ that our particular $H(m)$ experiences a polarity change of its real part while its imaginary part remains negative—this induces a true phase angle discontinuity that really is a constituent of $H(m)$ at $m = 17$. (Additional phase discontinuities occur each time the real part of $H(m)$ reverses polarity, as shown in Figure 5-35(c).) The reader may wonder why we care about the linear phase response of $H(m)$. The answer, an important one, requires us to introduce the notion of group delay.

Group delay is defined as the derivative of the phase with respect to frequency, or $G = \Delta\varnothing/\Delta f$. For FIR filters, then, group delay is the slope of the $H_\varnothing(m)$ response curve. When the group delay is constant, as it is over the

---

[†] When plotting filter phase responses, if we encounter a phase angle sample $\varnothing$ that looks like an unusual discontinuity, it's a good idea to add $360°$ to $\varnothing$ when $\varnothing$ is negative, or $-360°$ when $\varnothing$ is positive, to see if that compensates for any software anomalies.

passband of all FIR filters having symmetrical coefficients, all frequency components of the filter input signal are delayed by an equal amount of time G before they reach the filter's output. This means that no phase distortion is induced in the filter's desired output signal, and this is crucial in communications signals. For amplitude modulation (AM) signals, constant group delay preserves the time waveform shape of the signal's modulation envelope. That's important because the modulation portion of an AM signal contains the signal's information. Conversely, a nonlinear phase will distort the audio of AM broadcast signals, blur the edges of television video images, blunt the sharp edges of received radar pulses, and increase data errors in digital communication signals. (Group delay is sometimes called *envelope delay* because group delay was originally the subject of analysis due to its affect on the envelope, or modulation signal, of amplitude modulation AM systems.) Of course we're not really concerned with the group delay outside the passband because signal energy outside the passband is what we're trying to eliminate through filtering.

Over the passband frequency range for an S-tap FIR digital filter, group delay has been shown to be given by

$$G_{odd} = \frac{(S-1)t_s}{2} ,$$

(5-23)

when S is odd, and

$$G_{even} = \frac{S \cdot t_s}{2} ,$$

(5-23')

when S is even, where t_s is the sample period $(1/f_s)$.[†]

Although we used a 128-point DFT to obtain the frequency responses in Figures 5-34 and 5-35, we could just as well have used $N = 32$-point or $N = 64$-point DFTs. These smaller DFTs give us the phase response curves shown in Figure 5-36(a) and 5-36(b). Notice how different the phase response curves are when $N = 32$ in Figure 5-36(a) compared to when $N = 128$ in Figure 5-36(c). The phase angle resolution is much finer in Figure 5-36(c). The passband phase angle resolution, or increment $\Delta\phi$, is given by

$$\Delta\phi = \frac{-G \cdot 360°}{N} ,$$

(5-24)

[†] As derived in Section 3.4 of reference [15], and Section 5.7.3 of reference [19].

where N is the number of points in the DFT. So, for our $S = 25$-tap filter in Figure 5-34(a), $G = 12$, and $\Delta\phi$ is equal to $-12 \cdot 360°/32 = -135°$ in Figure 5-36(a), and $\Delta\phi$ is $-33.75°$ in Figure 5-36(c). If we look carefully at the sample values in Figure 5-36(a), we'll see that they're all included within the samples in Figures 5-36(b) and 5-36(c).

Let's conclude this FIR phase discussion by reiterating the meaning of phase response. The phase, or phase delay, at the output of an FIR filter is

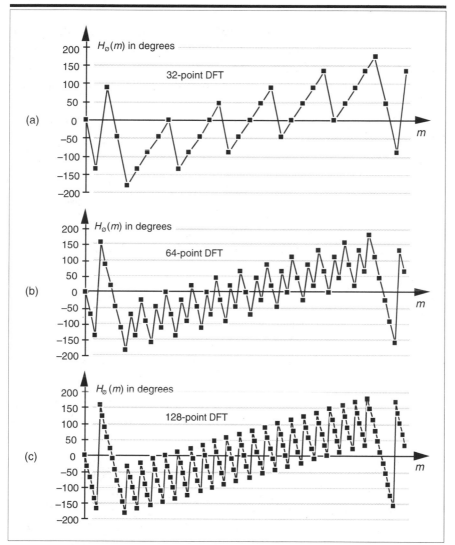

Figure 5-36 FIR filter phase response $H_\phi(m)$ in degrees: (a) calculated using a 32-point DFT; (b) using a 64-point DFT; (c) using a 128-point DFT.

Table 5-2 Values Used in Eq. (5-25) for the Frequency $f_s/32$

DFT size, N	Index m	$H_\emptyset(mf_s/N)$
32	1	-135°
64	2	-135°
128	4	-135°

the phase of the first output sample relative to the phase of the filter's first input sample. Over the passband, that phase shift, of course, is a linear function of frequency. This will be true only as long as the filter has symmetrical coefficients. Figure 5-10 is a good illustration of an FIR filter's output phase delay.

For FIR filters, the output phase shift measured in degrees, for the passband frequency $f = mf_s/N$, is expressed as

$$\text{phase delay} = H_\emptyset(mf_s/N) = m \cdot \Delta\emptyset = \frac{-m \cdot G \cdot 360^\circ}{N}. \tag{5-25}$$

We can illustrate Eq. (5-25) and show the relationship between the phase responses in Figure 5-36, by considering the phase delay associated with the frequency of $f_s/32$ in Table 5-2. The subject of group delay is described further in Appendix F where an example of envelope delay distortion, due to a filter's nonlinear phase, is illustrated.

5.9 A Generic Description of Discrete Convolution

Although convolution was originally an analysis tool used to prove continuous signal processing theorems, we now know that convolution affects every aspect of digital signal processing. Convolution influences our results whenever we analyze or filter any finite set of data samples from a linear time-invariant system. Convolution not only constrains DFTs to be just approximations of the continuous Fourier transform; it is the reason that discrete spectra are periodic in the frequency domain. It's interesting to note that, although we use the process of convolution to implement FIR digital filters, convolution effects induce frequency response ripple preventing us from ever building a perfect digital filter. Its influence is so pervasive that, to repeal the law of convolution, quoting a

$$Y_j = \sum_{k=0}^{N-1} P_k \cdot Q_{j-k}, \quad \text{or}$$

$$
\begin{bmatrix} Y \\ Y \\ Y \\ \cdot \\ \cdot \\ \cdot \\ Y_{N-1} \end{bmatrix}
=
\begin{bmatrix}
Q_0 & Q_{N-1} & Q_{N-2} & \cdots & Q_1 \\
Q_1 & Q_0 & Q_{N-1} & \cdots & Q_2 \\
Q_2 & Q_1 & Q_0 & \cdots & Q_3 \\
\cdot & \cdot & \cdot & \cdot & \cdot \\
\cdot & \cdot & \cdot & \cdot & \cdot \\
\cdot & \cdot & \cdot & \cdot & \cdot \\
Q_{N-1} & Q_{N-2} & Q_{N-3} & \cdots & Q_0
\end{bmatrix}
\cdot
\begin{bmatrix} P_0 \\ P_1 \\ P_2 \\ \cdot \\ \cdot \\ \cdot \\ P_{N-1} \end{bmatrix}
$$

Theorem: if

$$P_j \leftarrow \text{DFT} \rightarrow A_n \, ,$$

$$Q_j \leftarrow \text{DFT} \rightarrow B_n \, ,$$

$$Y_j \leftarrow \text{DFT} \rightarrow C_n \, , \quad \text{and}$$

then,

$$C_n = N \cdot A_n \cdot B_n$$

Figure 5-37 One very efficient, but perplexing, way of defining convolution.

phrase from Dr. Who, would "unravel the entire causal nexus" of digital signal processing.

Convolution has always been a somewhat difficult concept for the beginner to grasp. That's not too surprising for several reasons. Convolution's effect on discrete signal processing is not intuitively obvious for those without experience working with discrete signals, and the mathematics of convolution does seem a little puzzling at first. Moreover, in their sometimes justified haste, many authors present the convolution equation and abruptly start using it as an analysis tool without explaining its origin and meaning. For example, this author once encountered what was called a *tutorial* article on the FFT in a professional journal that proceeded to define *convolution* merely by presenting something like that shown in Figure 5-37 with no further explanation!

Unfortunately, few beginners can gain an understanding of the convolution process from Figure 5-37 alone. Here, we avoid this dilemma by defining the process of convolution and gently proceed through a couple of simple convolution examples. We conclude this chapter with a discussion of the powerful convolution theorem and show why it's so useful as a qualitative tool in discrete system analysis.

5.9.1 Discrete Convolution in the Time-Domain

Discrete convolution is a process, whose input is two sequences, that provides a single output sequence. Convolution inputs can be two time-domain sequences giving a time-domain output, or two frequency-domain input sequences providing a frequency-domain result. (Although the two input sequences must both be in the same domain for the process of convolution to have any practical meaning, their sequence lengths need not be the same.) Let's say we have two input sequences $h(k)$ of length P and $x(k)$ of length Q in the time domain. The output sequence $y(n)$ of the convolution of the two inputs is defined mathematically as

$$y(n) = \sum_{k=0}^{P+Q-2} h(k)x(n-k). \qquad (5\text{-}26)$$

Let's examine Eq. (5-26) by way of example using the $h(k)$ and $x(k)$ sequences shown in Figure 5-38. In this example, we can write the terms for each $y(n)$ in Eq. (5-26) as

$y(0) = h(0)x(0-0) + h(1)x(0-1) + h(2)x(0-2) + h(3)x(0-3) + h(4)x(0-4) + h(5)x(0-5),$
$y(1) = h(0)x(1-0) + h(1)x(1-1) + h(2)x(1-2) + h(3)x(1-3) + h(4)x(1-4) + h(5)x(1-5),$
$y(2) = h(0)x(2-0) + h(1)x(2-1) + h(2)x(2-2) + h(3)x(2-3) + h(4)x(2-4) + h(5)x(2-5),$
$y(3) = h(0)x(3-0) + h(1)x(3-1) + h(2)x(3-2) + h(3)x(3-3) + h(4)x(3-4) + h(5)x(3-5),$
$y(4) = h(0)x(4-0) + h(1)x(4-1) + h(2)x(4-2) + h(3)x(4-3) + h(4)x(4-4) + h(5)x(4-5),$

and

$y(5) = h(0)x(5-0) + h(1)x(5-1) + h(2)x(5-2) + h(3)x(5-3) + h(4)x(5-4) + h(5)x(5-5).$
$$(5\text{-}27)$$

With $P = 4$ and $Q = 3$, we need evaluate only $4 + 3 - 1 = 6$ individual $y(n)$ terms. Because $h(4)$ and $h(5)$ are zero, we can eliminate some of the terms in Eq. (5-27) and evaluate the remaining $x(n-k)$ indices giving the following expressions for $y(n)$ as

$y(0) = h(0)x(0) + h(1)x(-1) + h(2)x(-2) + h(3)x(-3) ,$
$y(1) = h(0)x(1) + h(1)x(0) + h(2)x(-1) + h(3)x(-2) ,$
$y(2) = h(0)x(2) + h(1)x(1) + h(2)x(0) + h(3)x(-1) ,$
$y(3) = h(0)x(3) + h(1)x(2) + h(2)x(1) + h(3)x(0) ,$
$y(4) = h(0)x(4) + h(1)x(3) + h(2)x(2) + h(3)x(1) ,$

and

$$y(5) = h(0)x(5) + h(1)x(4) + h(2)x(3) + h(3)x(2) . \qquad (5\text{-}28)$$

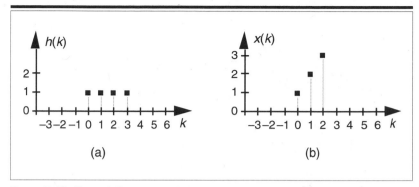

Figure 5-38 Convolution example input sequences: (a) first sequence *h(k)* of length *P* = 4; (b) second sequence *x(k)* of length *Q* = 3.

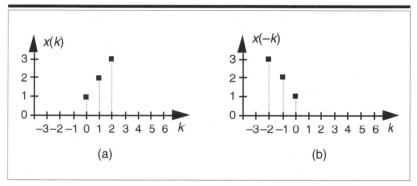

Figure 5-39 Convolution example input sequence: (a) second sequence *x(k)* of length 3; (b) reflection of the second sequence about the *k* = 0 index.

Looking at the indices of the $h(k)$ and $x(k)$ terms in Eq. (5-28), we see two very important things occurring. First, convolution is merely the summation of a series of products—so the process itself is not very complicated. Second, notice that, for a given $y(n)$, $h(k)$'s index is increasing as $x(k)$'s index is decreasing. This fact has led many authors to introduce a new sequence $x(-k)$, and use that new sequence to graphically illustrate the convolution process. The $x(-k)$ sequence is simply our original $x(k)$ reflected about the 0 index of the k axis as shown in Figure 5-39. Defining $x(-k)$ as such enables us to depict the products and summations of Eq. (5-28)'s convolution as in Figure 5-40; that is, we can now align the $x(-k)$ samples with the samples of $h(k)$ for a given n index to calculate $y(n)$. As shown in Figure 5-40(a), the alignment of $h(k)$ and $x(n-k)$, for $n = 0$, yields $y(0) = 1$. This is the result of the first line in Eq. (5-28) repeated on the right side of Figure 5-40(a). The calculation of $y(1)$, for $n = 1$, is depicted

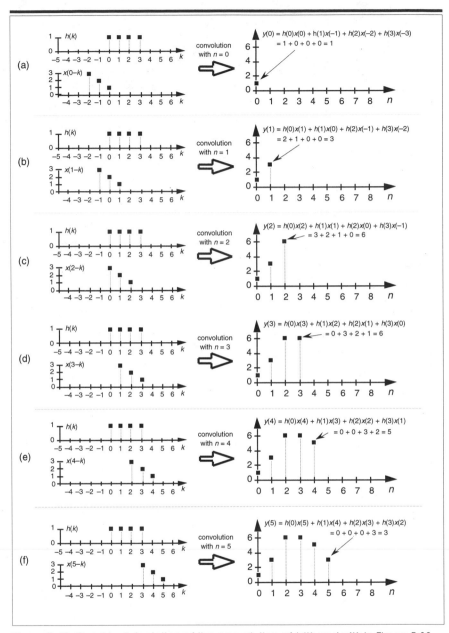

Figure 5-40 Graphical depiction of the convolution of $h(k)$ and $x(k)$ in Figure 5-38.

in Figure-34(b) where $x(n–k)$ is shifted one element to the right, resulting in $y(1) = 3$. We continue this $x(n–k)$ shifting and incrementing n until we arrive at the last nonzero convolution result of $y(5)$ shown in Figure 5-40(f). So, performing the convolution of $h(k)$ and $x(k)$ comprises

Step 1: plotting both $h(k)$ and $x(k)$,

Step 2: flipping the $x(k)$ sequence about the $k = 0$ value to get $x(-k)$,

Step 3: summing the products of $h(k)$ and $x(0-k)$ for all k to get $y(0)$,

Step 4: shifting $x(-k)$ one sample to the right,

Step 5: summing the products of $h(k)$ and $x(1-k)$ for all k to get $y(1)$, and

Step 6: shifting and summing products until there's no overlap of $h(k)$ and the shifted $x(n-k)$, in which case all further $y(n)$s are zero and we're done.

The full convolution of our $h(k)$ and $x(k)$ is the $y(n)$ sequence on the right side of Figure 5-40(f). We've scanned the $x(-k)$ sequence across the $h(k)$ sequence and summed the products where the sequences overlap. By the way, notice that the $y(n)$ sequence in Figure 5-40(f) has six elements where $h(k)$ had a length of four and $x(k)$ was of length three. In the general case, if $h(k)$ is of length P and $x(k)$ is of length Q, the length of $y(n)$ will have a sequence length of L, where

$$L = (P + Q - 1) . \tag{5-29}$$

At this point, it's fair for the beginner to ask, "OK, so what? What does this *strange* convolution process have to do with digital signal processing?" The answer to that question lies in understanding the effects of the convolution theorem.

5.9.2 The Convolution Theorem

The convolution theorem is a fundamental constituent of digital signal processing. It impacts our results anytime we filter or Fourier transform discrete data. To see why this is true, let's simplify the notation of Eq. (5-26) and use the abbreviated form

$$y(n) = h(k) * x(k) , \tag{5-30}$$

where, again, the "*" symbol means convolution. The convolution theorem may be stated as follows: If two time-domain sequences $h(k)$ and $x(k)$ have DFTs of $H(m)$ and $X(m)$, respectively, then the DFT of $h(k) * x(k)$ is the product $H(m) \cdot X(m)$. Likewise, the inverse DFT of $H(m) \cdot X(m)$ is $h(k) * x(k)$. We can represent this relationship with the expression

$$h(k) * x(k) \xrightarrow[\text{IDFT}]{\text{DFT}} H(m) \cdot X(m). \tag{5-31}$$

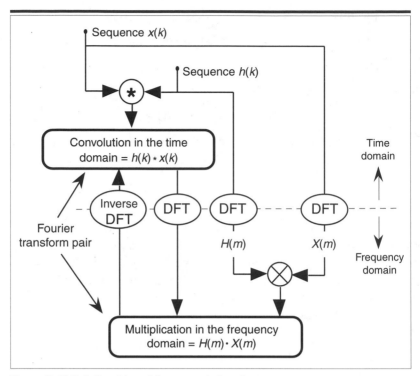

Figure 5-41 Relationships of the convolution theorem.

Equation (5-31) tells us that two sequences resulting from $h(k) * x(k)$ and $H(m) \cdot X(m)$ are Fourier transform pairs. So, taking the DFT of $h(k) * x(k)$ always gives us $H(m) \cdot X(m)$. Likewise, we can determine $h(k) * x(k)$ by taking the inverse DFT of $H(m) \cdot X(m)$. The important point to learn from Eq. (5-31) is that convolution in the time domain is equivalent to multiplication in the frequency domain. (We won't derive the convolution theorem here because it's derivation is readily available to the interested reader[26–29].) To help us appreciate this principle, Figure 5-41 sketches the relationship between convolution in the time domain and multiplication in the frequency domain.

We can easily illustrate the convolution theorem by taking 8-point DFTs of $h(k)$ and $x(k)$ to get $H(m)$ and $X(m)$, respectively, and listing these values as in Table 5-3. (Of course, we have to pad $h(k)$ and $x(k)$ with zeros, so they both have lengths of 8 to take 8-point DFTs.) Tabulating the inverse DFT of the product $H(m) \cdot X(m)$ allows us to verify Eq. (5-31), as listed in the last two columns of Table 5-3, where the acronym IDFT again means inverse DFT. The values from Table 5-3 are shown in Figure 5-42. (For simplicity, only the magnitudes of $H(m)$, $X(m)$, and $H(m) \cdot X(m)$ are shown in the figure.) We need to become comfortable with convolution in the time domain because,

Table 5-3 Convolution Values of $h(k)$ and $x(k)$ from Figure 5-38.

Index k or m	h(k)	x(k)	DFT of h(k) = H(m)	DFT of x(k) = X(m)	H(m) · X(m)	IDFT of H(m) · X(m)	h(k)∗x(k)
0	1	1	4.0 + j0.0	6.0 + j0.0	24.0 + j0.0	1.0 + j0.0	1
1	1	2	1.00 − j2.41	2.41 − j4.41	−8.24 − j10.24	3.0 + j0.0	3
2	1	3	0	−2.0 − j2.0	0	6.0 + j0.0	6
3	1	0	1.00 − j0.41	−0.41 + j1.58	0.24 + j1.75	6.0 + j0.0	6
4	0	0	0	2.0 + j0.0	0	5.0 + j0.0	5
5	0	0	1.00 + j0.41	−0.41 − j1.58	0.24 − j1.75	3.0 + j0.0	3
6	0	0	0	−2.00 + j2.00	0	0.0 + j0.0	0
7	0	0	1.00 + j2.41	2.41 + j4.41	−8.24 + j10.24	0.0 + j0.0	0

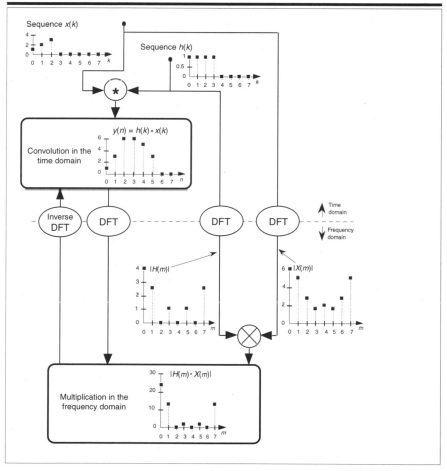

Figure 5-42 Convolution relationships of $h(k)$, $x(k)$, $H(m)$, and $X(m)$ from Figure 5-38.

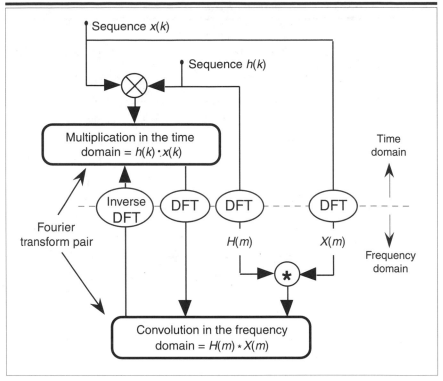

Figure 5-43 Relationships of the convolution theorem related to multiplication in the time domain.

as we've learned, it's the process used in FIR filters. As detailed in Section 5.2, we perform discrete time-domain FIR filtering by convolving an input sequence, $x(n)$ say, with the impulse response $x(k)$ of a filter, and, for FIR filters that impulse response happens to also be the filter's coefficients.[†] The result of that convolution is a filtered time-domain sequence whose spectrum is modified (multiplied) by the filter's frequency response $X(m)$. Section 10.10 describes a clever scheme to perform FIR filtering efficiently using the FFT algorithm to implement convolution.

Because of the duality of the convolution theorem, we could have swapped the time and frequency domains in our discussion of convolution and multiplication being a Fourier transform pair. This means that, similar to Eq. (5-31), we can also write

---

[†] As we'll see in Chapter 6, the coefficients used for an infinite impulse response (IIR) filter are not equal to that filter's impulse response.

$$h(k) \cdot x(k) \underset{\text{IDFT}}{\overset{\text{DFT}}{\rightleftharpoons}} H(m) * X(m) \ . \tag{5-32}$$

So the convolution theorem can be stated more generally as *Convolution in one domain is equivalent to multiplication in the other domain*. Figure 5-43 shows the relationship between multiplication in the time domain and convolution in the frequency domain. Equation (5-32) is the fundamental relationship used in the process of windowing time-domain data to reduce DFT leakage, as discussed in Section 3.9.

5.9.3 Applying the Convolution Theorem

The convolution theorem is useful as a qualitative tool in predicting the affects of different operations in discrete linear time-invariant systems. For example, many authors use the convolution theorem to show why periodic sampling of continuous signals results in discrete samples whose spectra are periodic in the frequency domain. Consider the real continuous time-domain waveform in Figure 5-44(a), with the one-sided spectrum of bandwidth B. Being a real signal, of course, its spectrum is symmetrical about 0 Hz. (In Figure 5-44, the large right-pointing shaded arrows represent Fourier transform operations.) Sampling this waveform is equivalent to multiplying it by a sequence of periodically spaced impulses, Figure 5-44(b), whose values are unity. If we say that the sampling rate is f_s samples/second, then the sample period $t_s = 1/f_s$ seconds. The result of this multiplication is the sequence of discrete time-domain impulses shown in Figure 5-44(c). We can use the convolution theorem to help us predict what the frequency-domain effect is of this multiplication in the time domain. From our theorem, we now realize that the spectrum of the time-domain product must be the convolution of the original spectra. Well, we know what the spectrum of the original continuous waveform is. What about the spectrum of the time-domain impulses? It has been shown that the spectrum of periodic impulses, whose period is t_s seconds, is also periodic impulses in the frequency domain with a spacing of f_s Hz as shown in Figure 5-44(b) [30].

Now, all we have to do is convolve the two spectra. In this case, convolution is straightforward because both of the frequency-domain functions are symmetrical about the zero-Hz point, and flipping one of them about zero Hz is superfluous. So we merely slide one of the functions across the other and plot the product of the two. The convolution of the original waveform spectrum and the spectral impulses results in replications of the waveform spectrum every f_s Hz, as shown in Figure 5-44(c).

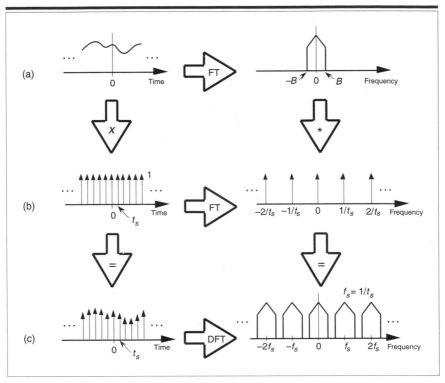

Figure 5-44 Using convolution to predict the spectral replication effects of periodic sampling.

This discussion reiterates the fact that the DFT is always periodic with a period of f_s Hz.

Here's another example of how the convolution theorem can come in handy when we try to understand digital signal processing operations. This author once used the theorem to resolve the puzzling result, at the time, of a triangular window function having its first frequency response null at twice the frequency of the first null of a rectangular window function. The question was "If a rectangular time-domain function of width T has its first spectral null at $1/T$ Hz, why does a triangular time-domain function of width T have its first spectral null at $2/T$ Hz?" We can answer this question by considering convolution in the time domain.

Look at the two rectangular time-domain functions shown in Figures 5-45(a) and 5-45(b). If their widths are each T seconds, their spectra are shown to have nulls at $1/T$ Hz as depicted in the frequency-domain functions in Figures 5-45(a) and 5-45(b). We know that the frequency magnitude responses will be the absolute value of the classic $\sin(x)/x$ function.[†] If we convolve those two rectangular time-domain functions of width T, we'll get the triangular function shown in Figure 5-45(c). Again, in this case, flipping one rectangular function about the zero time axis is unnecessary. To convolve them, we need only scan one function across the other and determine

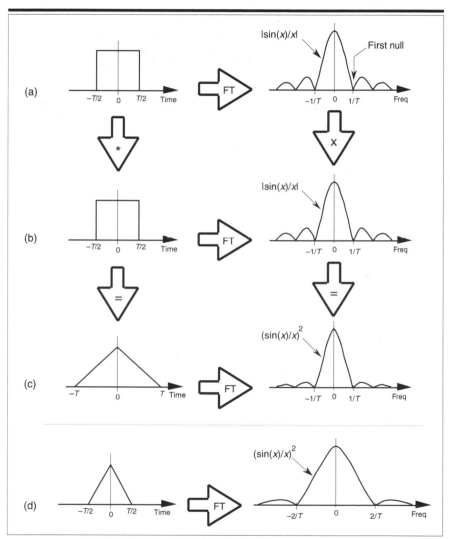

Figure 5-45 Using convolution to show that the Fourier transform of a triangular function has its first null at twice the frequency of the Fourier transform of a rectangular function.

[†] The $\sin(x)/x$ function was introduced in our discussion of window functions in Section 3.9 and is covered in greater detail in Section 3.13.

the area of their overlap. The time shift where they overlap the most happens to be a zero time shift. Thus, our resultant convolution has a peak at a time shift of zero seconds because there's 100 percent overlap. If we slide one of the rectangular functions in either direction, the convolution decreases linearly toward zero. When the time shift is $T/2$ seconds, the rectangular functions have a 50 percent overlap. The convolution is zero when the time shift is T seconds—that's when the two rectangular functions cease to overlap.

Notice that the triangular convolution result has a width of $2T$, and that's really the key to answering our question. Because convolution in the time domain is equivalent to multiplication in the frequency domain, the Fourier transform magnitude of our $2T$ width triangular function is the $|\sin(x)/x|$ in Figure 5-45(a) times the $|\sin(x)/x|$ in Figure 5-45(b), or the $(\sin(x)/x)^2$ function in Figure 5-45(c). If a triangular function of width 2T has its first frequency-domain null at $1/T$ Hz, then the same function of width T must have its first frequency null at $2/T$ Hz as shown in Figure 5-45(d), and that's what we set out to show. Comparison of Figure 5-45(c) and Figure 5-45(d) illustrates a fundamental Fourier transform property that compressing a function in the time domain results in an expansion of its corresponding frequency-domain representation.

References

[1] Shynk, J. J. "Adaptive IIR Filtering," *IEEE ASSP Magazine,* April 1989.

[2] Laundrie, A. "Adaptive Filters Enable Systems to Track Variations," *Microwaves & RF,* September 1989.

[3] Bullock, S. R. "High Frequency Adaptive Filter," *Microwave Journal,* September 1990.

[4] Haykin, S. S. *Adaptive Filter Theory,* Prentice-Hall, Englewood Cliffs, New Jersey, 1986.

[5] Goodwin, G. C., and Sin, K. S. *Adaptive Filtering Prediction and Control,* Prentice-Hall, Englewood Cliffs, New Jersey, 1984.

[6] Gibbs, J. W. *Nature,* Vol. 59, 1899, pp. 606.

[7] Stockham, T. G. "High-Speed Convolution and Correlation with Applications to Digital Filtering," Chapter 7 in *Digital Processing of Signals,* Ed. by B. Gold et al., McGraw-Hill, New York, 1969, pp. 203–232.

[8] Wait, J. V. "Digital Filters," in *Active Filters: Lumped, Distributed, Integrated, Digital, and Parametric.* Ed. by L. P. Huelsman. McGraw-Hill, New York, 1970, pp. 200–277.

[9] Dolph, C. L. "A Current Distribution for Broadside Arrays Which Optimizes the Relationship between Beam Width and Side-Lobe Level," *Proceedings of the IRE,* Vol. 35, June 1946.

[10] Barbiere, D. "A Method for Calculating the Current Distribution of Chebyshev Arrays," *Proceedings of the IRE,* Vol. 40, January 1952.

[11] Cook, C. E., and Bernfeld, M. *Radar Signals,* Academic Press, New York, 1967, pp. 178–180.

[12] Kaiser, J. F. "Digital Filters," in *System Analysis by Digital Computer.* Ed. by F. F. Kuo and J. F. Kaiser, John Wiley and Sons, New York, 1966, pp. 218–277.

[13] Williams, C. S. *Designing Digital Filters,* Prentice-Hall, Englewood Cliffs, New Jersey, 1986, pp. 117.

[14] Harris, F. J. "On the Use of Windows for Harmonic Analysis with the Discrete Fourier Transform," *Proceedings of the IEEE,* Vol. 66, No. 1, January 1978.

[15] McClellan, J. H., Parks, T. W., and Rabiner, L. R. "A Computer Program for Designing Optimum FIR Linear Phase Digital Filters," *IEEE Trans. on Audio and Electroacoustics,* Vol. AU-21, No. 6, December 1973, pp. 515.

[16] Rabiner, L. R. and Gold, B. *Theory and Application of Digital Signal Processing,* Prentice-Hall, Englewood Cliffs, New Jersey, 1975, pp. 136.

[17] Parks, T. W., and McClellan, J. H. "Chebyshev Approximation For Nonrecursive Digital Filters with Linear Phase," *IEEE Trans. on Circuit Theory,* Vol. CT-19, March 1972.

[18] McClellan, J. H., and Parks, T. W. "A Unified Approach to the Design of Optimum FIR Linear Phase Digital Filters," *IEEE Trans. on Circuit Theory,* Vol. CT-20, November 1973.

[19] Rabiner, L. R., McClellan, J. H., and Parks, T. W. "FIR Digital Filter Design Techniques Using Weighted Chebyshev Approximation," *Proc. IEEE,* Vol. 63, No. 4, April 1975.

[20] Oppenheim, A. V., and Schafer, R. W. *Discrete-Time Signal Processing,* Prentice-Hall, Englewood Cliffs, New Jersey, 1989, pp. 478.

[21] Funderburk, D. M., and Park, S. "Implementation of a C-QUAM AM-Stereo Receiver Using a General Purpose DSP Device," *RF Design,* June 1993.

[22] Harris Semiconductor Inc. "A Digital, 16-Bit, 52 Msps Halfband Filter," *Microwave Journal,* September 1993.

[23] Ballanger, M. G. "Computation Rate and Storage Estimation in Multirate Digital Filtering with Half-Band Filters," *IEEE Trans. on Acoust. Speech, and Signal Proc.,* Vol. ASSP-25, No. 4, August 1977.

[24] Crochiere, R. E., and Rabiner, L. R. "Decimation and Interpolation of Digital Signals—A Tutorial Review," *Proceedings of the IEEE,* Vol. 69, No. 3, March 1981, pp. 318.

[25] Ballanger, M. G., Daguet, J. L., and Lepagnol, G. P. "Interpolation, Extrapolation, and Reduction of Computational Speed in Digital Filters," *IEEE Trans. on Acoust. Speech, and Signal Proc.,* Vol. ASSP-22, No. 4, August 1974.

[26] Oppenheim, A. V., Willsky, A. S., and Young, I. T. *Signals and Systems,* Prentice-Hall, Englewood Cliffs, New Jersey, 1983, pp. 212.

[27] Stearns, S. *Digital Signal Analysis,* Hayden Book Co., Rochelle Park, New Jersey, 1975, pp. 93.

[28] Oppenheim, A. V., and Schafer, R. W. *Discrete-Time Signal Processing,* Prentice-Hall, Englewood Cliffs, New Jersey, 1989, pp. 58.

[29] Rabiner, L. R., and Gold, B. *Theory and Application of Digital Signal Processing,* Prentice-Hall, Englewood Cliffs, New Jersey, 1975, pp. 59.

[30] Oppenheim, A. V., Willsky, A. S., and Young, I. T. *Signals and Systems,* Prentice-Hall, Englewood Cliffs, New Jersey, 1983, pp. 201.

CHAPTER SIX

Infinite Impulse Response Filters

Infinite impulse response (IIR) digital filters are fundamentally different from FIR filters because practical IIR filters always require feedback. Where FIR filter output samples depend only on past input samples, each IIR filter output sample depends on previous input samples *and* previous filter output samples. IIR filters' *memory* of past outputs is both a blessing and a curse. Like all feedback systems, perturbations at the IIR filter input could, depending on the design, cause the filter output to become unstable and oscillate indefinitely. This characteristic of possibly having an infinite duration of nonzero output samples, even if the input becomes all zeros, is the origin of the phrase *infinite impulse response.* It's interesting at this point to know that, relative to FIR filters, IIR filters have more complicated structures (block diagrams), are harder to design and analyze, and do not have linear phase responses. Why in the world, then, would anyone use an IIR filter? Because they are very efficient. IIR filters require far fewer multiplications per filter output sample to achieve a given frequency magnitude response. From a hardware standpoint, this means that IIR filters can be very fast, allowing us to build real-time IIR filters that operate over much higher sample rates than FIR filters.[†]

To illustrate the utility of IIR filters, Figure 6-1 contrasts the frequency magnitude responses of what's called a fourth-order low-pass IIR filter and the 19-tap FIR filter of Figure 5-19(b) from Chapter 5. Where the 19-tap FIR filter in Figure 6-1 requires 19 multiplications per filter output sample, the fourth-order IIR filter requires only 9 multiplications for each filter output sample. Not only does the IIR filter give us reduced

---

[†] At the end of this chapter, we briefly compare the advantages and disadvantages of IIR filters relative to FIR filters.

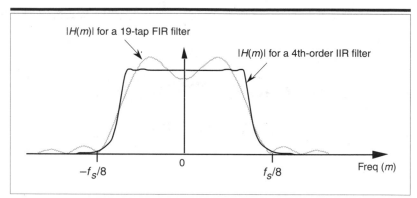

Figure 6-1 Comparison of the frequency magnitude responses of a 19-tap low-pass FIR filter and a 4th-order low-pass IIR filter.

passband ripple and a sharper filter roll-off; it does so with less than half the multiplication workload of the FIR filter.

Recall from Section 5.3 that, to force an FIR filter's frequency response to have very steep transition regions, we had to design an FIR filter with a very long impulse response. The longer the impulse response, the more ideal our filter frequency response will become. From a hardware standpoint, the maximum number of FIR filter taps we can have (the length of the impulse response) depends on how fast our hardware can perform the required number of multiplications and additions to get a filter output value before the next filter input sample arrives. IIR filters, however, can be designed to have impulse responses that are longer than their number of taps! Thus, IIR filters can give us much better filtering for a given number of multiplications per output sample than FIR filters. With this in mind, let's take a deep breath, flex our mathematical muscles, and learn about IIR filters.

6.1 An Introduction to Infinite Impulse Response Filters

IIR filters get their name from the fact that some of the filter's previous output samples are used to calculate the current output sample. Given a finite duration of nonzero input values, the effect is that an IIR filter could have a infinite duration of nonzero output samples. So, if the IIR filter's input suddenly becomes a sequence of all zeros, the filter's output could conceivably remain nonzero forever. This peculiar attribute of IIR filters comes about because of the way they're realized, i.e., the feedback structure of their delay units, multipliers, and adders. Understanding IIR filter structures is straightforward if we start by recalling the building blocks of an FIR filter. Figure 6-2(a) shows the now familiar structure of a 4-tap FIR digital filter that implements the time-domain FIR equation

$$y(n) = h(0)x(n) + h(1)x(n-1) + h(2)x(n-2) + h(3)x(n-3) + h(4)x(n-4) . \quad (6\text{-}1)$$

Although not specifically called out as such in Chapter 5, Eq. (6-1) is known as a *difference equation*. To appreciate how past filter output samples are used in the structure of IIR filters, let's begin by reorienting our FIR structure in Figure 6-2(a) to that of Figure 6-2(b). Notice how the structures in Figure 6-2 are computationally identical, and both are implementations, or realizations, of Eq. (6-1).

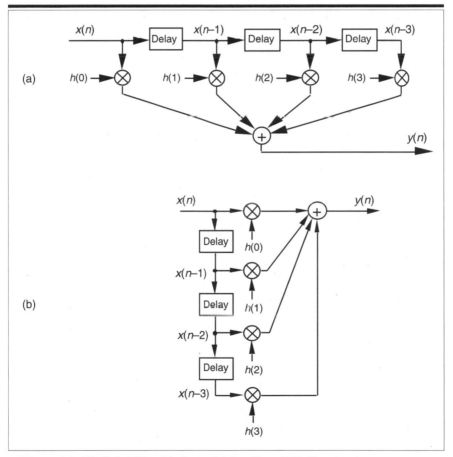

Figure 6-2 FIR digital filter structures: (a) traditional FIR filter structure; (b) rearranged, but equivalent, FIR filter structure.

We can now show how past filter output samples are combined with past input samples by using the IIR filter structure in Figure 6-3. Because IIR filters have two sets of coefficients, we'll use the standard notation of the variables $b(k)$ to denote the feed forward coefficients and the variables $a(k)$ to indicate the feedback coefficients in Figure 6-3. OK, the difference equation describing the IIR filter in Figure 6-3 is

$$y(n) = b(0)x(n) + b(1)x(n-1) + b(2)x(n-2) + b(3)x(n-3)$$

$$+ a(1)y(n-1) + a(2)y(n-2) + a(3)y(n-3) \ . \qquad (6\text{-}2)$$

Look at Figure 6-3 and Eq. (6-2) carefully. It's important to convince ourselves that Figure 6-3 really is a valid implementation of Eq. (6-2) and that, conversely, difference equation Eq. (6-2) fully describes the IIR filter structure in Figure 6-3. Keep in mind now that the sequence $y(n)$ in Figure 6-3 is not the same $y(n)$ sequence that's shown in Figure 6-2. The $d(n)$ sequence in Figure 6-3 is equal to the $y(n)$ sequence in Figure 6-2.

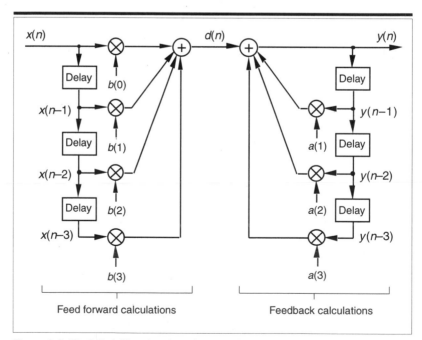

Figure 6-3 IIR digital filter structure showing feed forward and feedback calculations.

By now you're probably wondering, "Just how do we determine those $a(k)$ and $b(k)$ IIR filter coefficients if we actually want to design an IIR filter?" Well, fasten your seat belt because this is where we get serious about understanding IIR filters. Recall from the last chapter concerning the window method of low-pass FIR filter design that we defined the frequency response of our desired FIR filter, took the inverse Fourier transform of that frequency response, and then shifted that transform result to get the filter's time-domain impulse response. Happily, due to the nature of transversal FIR filters, the desired $h(k)$ filter coefficients turned out to be exactly equal to the impulse response sequence. Following that same procedure with IIR filters, we could define the desired frequency response of our IIR filter and then take the inverse Fourier transform of that response to yield the filter's time-domain impulse response. The bad news is that there's no direct method for computing the IIR filter's $a(k)$ and $b(k)$ coefficients from the impulse response! Unfortunately, the FIR filter design techniques that we've learned so far simply cannot be used to design IIR filters. Fortunately for us, this wrinkle can be ironed out by using one of several available methods of designing IIR filters.

Standard IIR filter design techniques fall into three basic classes: the impulse invariance, bilinear transform, and optimization methods. These design methods use a discrete sequence, mathematical transformation process known as the z-transform whose origin is the Laplace transform traditionally used in the analyzing of continuous systems. With that in mind, let's start this IIR filter analysis and design discussion by briefly reacquainting ourselves with the fundamentals of the Laplace transform.

6.2 The Laplace Transform

The Laplace transform is a mathematical method of solving linear differential equations that has proved very useful in the fields of engineering and physics. This transform technique, as it's used today, originated from the work of the brilliant English physicist Oliver Heaviside.[†] The fundamental process of using the Laplace transform goes something like the following:

[†] Heaviside (1850–1925), who was interested in electrical phenomena, developed an efficient algebraic process of solving differential equations. He initially took a lot of heat from his contemporaries because they thought his work was not sufficiently justified from a mathematical standpoint. However, the discovered correlation of Heaviside's methods with the rigorous mathematical treatment of the French mathematician Marquis Pierre Simon de Laplace's (1749–1827) operational calculus verified the validity of Heaviside's techniques.

Step 1: A time-domain differential equation is written that describes the input/output relationship of a physical system (and we want to find the output function that satisfies that equation with a given input).

Step 2: The differential equation is Laplace transformed, converting it to an algebraic equation.

Step 3: Standard algebraic techniques are used to determine the desired output function's equation in the Laplace domain.

Step 4: The desired Laplace output equation is, then, inverse Laplace transformed to yield the desired time-domain output function's equation.

This procedure, at first, seems cumbersome because it forces us to *go the long way around*, instead of just solving a differential equation directly. The justification for using the Laplace transform is that although solving differential equations by classical methods is a very powerful analysis technique for all but the most simple systems, it can be tedious and (for some of us) error-prone. The reduced complexity of using algebra outweighs the extra effort needed to perform the required forward and inverse Laplace transformations. This is especially true now that tables of forward and inverse Laplace transforms exist for most of the commonly encountered time functions. Well-known properties of the Laplace transform also allow practitioners to decompose complicated time functions into combinations of simpler functions and, then, use the tables. (Tables of Laplace transforms allow us to translate quickly back and forth between a time function and its Laplace transform—analogous to, say, a German-English dictionary if we were studying the German language.[†]) Let's briefly look at a few of the more important characteristics of the Laplace transform that will prove useful as we make our way toward the discrete z-transform used to design and analyze IIR digital filters.

The Laplace transform of a continuous time-domain function $f(t)$, where $f(t)$ is defined only for positive time ($t > 0$), is expressed mathematically as

$$F(s) = \int_0^\infty f(t)e^{-st}dt \; . \tag{6-3}$$

[†] Although tables of commonly encountered Laplace transforms are included in almost every system analysis textbook, very comprehensive tables are also available[1–3].

$F(s)$ is called "the Laplace transform of $f(t)$," and the variable s is the complex number

$$s = \sigma + j\omega . \tag{6-4}$$

A more general expression for the Laplace transform, called the bilateral or two-sided transform, uses negative infinity ($-\infty$) as the lower limit of integration. However, for the systems that we'll be interested in, where system conditions for negative time ($t < 0$) are not needed in our analysis, the *one-sided* Eq. (6-3) applies. Those systems, often referred to as causal systems, may have initial conditions at $t = 0$ that must be taken into account (velocity of a mass, charge on a capacitor, temperature of a body, etc.) but we don't need to know what the system was doing prior to $t = 0$.

In equation (6-4), σ is a real number and ω is frequency in radians/second. Because e^{-st} is dimensionless, the exponent term s must have the dimension of 1/time, or frequency. That's why the Laplace variable s is often called a complex frequency.

To put Eq. (6-3) into words, we can say that it requires us to multiply, point for point, the function $f(t)$ by the complex function e^{-st} for a given value of s. (We'll soon see that using the function e^{-st} here is not accidental; e^{-st} is used because it's the general form for the solution of linear differential equations.) After the point-for-point multiplications, we find the area under the curve of the function $f(t)e^{-st}$ by summing all the products. That area, a complex number, represents the value of the Laplace transform for the particular value of $s = \sigma + j\omega$ chosen for the original multiplications. If we were to go through this process for all values of s, we'd have a full description of $F(s)$ for every value of s.

I like to think of the Laplace transform as a continuous function, where the complex value of that function for a particular value of s is a correlation of $f(t)$ and a damped complex e^{-st} sinusoid whose frequency is ω and whose damping factor is σ. What do these complex sinusoids look like? Well, they are rotating phasors described by

$$e^{-st} = e^{-(\sigma+j\omega)t} = e^{-\sigma t}e^{-j\omega t} = \frac{e^{-j\omega t}}{e^{\sigma t}} . \tag{6-5}$$

From our knowledge of complex numbers, we know that $e^{-j\omega t}$ is a unity-magnitude phasor rotating clockwise around the origin of a complex plane at a frequency of ω radians per second. The denominator of Eq. (6-5) is a real number whose value is one at time $t = 0$. As t increases, the denominator $e^{\sigma t}$ gets larger (when σ is positive), and the complex e^{-st} phasor's magnitude gets smaller as the phasor rotates on the complex plane. The tip of that phasor traces out a curve spiraling in toward the origin of the complex plane. One way to visualize a complex sinusoid is to consider its real and imaginary parts individually. We do this by expressing the complex e^{-st} sinusoid from Eq. (6-5) in rectangular form as

$$e^{-st} = \frac{e^{-j\omega t}}{e^{\sigma t}} = \frac{\cos(\omega t)}{e^{\sigma t}} - j\frac{\sin(\omega t)}{e^{\sigma t}}. \tag{6-5'}$$

Figure 6-4 shows the real parts (cosine) of several complex sinusoids with different frequencies and different damping factors. In Figure 6-4(a), the complex sinusoid's frequency is the arbitrary ω', and the damping factor is the arbitrary σ'. So the real part of $F(s)$, at $s = \sigma' + j\omega'$, is equal to the correlation of $f(t)$ and the wave in Figure 6-4(a). For different values of s, we'll correlate $f(t)$ with different complex sinusoids as shown in Figure 6-4. (As we'll see, this correlation is very much like the correlation of $f(t)$ with various sine and cosine waves when we were calculating the discrete Fourier transform.) Again, the real part of $F(s)$, for a particular value of s, is the correlation of $f(t)$ with a cosine wave of frequency ω and a damping factor of σ, and the imaginary part of $F(s)$ is the correlation of $f(t)$ with a sinewave of frequency ω and a damping factor of σ.

Now, if we associate each of the different values of the complex s variable with a point on a complex plane, rightfully called the s-plane, we could plot the real part of the $F(s)$ correlation as a surface above (or below) that s-plane and generate a second plot of the imaginary part of the $F(s)$ correlation as a surface above (or below) the s-plane. We can't plot the full complex $F(s)$ surface on paper because that would require four dimensions. That's because s is complex, requiring two dimensions, and $F(s)$ is itself complex and also requires two dimensions. What we can do, however, is graph the magnitude $|F(s)|$ as a function of s because this graph requires only three dimensions. Let's do that as we demonstrate this notion of an $|F(s)|$ surface by illustrating the Laplace transform in a tangible way.

Say, for example, that we have the linear system shown in Figure 6-5. Also, let's assume that we can relate the $x(t)$ input and the $y(t)$ output of the linear time invariant physical system in Figure 6-5 with the following messy homogeneous constant-coefficient differential equation

$$a_2\frac{d^2y(t)}{dt^2} + a_1\frac{dy(t)}{dt} + a_0y(t) = b_1\frac{dx(t)}{dt} + b_0x(t) . \tag{6-6}$$

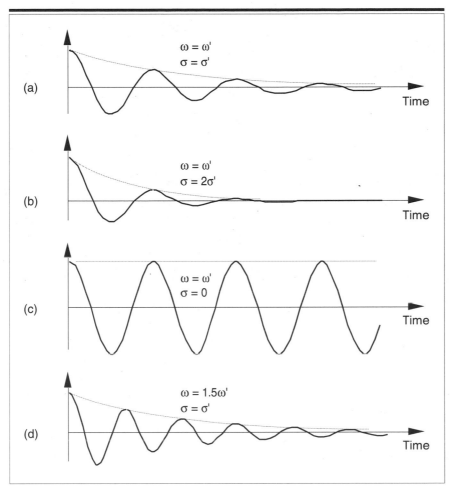

Figure 6-4 Real part (cosine) of various e^{-st} functions, where $s = \sigma + j\omega$, to be correlated with $f(t)$.

We'll use the Laplace transform toward our goal of figuring out how the system will behave when various types of input functions are applied, i.e., what will the $y(t)$ output be for any given $x(t)$ input.

Let's slow down here and see exactly what Figure 6-5 and Eq. (6-6) are telling us. First, if the system is time invariant, then the a_n and b_n coefficients in Eq. (6-6) are constant. They may be positive or negative, zero, real or complex, but they do not change with time. If the system is electrical, the coefficients might be related to capacitance, inductance, and resistance. If the system is mechanical with masses and springs, the coefficients could be related to mass, coefficient of damping, and coefficient of

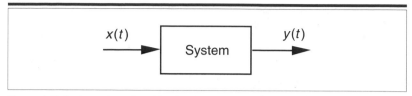

Figure 6-5 System described by Eq. (6-6). The system's input and output are the continuous time functions $x(t)$ and $y(t)$ respectively.

resilience. Then, again, if the system is thermal with masses and insulators, the coefficients would be related to thermal capacity and thermal conductance. To keep this discussion general, though, we don't really care what the coefficients actually represent.

OK, Eq. (6-6) also indicates that, ignoring the coefficients for the moment, the sum of the $y(t)$ output plus derivatives of that output is equal to the sum of the $x(t)$ input plus the derivative of that input. Our problem is to determine exactly what input and output functions satisfy the elaborate relationship in Eq. (6-6). (For the stout hearted, classical methods of solving differential equations could be used here, but the Laplace transform makes the problem much simpler for our purposes.) Thanks to Laplace, the complex exponential time function of e^{st} is the one we'll use. It has the beautiful property that it can be differentiated any number of times without destroying its original form. That is

$$\frac{d(e^{st})}{dt} = se^{st}, \quad \frac{d^2(e^{st})}{dt^2} = s^2 e^{st}, \quad \frac{d^3(e^{st})}{dt^3} = s^3 e^{st}, \dots, \quad \frac{d^n(e^{st})}{dt^n} = s^n e^{st} . \quad (6\text{-}7)$$

If we let $x(t)$ and $y(t)$ be functions of e^{st}, $x(e^{st})$ and $y(e^{st})$, and use the properties shown in Eq. (6-7), Eq. (6-6) becomes

$$a_2 s^2 y(e^{st}) + a_1 s y(e^{st}) + a_0 y(e^{st}) = b_1 s x(e^{st}) + b_0 x(e^{st}),$$

or

$$(a_2 s^2 + a_1 s + a_0) y(e^{st}) = (b_1 s + b_0) x(e^{st}) . \quad (6\text{-}8)$$

Although it's simpler than Eq. (6-6), we can further simplify the relationship in the last line in Eq. (6-8) by considering the ratio of $y(e^{st})$ over $x(e^{st})$ as the Laplace transfer function of our system in Figure 6-5. If we call that ratio of polynomials the transfer function $H(s)$,

$$H(s) = \frac{y(e^{st})}{x(e^{st})} = \frac{b_1 s + b_0}{a_2 s^2 + a_1 s + a_0} \quad , \tag{6-9}$$

To indicate that the original $x(t)$ and $y(t)$ have the identical functional form of e^{st}, we can follow the standard Laplace notation of capital letters and show the transfer function as

$$H(s) = \frac{Y(s)}{X(s)} = \frac{b_1 s + b_0}{a_2 s^2 + a_1 s + a_0} \quad , \tag{6-10}$$

where the output $Y(s)$ is given by

$$Y(s) = X(s) \frac{b_1 s + b_0}{a_2 s^2 + a_1 s + a_0} = X(s)H(s) \quad . \tag{6-11}$$

Equation (6-11) leads us to redraw the original system diagram in a form that highlights the definition of the transfer function $H(s)$ as shown in Figure 6-6.

The cautious reader may be wondering, "Is it really valid to use this Laplace analysis technique when it's strictly based on the system's $x(t)$ input being some function of e^{st}, or $x(e^{st})$?" The answer is that the Laplace analysis technique, based on the complex exponential $x(e^{st})$, is valid because all practical $x(t)$ input functions can be represented with complex exponentials. For example,

- a constant, $c = ce^{0t}$,

- sinusoids, $\sin(\omega t) = (e^{j\omega t} - e^{-j\omega t})/2j$ or $\cos(\omega t) = (e^{j\omega t} + e^{-j\omega t})/2$,

- a monotonic exponential, e^{at}, and

- an exponentially varying sinusoid, $e^{at}\cos(\omega t)$.

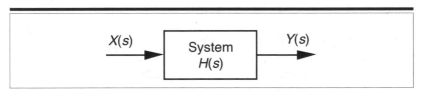

Figure 6-6 Linear system described by Eqs. (6-10) and (6-11). The system's input is the Laplace function $X(s)$, its output is the Laplace function $Y(s)$, and the system transfer function is $H(s)$.

With that said, if we know a system's transfer function $H(s)$, we can take the Laplace transform of any $x(t)$ input to determine $X(s)$, multiply that $X(s)$ by $H(s)$ to get $Y(s)$, and then inverse Laplace transform $Y(s)$ to yield the time-domain expression for the output $y(t)$. In practical situations, however, we usually don't go through all those analytical steps because it's the system's transfer function $H(s)$ in which we're most interested. Being able to express $H(s)$ mathematically or graph the surface $|H(s)|$ as a function of s will tell us the two most important properties we need to know about the system under analysis: Is the system stable, and if so, what is its frequency response?

"But wait a minute," you say. "Equations (6-10) and (6-11) indicate that we have to know the $Y(s)$ output before we can determine $H(s)$!" Not really. All we really need to know is the time-domain differential equation like that in Eq. (6-6). Next we take the Laplace transform of that differential equation and rearrange the terms to get the $H(s)$ ratio in the form of Eq. (6-10). With practice, systems designers can look at a diagram (block, circuit, mechanical, whatever) of their system and promptly write the Laplace expression for $H(s)$. Let's use the concept of the Laplace transfer function $H(s)$ to determine the stability and frequency response of simple continuous systems.

6.2.1 Poles and Zeros on the *s*-plane and Stability

One of the most important characteristics of any system involves the concept of stability. We can think of a system as stable if, given any bounded input, the output will always be bounded. This sounds like an easy condition to achieve because most systems we encounter in our daily lives are indeed stable. Nevertheless, we have all experienced instability in a system containing feedback. Recall the annoying *howl* when a public address system's microphone is placed too close to the loudspeaker. A sensational example of an unstable system occurred in western Washington when the first Tacoma Narrows Bridge began oscillating on the afternoon of November 7th, 1940. Those oscillations, caused by 42 mph winds, grew in amplitude until the bridge destroyed itself. For IIR digital filters with their built-in feedback, instability would result in a filter output that's not at all representative of the filter input; that is, our filter output samples would not be a filtered version of the input; they'd be some strange oscillating or pseudorandom values. A situation we'd like to avoid if we can, right? Let's see how.

We can determine a continuous system's stability by examining several different examples of $H(s)$ transfer functions associated with linear time-invariant systems. Assume that we have a system whose Laplace transfer function is of the form of Eq. (6-10), the coefficients are all real, and the coefficients b_1 and a_2 are equal to zero. We'll call that Laplace transfer function $H_1(s)$, where

$$H_1(s) = \frac{b_0}{a_1 s + a_0} = \frac{b_0 / a_1}{s + a_0 / a_1} . \qquad (6\text{-}12)$$

Notice that if $s = -a_0/a_1$, the denominator in Eq. (6-12) equals zero and $H_1(s)$ would have an infinite magnitude. This $s = -a_0/a_1$ point on the s-plane is called a pole, and that pole's location is shown by the "*x*" in Figure 6-7(a). Notice that the pole is located exactly on the negative portion of the real σ axis. If the system described by H_1 were at rest and we disturbed it with an impulse like $x(t)$ input at time $t = 0$, its continuous time-domain $y(t)$ output would be the damped exponential curve shown in Figure 6-7(b). We can see that $H_1(s)$ is stable because its $y(t)$ output approaches zero as time passes. By the way, the distance of the pole from the $\sigma = 0$ axis, a_0/a_1 for our $H_1(s)$, gives the decay rate of the $y(t)$ impulse response. To illustrate why the term *pole* is appropriate, Figure 6-8(b) depicts the three-dimensional surface of $|H_1(s)|$ above the s-plane. Look at Figure 6-8(b) carefully and see how we've reoriented the s-plane axis. This new axis orientation allows us to see how the $H_1(s)$ system's frequency magnitude response can be determined from its three-dimensional s-plane surface. If we examine the $|H_1(s)|$ surface at $\sigma = 0$, we get the bold curve in Figure 6-8(b). That bold curve, the intersection of the vertical $\sigma = 0$ plane (the $j\omega$ axis plane) and the $|H_1(s)|$ surface, gives us the frequency magnitude response $|H_1(\omega)|$ of the system—and that's one of the things we're after here. The bold $|H_1(\omega)|$

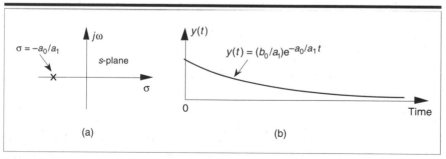

Figure 6-7 Descriptions of $H_1(s)$: (a) pole located at $s = \sigma + j\omega = -a_0/a_1 + j0$ on the s-plane; (b) time-domain $y(t)$ impulse response of the system.

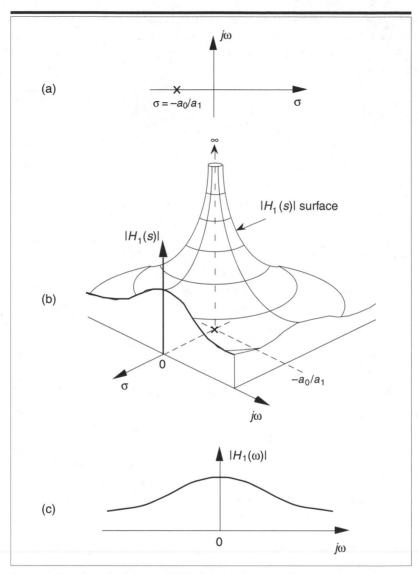

Figure 6-8 Further depictions of $H_1(s)$: (a) pole located at $\sigma = -a_0/a_1$ on the
s-plane; (b) $|H_1(s)|$ surface; (c) curve showing the intersection of the
$|H_1(s)|$ surface and the vertical $\sigma = 0$ plane. This is the conventional
depiction of the $|H_1(\omega)|$ frequency magnitude response.

curve in Figure 6-8(b) is shown in a more conventional way in Figure
6-8(c). Figures 6-8(b) and 6-8(c) highlight the very important property
that the Laplace transform is a more general case of the Fourier
transform because if $\sigma = 0$, then $s = j\omega$. In this case, the $|H_1(s)|$ curve for

$\sigma = 0$ above the s-plane becomes the $|H_1(\omega)|$ curve above the $j\omega$ axis in Figure 6-8(c).

Another common system transfer function leads to an impulse response that oscillates. Let's think about an alternate system whose Laplace transfer function is of the form of Eq. (6-10), the coefficient b_0 equals zero, and the coefficients lead to complex terms when the denominator polynomial is factored. We'll call this particular second-order transfer function $H_2(s)$, where

$$H_2(s) = \frac{b_1 s}{a_2 s^2 + a_1 s + a_0} = \frac{(b_1 / a_2)s}{s^2 + (a_1 / a_2)s + a_0 / a_2} \ . \tag{6-13}$$

(By the way, when a transfer function has the Laplace variable s in both the numerator and denominator, the *order* of the overall function is defined by the largest exponential order of s in the denominator polynomial. So our $H_2(s)$ is a second-order transfer function.) To keep the following equations from becoming too messy, let's factor its denominator and rewrite Eq. (6-13) as

$$H_2(s) = \frac{As}{(s+p)(s+p^*)} \ . \tag{6-14}$$

where $A = b_1/a_2$, $p = p_{\text{real}} + jp_{\text{imag}}$, and $p^* = p_{\text{real}} - jp_{\text{imag}}$ (complex conjugate of p). Notice that, if s is equal to $-p$ or $-p^*$, one of the polynomial roots in the denominator of Eq. (6-14) will equal zero, and $H_2(s)$ will have an infinite magnitude. Those two complex poles, shown in Figure 6-9(a), are located off the negative portion of the real σ axis. If the H_2 system were at rest and we disturbed it with an impulselike $x(t)$ input at time $t = 0$, its continuous time-domain $y(t)$ output would be the damped sinusoidal curve shown in Figure 6-9(b). We see that $H_2(s)$ is stable because its oscillating $y(t)$ output, like a plucked guitar string, approaches zero as time increases. Again, the distance of the poles from the $\sigma = 0$ axis ($-p_{\text{real}}$) gives the decay rate of the sinusoidal $y(t)$ impulse response. Likewise, the distance of the poles from the $j\omega = 0$ axis ($\pm p_{\text{imag}}$) gives the frequency of the sinusoidal $y(t)$ impulse response. Notice something new in Figure 6-9(a). When $s = 0$, the numerator of Eq. (6-14) is zero, making the transfer function $H_2(s)$ equal to zero. Any value of s where $H_2(s) = 0$ is sometimes of interest and is usually plotted on the s-plane as the little circle, called a "zero," shown in Figure 6-9(a). At this point we're not very interested in knowing exactly what p and p^* are in

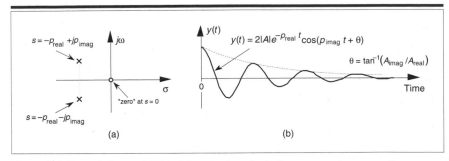

Figure 6-9 Descriptions of $H_2(s)$: (a) poles located at $s = p_{real} \pm jp_{imag}$ on the s-plane; (b) time domain $y(t)$ impulse response of the system.

terms of the coefficients in the denominator of Eq. (6-13). However, an energetic reader could determine the values of p and p^* in terms of a_0, a_1, and a_2 by using the following well-known quadratic factorization formula: Given the second-order polynomial $f(s) = as^2 + bs + c$, then $f(s)$ can be factored as

$$f(s) = as^2 + bs + c = \left(s + \frac{b}{2a} + \sqrt{\frac{b^2 - 4ac}{4a^2}} \right) \cdot \left(s + \frac{b}{2a} - \sqrt{\frac{b^2 - 4ac}{4a^2}} \right) . \quad (6-15)$$

Figure 6-10(b) illustrates the $|H_2(s)|$ surface above the s-plane. Again, the bold $|H_2(\omega)|$ curve in Figure 6-10(b) is shown in the conventional way in Figure 6-10(c) to indicate the frequency magnitude response of the system described by Eq. (6-13). Although the three-dimensional surfaces in Figures 6-8(b) and 6-10(b) are informative, they're also unwieldy and unnecessary. We can determine a system's stability merely by looking at the locations of the poles on the two-dimensional s-plane.

To further illustrate the concept of system stability, Figure 6-11 shows the s-plane pole locations of several example Laplace transfer functions and their corresponding time-domain impulse responses. We recognize Figures 6-11(a) and 6-11(b), from our previous discussion, as indicative of stable systems. When disturbed from their at-rest condition they respond and, at some later time, return to that initial condition. The single pole location at $s = 0$ in Figure 6-11(c) is indicative of the $1/s$ transfer function of a single element of a linear system. In an electrical system, this $1/s$ transfer function could be a capacitor that was charged with an impulse of current, and there's no discharge path in the circuit.

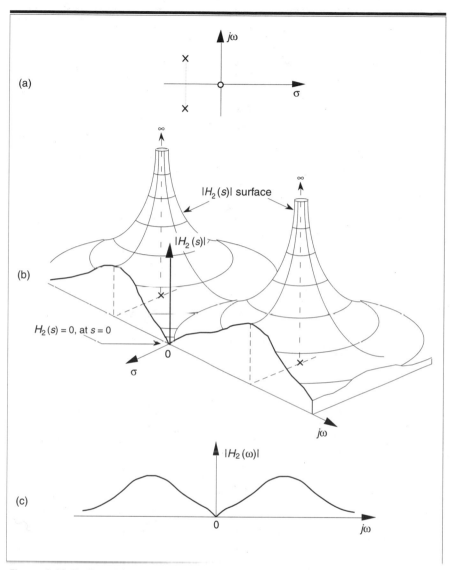

Figure 6-10 Further depictions of $H_2(s)$: (a) poles and zero locations on the s-plane; (b) $|H_2(s)|$ surface; (c) $|H_2(\omega)|$ frequency magnitude response curve.

For a mechanical system, Figure 6-11(c) would describe a kind of spring that's compressed with an impulse of force and, for some reason, remains under compression. Notice, in Figure 6-11(d), that, if an $H(s)$ transfer function has conjugate poles located exactly on the $j\omega$ axis

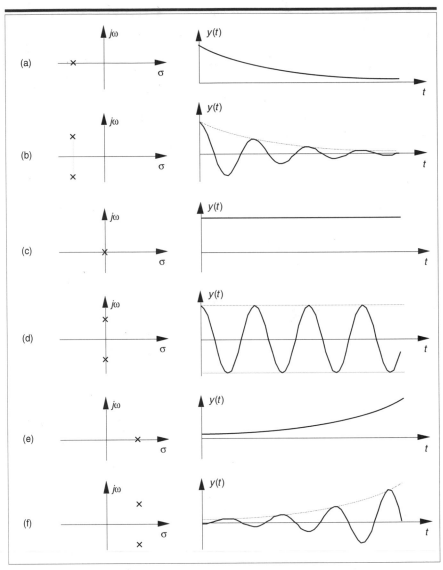

Figure 6-11 Various *H(s)* pole locations and their time-domain impulse responses: (a) single pole at σ < 0; (b) conjugate poles at σ < 0; (c) single pole located at σ = 0; (d) conjugate poles located at σ = 0; (e) single pole at σ > 0; (f) conjugate poles at σ > 0.

(σ = 0), the system will go into oscillation when disturbed from its initial condition. This situation, called conditional stability, happens to describe the intentional transfer function of electronic oscillators. Instability is indicated in Figures 6-11(e) and 6-11(f). Here, the poles lie to the right of the *jω* axis. When disturbed from their initial at-rest condition

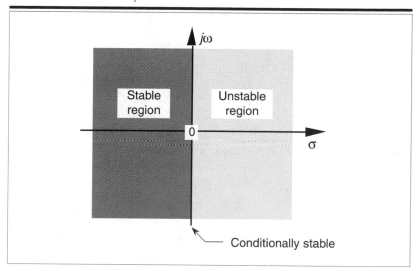

Figure 6-12 The Laplace *s*–plane showing the regions of stability and instability for pole locations for linear continuous systems.

by an impulse input, their outputs grow without bound.[†] See how the value of σ, the real part of s, for the pole locations is the key here? When $\sigma < 0$, the system is well behaved and stable; when $\sigma = 0$, the system is conditionally stable; and when $\sigma > 0$ the system is unstable. So we can say that, when σ is located on the right half of the *s*-plane, the system is unstable. We show this characteristic of linear continuous systems in Figure 6-12. Keep in mind that real-world systems often have more than two poles, and a system is only as stable as its least stable pole. For a system to be stable, all of its transfer-function poles must lie on the left half of the *s*-plane.

To consolidate what we've learned so far: $H(s)$ is determined by writing a linear system's time-domain differential equation and taking the Laplace transform of that equation to obtain a Laplace expression in terms of $X(s)$, $Y(s)$, s, and the system's coefficients. Next we rearrange the Laplace expression terms to get the $H(s)$ ratio in the form of Eq. (6-10). (The really slick part is that we do not have to know what the time-domain $x(t)$ input

---

[†] Impulse response testing in a laboratory can be an important part of the system design process. The difficult part is generating a true impulselike input. If the system is electrical, for example, although somewhat difficult to implement, the input $x(t)$ impulse would be a very short duration voltage or current pulse. If, however, the system were mechanical, a whack with a hammer would suffice as an $x(t)$ impulse input. For digital systems, on the other hand, an impulse input is easy to generate; it's a single unity-valued sample preceded and followed by all zero-valued samples.

is to analyze a linear system!) We can get the expression for the continuous frequency response of a system just by substituting $j\omega$ for s in the $H(s)$ equation. To determine system stability, the denominator polynomial of $H(s)$ is factored to find each of its roots. Each root is set equal to zero and solved for s to find the location of the system poles on the s-plane. Any pole located to the right of the $j\omega$ axis on the s-plane will indicate an unstable system.

OK, returning to our original goal of understanding the z-transform, the process of analyzing IIR filter systems requires us to replace the Laplace transform with the z-transform and to replace the s-plane with a z-plane. Let's introduce the z-transform, determine what this new z-plane is, discuss the stability of IIR filters, and design and analyze a few simple IIR filters.

6.3 The z-Transform

The z-transform is the discrete-time cousin of the continuous Laplace transform.[†] While the Laplace transform is used to simplify the analysis of continuous differential equations, the z-transform facilitates the analysis of discrete difference equations. Let's define the z-transform and explore its important characteristics to see how it's used in analyzing IIR digital filters.

The z-transform of a discrete sequence $h(n)$, expressed as $H(z)$, is defined as

$$H(z) = \sum_{n=-\infty}^{\infty} h(n)z^{-n} , \tag{6-16}$$

where the variable z is complex. Where Eq. (6-3) allowed us to take the Laplace transform of a continuous signal, the z-transform is performed on a discrete $h(n)$ sequence, converting that sequence into a continuous function $H(z)$ of the continuous complex variable z. Similarly, as the function e^{-st} is the general form for the solution of linear differential equations, z^{-n} is the general form for the solution of linear difference equations. Moreover, as a Laplace function $F(s)$ is a continuous surface

---

[†] In the early 1960s, James Kaiser, after whom the Kaiser window function is named, consolidated the theory of digital filters using a mathematical description known as the z-transform[4,5]. Until that time, the use of the z-transform had generally been restricted to the field of discrete control systems[6–9].

above the s-plane, the z-transform function $H(z)$ is a continuous surface above a z-plane. To whet your appetite, we'll now state that, if $H(z)$ represents an IIR filter's z-domain transfer function, evaluating the $H(z)$ surface will give us the filter's frequency magnitude response, and $H(z)$'s pole and zero locations will determine the stability of the filter.

We can determine the frequency response of an IIR digital filter by expressing z in polar form as $z = re^{j\omega}$, where r is a magnitude and ω is the angle. In this form, the z-transform equation becomes

$$H(z) = H(re^{j\omega}) = \sum_{n=-\infty}^{\infty} h(n)(re^{j\omega})^{-n} = \sum_{n=-\infty}^{\infty} h(n)r^{-n}(e^{-j\omega n}) \ . \qquad (6\text{-}17)$$

Equation (6-17) can be interpreted as the Fourier transform of the product of the original sequence $h(n)$ and the exponential sequence r^{-n}. When r equals one, Eq. (6-17) simplifies to the Fourier transform. Thus on the z-plane, the contour of the $H(z)$ surface for those values where $|z| = 1$ is the Fourier transform of $h(n)$. If $h(n)$ represents a filter impulse response sequence, evaluating the $H(z)$ transfer function for $|z| = 1$ yields the frequency response of the filter. So where on the z-plane is $|z| = 1$? It's a circle with a radius of one, centered about the $z = 0$ point. This circle, so important that it's been given the name *unit circle*, is shown in Figure 6-13. Recall that the $j\omega$ frequency axis on the continuous Laplace s-plane was linear and ranged from $-\infty$ to $+\infty$ radians/s. The ω frequency axis on the complex z-plane, however, spans only the range from $-\pi$ to $+\pi$ radians. With this relationship between the $j\omega$ axis on the Laplace s-plane and the unit circle on the z-plane, we can see that the z-plane frequency axis is

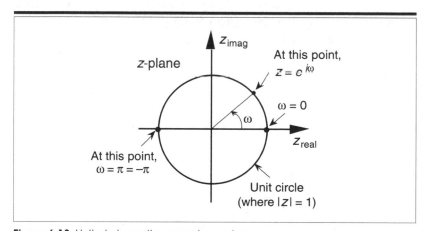

Figure 6-13 Unit circle on the complex z-plane.

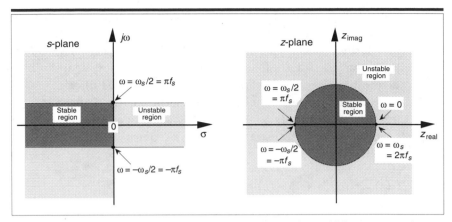

Figure 6-14 Mapping of the Laplace s–plane to the z–plane. All frequency values are in radians/s.

equivalent to coiling the s-plane's jω axis about the unit circle on the z-plane as shown in Figure 6-14.

Then, frequency ω on the z-plane, is not a distance along a straight line, but rather an angle around a circle. With ω in Figure 6-13 being a general normalized angle in radians ranging from $-\pi$ to $+\pi$, we can relate ω to an equivalent f_s sampling rate by defining a new frequency variable $\omega_s = 2\pi f_s$ in radians/s. The periodicity of discrete frequency representations, with a period of $\omega_s = 2\pi f_s$ radians/s or f_s Hz, is indicated for the z-plane in Figure 6-14. Where a walk along the jω frequency axis on the s-plane could take us to infinity in either direction, a trip on the ω frequency path on the z-plane leads us in circles (on the unit circle). Figure 6-14 shows us that only the $-\pi f_s$ to $+\pi f_s$ radians/s frequency range for ω can be accounted for on the z-plane, and this is another example of the universal periodicity of the discrete frequency domain. (Of course the $-\pi f_s$ to $+\pi f_s$ radians/s range corresponds to a cyclic frequency range of $-f_s/2$ to $+f_s/2$.) With the perimeter of the unit circle being $z = e^{j\omega}$, later, we'll show exactly how to substitute $e^{j\omega}$ for z in a filter's H(z) transfer function, giving us the filter's frequency response.

6.3.1 Poles and Zeros on the *z–plane* and Stability

One of the most important characteristics of the z-plane is that the region of filter stability is mapped to the inside of the unit circle on the z-plane. Given the H(z) transfer function of a digital filter, we can examine that function's pole locations to determine filter stability. If all poles are located inside the unit circle, the filter will be stable. On the other hand, if

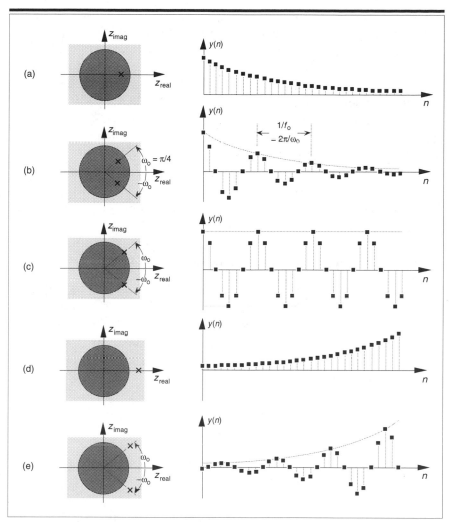

Figure 6-15 Various *H(z)* pole locations and their discrete time-domain impulse
responses: (a) single pole inside the unit circle; (b) conjugate poles
located inside the unit circle; (c) conjugate poles located on the
unit circle; (d) single pole outside the unit circle; (e) conjugate
poles located outside the unit circle.

any pole is located outside the unit circle, the filter will be unstable. Figure
6-15 shows the *z*-plane pole locations of several example *z*-domain transfer functions and their corresponding discrete time-domain impulse
responses. It's a good idea for the reader to compare the *z*-plane and
discrete-time responses of Figure 6-15 with the *s*-plane and the continuous
time responses of Figure 6-11. The $y(n)$ outputs in Figures 6-15(d) and

(e) show examples of how unstable filter outputs increase in amplitude, as time passes, whenever their $x(n)$ inputs are nonzero. To avoid this situation, any IIR digital filter that we design should have an $H(z)$ transfer function with all of its individual poles inside the unit circle. Like a chain that's only as strong as its weakest link, an IIR filter is only as stable as its least stable pole.

The ω_o oscillation frequency of the impulse responses in Figures 6-15(c) and (e) is, of course, proportional to the angle of the conjugate pole pairs from the z_{real} axis, or ω_o radians/s corresponding to $f_o = \omega_o/2\pi$ Hz. Because the intersection of the $-z_{real}$ axis and the unit circle, point $z = -1$, corresponds to π radians (or πf_s radians/s $= f_s/2$ Hz), the ω_o angle of $\pi/4$ in Figure 6-15 means that $f_o = f_s/8$ and our $y(n)$ will have eight samples per cycle of f_o.

6.3.2 Using the z-transform to Analyze IIR Filters

We have one last concept to consider before we can add the z-transform to our collection of digital signal processing tools. We need to determine what the time delay operation in Figure 6-3 is relative to the z-transform. To do this, assume we have a sequence $x(n)$ whose z-transform is $X(z)$ and a sequence $y(n) = x(n-1)$ whose z-transform is $Y(z)$ as shown in Figure 6-16. The z-transform of $y(n)$ is, by definition,

$$Y(z) = \sum_{n=-\infty}^{\infty} y(n)z^{-n} = \sum_{n=-\infty}^{\infty} x(n-1)z^{-n} \ . \tag{6-18}$$

Now if we let $k = n-1$, then, $Y(z)$ becomes

$$Y(z) = \sum_{k=-\infty}^{\infty} x(k)z^{-(k+1)} = \sum_{k=-\infty}^{\infty} x(k)z^{-k}z^{-1} \ , \tag{6-19}$$

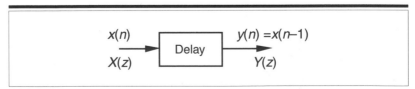

Figure 6-16 Output sequence $y(n)$ equal to a *unit delayed* version of the input $x(n)$ sequence.

which we can write as

$$Y(z) = z^{-1} \sum_{k=-\infty}^{\infty} x(k)z^{(-k)} = z^{-1}[X(z)] \ . \tag{6-20}$$

Thus, the effect of a single unit of time delay is to multiply the z-transform by z^{-1}.

Interpreting a unit time delay to be equivalent to the z^{-1} operator leads us to the relationship shown in Figure 6-17, where we can say $X(z)z^0 = X(z)$ is the z-transform of $x(n)$, $X(z)z^{-1}$ is the z-transform of $x(n)$ delayed by one sample, $X(z)z^{-2}$ is the z-transform of $x(n)$ delayed by two samples, and $X(z)z^{-k}$ is the z-transform of $x(n)$ delayed by k samples. So a transfer function of z^{-k} is equivalent to a delay of kt_s seconds from the instant when $t = 0$, where t_s is the period between data samples, or one over the sample rate. Specifically, $t_s = 1/f_s$. Because a delay of one sample is equivalent to the factor z^{-1}, the unit time delay symbol used in Figures 6 2 and 6-3 is usually indicated by the z^{-1} operator.

Let's pause for a moment and consider where we stand so far. Our acquaintance with the Laplace transform with its s-plane, the concept of stability based on $H(s)$ pole locations, the introduction of the z-transform with its z-plane poles, and the concept of the z^{-1} operator signifying a single unit of time delay has led us to our goal: the ability to inspect an IIR filter difference equation or filter structure and immediately write the filter's z-domain transfer function $H(z)$. Accordingly, by evaluating an IIR filter's $H(z)$ transfer function appropriately, we can determine the filter's frequency response and its stability. With those ambitious thoughts in mind, let's develop the z-domain equations we need to analyze IIR filters.

Using the relationships of Figure 6-17, we redraw Figure 6-3 as a general Mth-order IIR filter using the z^{-1} operator as shown in Figure 6-18. (In hardware, those z^{-1} operations are shift registers holding successive filter

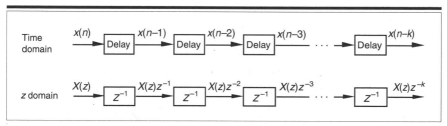

Figure 6-17 Correspondence of the delay operation in the time domain and the z^{-k} operation in the z-domain.

input and output sample values. When implementing an IIR filter in a software routine, the z^{-1} operation merely indicates sequential memory locations where input and output sequences are stored.) The IIR filter structure in Figure 6-18 is often called the *Direct Form I* structure.

The time-domain difference equation describing the general Mth-order IIR filter, having N feed forward stages and M feedback stages, in Figure 6-18 is

$$y(n) = b(0)x(n) + b(1)x(n-1) + b(2)x(n-2) + \ldots + b(N)x(n-N)$$
$$\quad + a(1)y(n-1) + a(2)y(n-2) + \ldots + a(M)y(n-M) \; . \tag{6-21}$$

Time domain expression for an Mth-order IIR filter

In the z-domain, that IIR filter's output can be expressed by

$$Y(z) = b(0)X(z) + b(1)X(z)z^{-1} + b(2)X(z)z^{-2} + \ldots + b(N)X(z)z^{-N}$$
$$\quad + a(1)Y(z)z^{-1} + a(2)Y(z)z^{-2} + \ldots + a(M)Y(z)z^{-M} \; , \tag{6-22}$$

where $Y(z)$ and $X(z)$ represent the z-transform of $y(n)$ and $x(n)$. Look Eqs. (6-21) and (6-22) over carefully and see how the unit time delays translate

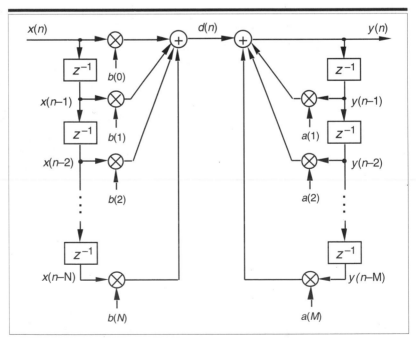

Figure 6-18 General (Direct Form I) structure of an Mth-order IIR filter, having N feed forward stages and M feedback stages, with the z^{-1} operator indicating a unit time delay.

to negative powers of z in the z-domain expression. A more compact form for $Y(z)$ is

z-domain expression for an Mth-order IIR filter: →
$$Y(z) = X(z)\sum_{k=0}^{N} b(k)z^{-k} + Y(z)\sum_{k=1}^{M} a(k)z^{-k} \quad . \quad (6\text{-}23)$$

OK, now we've arrived at the point where we can describe the transfer function of a general IIR filter. Rearranging Eq. (6-23) to collect like terms,

$$Y(z)\left[1 - \sum_{k=1}^{M} a(k)z^{-k}\right] = X(z)\sum_{k=0}^{N} b(k)z^{-k} \quad . \quad (6\text{-}24)$$

Finally, we define the filter's z-domain transfer function as $H(z) = Y(z)/X(z)$, where $H(z)$ is given by

z-domain transfer function of an Mth-order IIR filter: →
$$H(z) = \frac{Y(z)}{X(z)} = \frac{\displaystyle\sum_{k=0}^{N} b(k)z^{-k}}{1 - \displaystyle\sum_{k=1}^{M} a(k)z^{-k}} \quad . \quad (6\text{-}25)$$

(Just like Laplace transfer functions, the order of our z-domain transfer function is defined by the largest exponential order of z in the denominator, in this case M.) Equation (6-25) tells us all we need to know about an IIR filter. We can evaluate the denominator of Eq. (6-25) to determine the positions of the filter's poles on the z-plane indicating the filter's stability.

Remember, now, just as the Laplace transfer function $H(s)$ in Eq. (6-9) was a complex-valued surface on the s-plane, $H(z)$ is a complex-valued surface above, or below, the z-plane. The intersection of that $H(z)$ surface and the perimeter of a cylinder representing the $z = e^{j\omega}$ unit circle is the filter's complex frequency response. This means that substituting $e^{j\omega}$ for z in Eq. (6-25)'s transfer function gives us the expression for the filter's $H_{IIR}(\omega)$ frequency response as

Frequency response of an Mth-order IIR filter filter (exponential form): →
$$H_{IIR}(\omega) = H(z)\Big|_{z=e^{j\omega}} = \frac{\displaystyle\sum_{k=0}^{N} b(k)e^{-jk\omega}}{1 - \displaystyle\sum_{k=1}^{M} a(k)e^{-jk\omega}} \quad . \quad (6\text{-}26)$$

Let's alter the form of Eq. (6-26) to obtain more useful expressions for $H_{IIR}(\omega)$'s frequency magnitude and phase responses. Because a typical IIR filter frequency response $H_{IIR}(\omega)$ is the ratio of two complex functions, we can express $H_{IIR}(\omega)$ in its equivalent rectangular form as

$$H_{IIR}(\omega) = \frac{\sum\limits_{k=0}^{N} b(k) \cdot [\cos(k\omega) - j\sin(k\omega)]}{1 - \sum\limits_{k=1}^{M} a(k) \cdot [\cos(k\omega) - j\sin(k\omega)]} \,, \qquad (6\text{-}27)$$

or

Frequency response of an *M*th-order IIR filter (rectangular form): →

$$H_{IIR}(\omega) = \frac{\sum\limits_{k=0}^{N} b(k) \cdot \cos(k\omega) - j\sum\limits_{k=0}^{N} b(k) \cdot \sin(k\omega)}{1 - \sum\limits_{k=1}^{M} a(k) \cdot \cos(k\omega) + j\sum\limits_{k=1}^{M} a(k) \cdot \sin(k\omega)}$$

$$(6\text{-}28)$$

It's usually easier and, certainly, more useful, to consider the complex frequency response expression in terms of its magnitude and phase. Let's do this by representing the numerator and denominator in Eq. (6-28) as two complex functions of radian frequency ω. Calling the numerator of Eq. (6-28) $Num(\omega)$, then,

$$Num(\omega) = Num_{real}(\omega) + jNum_{imag}(\omega) \,, \qquad (6\text{-}29)$$

where

$$Num_{real}(\omega) = \sum\limits_{k=0}^{N} b(k) \cdot \cos(k\omega) \,,$$

and

$$Num_{imag}(\omega) = -\sum\limits_{k=0}^{N} b(k) \cdot \sin(k\omega) \,. \qquad (6\text{-}30)$$

Likewise, the denominator in Eq. (6-28) can be expressed as

$$Den(\omega) = Den_{real}(\omega) + jDen_{imag}(\omega) \,, \qquad (6\text{-}31)$$

where

$$Den_{real}(\omega) = 1 - \sum_{k=1}^{M} a(k) \cdot \cos(k\omega) \ ,$$

and

$$Den_{imag}(\omega) = \sum_{k=1}^{M} a(k) \cdot \sin(k\omega) \ . \tag{6-32}$$

These $Num(\omega)$ and $Den(\omega)$ definitions allow us to represent $H_{IIR}(\omega)$ in the more simple forms of

$$H_{IIR}(\omega) = \frac{Num(\omega)}{Den(\omega)} = \frac{Num_{real}(\omega) + jNum_{imag}(\omega)}{Den_{real}(\omega) + jDen_{imag}(\omega)} \tag{6-33}$$

$$= \frac{|Num(\omega)| \angle \emptyset_{Num}(\omega)}{|Den(\omega)| \angle \emptyset_{Den}(\omega)} \ . \tag{6-33'}$$

Given the form in Eq. (6-33) and the rules for dividing one complex number by another, provided by Eqs. (A-2) and (A-19') in Appendix A, the frequency magnitude response of a general IIR filter is

$$|H_{IIR}(\omega)| = \frac{|Num(\omega)|}{|Den(\omega)|} = \frac{\sqrt{[Num_{real}(\omega)]^2 + [Num_{imag}(\omega)]^2}}{\sqrt{[Den_{real}(\omega)]^2 + [Den_{imag}(\omega)]^2}} \ . \tag{6-34}$$

Furthermore, the filter's phase response $\emptyset_{IIR}(\omega)$ is the phase of the numerator minus the phase of the denominator, or

$$\emptyset_{IIR}(\omega) = \emptyset_{Num}(\omega) - \emptyset_{Den}(\omega)$$

$$= \tan^{-1}\left(\frac{Num_{imag}(\omega)}{Num_{real}(\omega)}\right) - \tan^{-1}\left(\frac{Den_{imag}(\omega)}{Den_{real}(\omega)}\right) \ . \tag{6-35}$$

To reiterate our intent here, we've gone through the above algebraic gymnastics to develop expressions for an IIR filter's frequency magnitude response $|H_{IIR}(\omega)|$ and phase response $\emptyset_{IIR}(\omega)$ in terms of the filter coefficients in Eqs. (6-30) and (6-32). Shortly, we'll use these expressions to analyze an actual IIR filter.

Pausing a moment to gather our thoughts, we realize that we can use Eqs. (6-34) and (6-35) to compute the magnitude and phase response of IIR filters as a function of the frequency ω. And again, just what is ω? It's the normalized radian frequency represented by the angle around the unit circle in Figure 6-13, having a range of $-\pi \leq \omega \leq +\pi$. In terms of a discrete sampling frequency ω_s measured in radians/s, from Figure 6-14, we see that ω covers the range $-\omega_s/2 \leq \omega \leq +\omega_s/2$. In terms of our old friend f_s Hz, Eqs. (6-34) and (6-35) apply over the equivalent frequency range of $-f_s/2$ to $+f_s/2$ Hz. So, for example, if digital data is arriving at the filter's input at a rate of $f_s = 1000$ samples/s, we could use Eq. (6-34) to plot the filter's frequency magnitude response over the frequency range of -500 Hz to $+500$ Hz.

Although the equations describing the transfer function $H_{IIR}(\omega)$, its magnitude response $|H_{IIR}(\omega)|$, and phase response $\emptyset_{IIR}(\omega)$ look somewhat complicated at first glance, let's illustrate their simplicity and utility by analyzing the simple second-order low-pass IIR filter in Figure 6-19 whose positive cutoff frequency is $\omega_s/10$. By inspection, we can write the filter's time-domain difference equation as

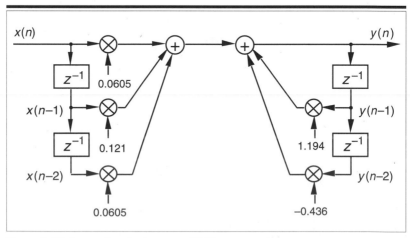

Figure 6-19 Second-order low-pass IIR filter example.

$$y(n) = 0.0605 \cdot x(n) + 0.121 \cdot x(n-1) + 0.0605 \cdot x(n-2)$$
$$+ 1.194 \cdot y(n-1) - 0.436 \cdot y(n-2) , \tag{6-36}$$

or the z-domain expression as

$$Y(z) = 0.0605 \cdot X(z) + 0.121 \cdot X(z)z^{-1} + 0.0605 \cdot X(z)z^{-2}$$
$$+ 1.194 \cdot Y(z)z^{-1} - 0.436 \cdot Y(z)z^{-2} . \tag{6-37}$$

Using Eq. (6-25), we write the z-domain transfer function $H(z)$ as

$$H(z) = \frac{Y(z)}{X(z)} = \frac{0.0605 \cdot z^0 + 0.121 \cdot z^{-1} + 0.0605 \cdot z^{-2}}{1 - 1.194 \cdot z^{-1} + 0.436 \cdot z^{-2}} . \tag{6-38}$$

Replacing z with $e^{j\omega}$, we see that the frequency response of our example IIR filter is

$$H_{IIR}(\omega) = \frac{0.0605 \cdot e^{-j0\omega} + 0.121 \cdot e^{-j1\omega} + 0.0605 \cdot e^{-j2\omega}}{1 - 1.194 \cdot e^{-j1\omega} + 0.436 \cdot e^{-j2\omega}} . \tag{6-39}$$

We're almost there. Remembering Euler's equations and that $\cos(0) = 1$ and $\sin(0) = 0$, we can write the rectangular form of $H_{IIR}(\omega)$ as

$$H_{IIR}(\omega) = \frac{0.0605 + 0.121 \cdot \cos(1\omega) + 0.0605 \cdot \cos(2\omega) - j[0.121 \cdot \sin(1\omega) + 0.0605 \cdot \sin(2\omega)]}{1 - 1.194 \cdot \cos(1\omega) + 0.436 \cdot \cos(2\omega) + j[1.194 \cdot \sin(1\omega) - 0.436 \cdot \sin(2\omega)]} . \tag{6-40}$$

Equation (6-40) is what we're after here, and if we calculate its magnitude over the frequency range of $-\pi \leq \omega \leq \pi$, we get the $|H_{IIR}(\omega)|$ shown as the solid curve in Figure 6-20(a). For comparison purposes we also show a 5-tap low-pass FIR filter magnitude response in Figure 6-20(a). Although both filters require the same computational workload, five multiplications per filter output sample, the low-pass IIR filter has the superior frequency magnitude response. Notice the steeper roll-off and lower sidelobes of the IIR filter relative to the FIR filter.[†]

[†] To make this IIR and FIR filter comparison valid, the coefficients used for both filters were chosen so that each filter would approximate the ideal low-pass frequency response shown in Figure 5-17(a).

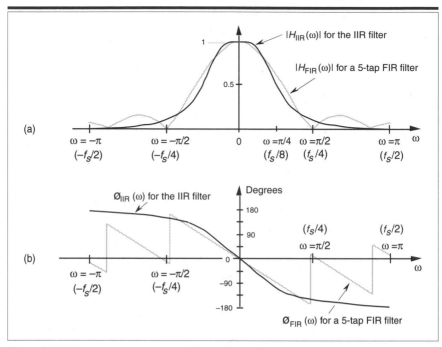

Figure 6-20 Frequency responses of the example IIR filter (solid line) in Figure 6-19 and a 5-tap FIR filter (dashed line): (a) magnitude responses; (b) phase responses.

A word of warning here. It's easy to reverse some of the signs for the terms in the denominator of Eq. (6-40), so be careful if you attempt these calculations at home. Some authors avoid this problem by showing the $a(k)$ coefficients in Figure 6-18 as negative values, so that the summation in the denominator of Eq. (6-25) is always positive. Moreover, some commercial software IIR design routines provide $a(k)$ coefficients whose signs must be reversed before they can be applied to the IIR structure in Figure 6-18. (If, while using software routines to design or analyze IIR filters, your results are very strange or unexpected, the first thing to do is reverse the signs of the $a(k)$ coefficients and see if that doesn't solve the problem.)

The solid curve in Figure 6-20(b) is our IIR filter's $\varnothing_{IIR}(\omega)$ phase response. Notice its nonlinearity relative to the FIR filter's phase response. (Remember now, we're only interested in the filter phase responses over the low-pass filters' passband. So those phase discontinuities for the FIR filter are of no consequence.) Phase nonlinearity is inherent in IIR filters and, based on the ill effects of nonlinear phase introduced in the group delay discussion of Section 5.8, we must carefully consider its implications whenever we decide to use an IIR filter instead of an FIR filter in any given application. The question any filter designer must ask and answer, is "How much phase distortion can I tolerate to realize the benefits of the reduced computational load and high data rates afforded by IIR filters?"

We've arrived at an appropriate point to illustrate an important relationship between IIR and FIR filters. We can use our IIR analysis equations to characterize FIR filters. Notice that, if there were no feedback coefficients in Figure 6-18, equivalent to setting $a(1)$ through $a(M)$ in Eq. (6-21) all to zero, we'd have an N-tap FIR filter. This simplifies Eqs. (6-25) through (6-28) because their denominators would be equal to one. In addition, $Den(\omega)$ in Eq. (6-33) and $|Den(\omega)|$ in Eq. (6-34) would both be unity, and $\phi_{Den}(\omega)$ would be zero in Eq. (6-35). Thus, for an FIR filter, with the $a(k)$ coefficients in Figure 6-18 being all zeros, we can rewrite Eqs. (6-26) and (6-28) as

Frequency response of an N-stage FIR filter (exponential form): →

$$H_{\text{FIR}}(\omega) = \sum_{k=0}^{N} b(k)e^{-jk\omega}, \qquad (6\text{-}41)$$

and

Frequency response of an N-stage FIR filter (rectangular form): →

$$H_{\text{FIR}}(\omega) = \sum_{k=0}^{N} b(k)\cdot\cos(k\omega) - j\sum_{k=0}^{N} b(k)\cdot\sin(k\omega) . \qquad (6\text{-}42)$$

Just as in Eqs. (6-34) and (6-35), the ω in Eqs. (6-41) and (6-42), is the normalized radian frequency represented by the angle around the unit circle in Figure 6-13, having a range of $-\pi \leq \omega \leq +\pi$, corresponding to the equivalent frequency range of $-f_s/2$ to $+f_s/2$ Hz.

6.3.3 An Improved IIR Filter Structure

The structure of the IIR filter in Figure 6-18 can be converted to a more efficient and popular form. It's easy to explore this idea by assuming that there are an equal number of feed forward and feedback stages, letting $M = N$ in Figure 6-18 and thinking of the feed forward and feedback stages as two separate filters. Specifically, the $b(k)$ coefficients are used to calculate the sequence $d(n)$ from $x(n)$, and the $a(k)$ coefficients are used to calculate the filter output $y(n)$ sequence from $d(n)$. Because the two halves of the filter are both linear, we can swap them, as shown in Figure 6-21(a), with no change in the final $y(n)$ output. Of course, after this reorientation,

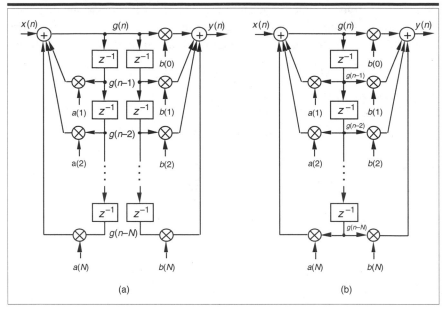

Figure 6-21 Rearranged Nth-order IIR filter structure: (a) feed forward and feedback paths swapped from the structure in Figure 6-3; (b) final simplified Direct Form II filter structure.

a new sequence that we'll call g(n) is generated. The two identical delay paths in Figure 6-21(a) provide the motivation for this reorientation. Because the sequence g(n) is being shifted down along both delay paths in Figure 6-21(a), we can eliminate one of the paths and arrive at the simplified *Direct Form II* filter structure shown in Figure 6-21(b). Notice that eliminating a delay path simplifies the structure, so that a hardware implementation would need only half the number of delay storage registers required by the original IIR structure in Figure 6-18.

Although the filter structure depicted in Figure 6-21(b) is very popular, in the literature we'll also encounter the equivalent representation shown in Figure 6-22. It's interesting to compare Figure 6-22's IIR structure with the FIR structure in Figure 6-2(a). If the feedback coefficients $a(k)$ in Figure 6-22 were all zeros, for example, then this IIR filter would be equivalent to the FIR filter in Figure 6-2(a) as long as the $b(k)$ IIR coefficients were equal to the $h(k)$ FIR coefficients. It's very important to keep in mind that Eq. (6-21), associated with the IIR structure in Figure 6-18, does not explicitly describe the filter structures in Figures 6-21(b) and 6-22. However, the frequency-domain expressions in Eqs. (6-25) through (6-35) can be used to analyze the performance of the popular filter structures in Figures 6-21(b) and 6-22.

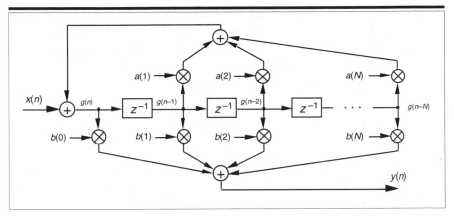

Figure 6-22 Alternate depiction of the IIR filter Direct Form II structure in Figure 6-21(b).

By the way, because of the feedback nature of IIR filters, they're some-times referred to as *recursive* filters. Similarly, FIR filters are often called *nonrecursive* filters. (The term *recursive* originally signified filters with both poles and zeros on the complex z-plane whereas *nonrecursive* signi-fied filters with only zeros on the z-plane.) We'll avoid the terms *recursive* and *nonrecursive*, however, because it's been shown that filters with only zeros can be designed with recursive structures and that filters with both poles and zeros on the z-plane can be designed with nonrecursive struc-tures[10]. So the terms *recursive* and *nonrecursive* more correctly apply to the structures used to implement a digital filter, and the terms *IIR* and *FIR* better describe whether a filter has only zeros (FIR) or both poles and zeros (IIR) on the z-plane[11].

Now that we have a feeling for the performance and implementation structures of IIR filters, let's briefly introduce three filter design tech-niques. These IIR design methods go by the impressive names of impulse *invariance*, *bilinear transform*, and *optimized methods*. The first two methods use *analytical* techniques to design digital filters that approximate con-tinuous analog filters.† The impulse invariance and bilinear transform IIR design methods both start with a closed-form Laplace equation describing the desired analog filter and yield z-transform expressions for the approximating IIR digital filter. The filter designer then solves the z-transform equations to obtain the $a(k)$ and $b(k)$ coefficients for use in the IIR structures in Figures 6-21(b) or 6-22. Because analog filter design

---

† Due to its popularity, throughout the rest of this chapter we'll use the phrase *analog filter* to designate those filters made up of resistors, capacitors, and inductors, designed to operate on continuous signals.

methods are very well understood, designers can take advantage of an abundant variety of analog filter design techniques to design, say, a Butterworth filter with its very flat passband response or, perhaps, go with a Chebyshev filter with its fluctuating passband response and sharper passband-to-stopband cutoff characteristics.

6.4 Impulse Invariance IIR Filter Design Method

The impulse invariance method of IIR filter design is based upon the notion that we can design a discrete filter whose time-domain impulse response is a sampled version of the impulse response of a continuous analog filter. If that analog filter (often called the *prototype filter*) has some desired frequency response, then our IIR filter will yield a discrete approximation of that desired response. The impulse response equivalence of this design method is depicted in Figure 6-23 where we use the conventional notation of δ to represent an impulse function and $h_c(t)$ is the analog filter's impulse response. We use the subscript "*c*" in Figure 6-23(a) to emphasize the continuous nature of the analog filter. Figure 6-23(b) illustrates the definition of the discrete filter's impulse response: the filter's time-domain output sequence when the input is a single unity-valued sample (impulse) preceded and followed by all zero-valued samples. Our goal is to design a digital filter whose impulse response is a sampled version of the analog filter's continuous impulse response. Implied in the correspondence of the continuous and discrete impulse responses is the property that we can map each pole on the *s*-plane for the analog filter's

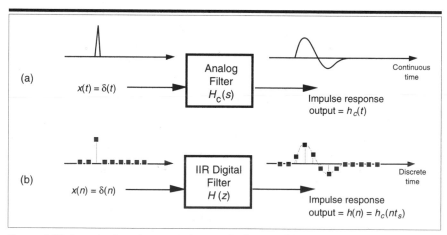

Figure 6-23 Impulse invariance design equivalence of (a) analog filter continuous impulse response; (b) digital filter discrete impulse response.

$H_c(s)$ transfer function to a pole on the z-plane for the discrete IIR filter's $H(z)$ transfer function. What designers have found is that the impulse invariance method does yield useful IIR filters, as long as the sampling rate is high relative to the bandwidth of the signal to be filtered. In other words, IIR filters designed using the impulse invariance method are susceptible to aliasing problems because practical analog filters cannot be perfectly band-limited. Aliasing will occur in an IIR filter's frequency response as shown in Figure 6-24.

From what we've learned in Chapter 2 about the spectral replicating effects of sampling, if Figure 6-4(a) is the spectrum of the continuous $h_c(t)$ impulse response, then the spectrum of the discrete $h_c(nt_s)$ sample sequence is the replicated spectra in Figure 6-24(b).

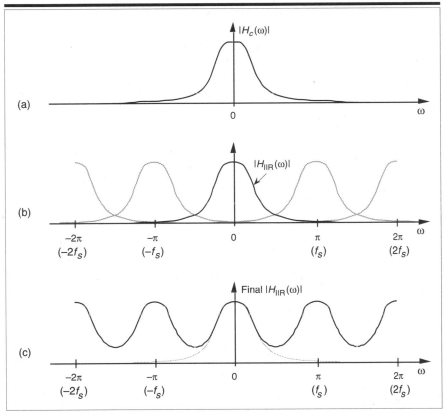

Figure 6-24 Aliasing in the impulse invariance design method: (a) prototype analog filter magnitude response; (b) replicated magnitude responses where $H_{IIR}(\omega)$ is the discrete Fourier transform of $h(n) = h_c(nt_s)$; (c) potential resultant IIR filter magnitude response with aliasing effects.

In Figure 6-24(c) we show the possible effect of aliasing where the dashed curve is a desired $H_{IIR}(\omega)$ frequency magnitude response. However, the actual frequency magnitude response, indicated by the solid curve, can occur when we use the impulse invariance design method. For this reason, we prefer to make the sample frequency f_s as large as possible to minimize the overlap between the primary frequency response curve and its replicated images spaced at multiples of $\pm f_s$ Hz. To see how aliasing can affect IIR filters designed with this method, let's list the necessary impulse invariance design steps and then go through a filter design example.

There are two different methods for designing IIR filters using impulse invariance. The first method, which we'll call Method 1, requires that an inverse Laplace transform as well as a z-transform be performed[12,13]. The second impulse invariance design technique, Method 2, uses a direct substitution process to avoid the inverse Laplace and z-transformations at the expense of needing partial fraction expansion algebra necessary to handle polynomials[14–17]. Both of these methods seem complicated when described in words, but they're really not as difficult as they sound. Let's compare the two methods by listing the steps required for each of them. The impulse invariance design Method 1 goes like this:

Method 1, Step 1:   Design (or have someone design for you) a prototype analog filter with the desired frequency response.†
The result of this step is a continuous Laplace transfer function $H_c(s)$ expressed as the ratio of two polynomials, such as,

$$H_c(s) = \frac{b(N)s^N + b(N-1)s^{N-1} + ... + b(1)s + b(0)}{a(M)s^M + a(M-1)s^{M-1} + ... + a(1)s + a(0)} = \frac{\displaystyle\sum_{k=0}^{N} b(k)s^k}{\displaystyle\sum_{k=0}^{M} a(k)s^k} \qquad (6\text{-}43)$$

which is the general form of Eq. (6-10) with $N < M$, and $a(k)$ and $b(k)$ are constants.

Method 1, Step 2:   Determine the analog filter's continuous time-domain impulse response $h_c(t)$ from the $H_c(s)$ Laplace

---

† In a low-pass filter design, for example, the filter type (Chebyshev, Butterworth, elliptic), filter order (number of poles), and the cutoff frequency are parameters to be defined in this step.

transfer function. I hope, this can be done using Laplace tables as opposed to actually evaluating an inverse Laplace transform equation.

Method 1, Step 3: Determine the digital filter's sampling frequency f_s, and calculate the sample period as $t_s = 1/f_s$. The f_s sampling rate is chosen based on the absolute frequency, in Hz, of the prototype analog filter. Because of the aliasing problems associated with this impulse invariance design method, later, we'll see why f_s should be made as large as practical.

Method 1, Step 4: Find the z-transform of the continuous $h_c(t)$ to obtain the IIR filter's z-domain transfer function $H(z)$ in the form of a ratio of polynomials in z.

Method 1, Step 5: Substitute the value (not the variable) t_s for the continuous variable t in the $H(z)$ transfer function obtained in Step 4. In performing this step, we are ensuring, as in Figure 6-23, that the IIR filter's discrete $h(n)$ impulse response is a sampled version of the continuous filter's $h_c(t)$ impulse response so that $h(n) = h_c(nt_s)$, for $0 \leq n \leq \infty$.

Method 1, Step 6: Our $H(z)$ from Step 5 will now be of the general form

$$H(z) = \frac{b(N)z^{-N} + b(N-1)z^{-(N-1)} + \ldots + b(1)z^{-1} + b(0)}{a(M)z^M + a(M-1)z^{-(M-1)} + \ldots + a(1)z^{-1} + a(0)} = \frac{\displaystyle\sum_{k=0}^{N} b(k)z^{-k}}{1 - \displaystyle\sum_{k=1}^{M} a(k)z^{-k}} . \quad (6\text{-}44)$$

Because the process of sampling the continuous impulse response results in a digital filter frequency response that's scaled by a factor of $1/t_s$, many filter designers find it appropriate to include the t_s factor in Eq. (6-44). So we can rewrite Eq. (6-44) as

$$H(z) = \frac{Y(z)}{X(z)} = \frac{t_s \displaystyle\sum_{k=0}^{N} b(k)z^{-k}}{1 - \displaystyle\sum_{k=1}^{M} a(k)z^{-k}} . \quad (6\text{-}45)$$

Incorporating the value of t_s in Eq. (6-45), then, makes the IIR filter time-response scaling independent of the sampling rate, and the discrete filter will have the same gain as the prototype analog filter.[†]

Method 1, Step 7:   Because Eq. (6-44) is in the form of Eq. (6-25), by inspection, we can express the filter's time-domain difference equation in the general form of Eq. (6-21) as

$$y(n) = b(0)x(n) + b(1)x(n-1) + b(2)x(n-2) + ... + b(N)x(n-N)$$
$$+ a(1)y(n-1) + a(2)y(n-2) + ... + a(M)y(n-M) \; . \qquad (6\text{-}46)$$

Choosing to incorporate t_s, as in Eq. (6-45), to make the digital filter's gain equal to the prototype analog filter's gain, by multiplying the $b(k)$ coefficients by the sample period t_s, leads to an IIR filter time-domain expression of the form

$$y(n) = t_s \cdot [b(0)x(n) + b(1)x(n-1) + b(2)x(n-2) + ... + b(N)x(n-N)]$$
$$+ a(1)y(n-1) + a(2)y(n-2) + ... + a(M)y(n-M). \qquad (6\text{-}47)$$

Notice how the signs changed for the $a(k)$ coefficients from Eqs. (6-44) and (6-45) to Eqs. (6-46) and (6-47). These sign changes always seem to cause problems for beginners, so watch out. Also, keep in mind that the time-domain expressions in Eq. (6-46) and Eq. (6-47) apply only to the filter structure in Figure 6-18. The $a(k)$ and $b(k)$, or $t_s \cdot b(k)$, coefficients, however, can be applied to the improved IIR structure shown in Figure 6-22 to complete our design.

Before we go through an actual example of this design process, let's discuss the other impulse invariance design method.

[†] Some authors have chosen to include the t_s factor in the discrete $h(n)$ impulse response in the above Step 4; that is, make $h(n) = t_s h_c(nt_s)$ [14, 18]. The final result of this, of course, is the same as that obtained by including t_s as described in Step 5.

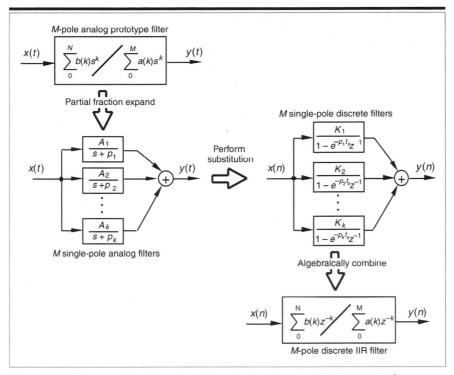

Figure 6-25 Mathematical flow of the impulse invariance design Method 2.

The impulse invariance Design Method 2, also called the standard z-transform method, takes a different approach. It mathematically partitions the prototype analog filter into multiple single-pole continuous filters and then approximates each one of those by a single-pole digital filter. The set of M single-pole digital filters is then algebraically combined to form an M-pole, Mth-ordered IIR filter. This process of breaking the analog filter to discrete filter approximation into manageable pieces is shown in Figure 6-25. The steps necessary to perform an impulse invariance Method 2 design are

Method 2, Step 1: Obtain the Laplace transfer function $H_c(s)$ for the prototype analog filter in the form of Eq. (6-43). (Same as Method 1, Step 1.)

Method 2, Step 2: Select an appropriate sampling frequency f_s and calculate the sample period as $t_s = 1/f_s$. (Same as Method 1, Step 3.)

Method 2, Step 3:   Express the analog filter's Laplace transfer function $H_c(s)$ as the sum of single-pole filters. This requires us to use partial fraction expansion methods to express the ratio of polynomials in Eq. (6-43) in the form of

$$H_c(s) = \frac{b(N)s^N + b(N-1)s^{N-1} + \ldots + b(1)s + b(0)}{a(M)s^M + a(M-1)s^{M-1} + \ldots + a(1)s + a(0)}$$

$$= \sum_{k=1}^{M} \frac{A_k}{s+p_k} = \frac{A_1}{s+p_1} + \frac{A_2}{s+p_2} + \ldots + \frac{A_M}{s+p_M} , \qquad (6\text{-}48)$$

where the individual A_k factors are constants and the kth pole is located at $-p_k$ on the s-plane. We'll denote the kth single-pole analog filter as $H_k(s)$, or

$$H_k(s) = \frac{A_k}{s+p_k} . \qquad (6\text{-}49)$$

Method 2, Step 4:   Substitute $1 - e^{-p_k t_s} z^{-1}$ for $s + p_k$ in Eq. (6-48). This mapping of each $H_k(s)$ pole, located at $s = -p_k$ on the s-plane, to the $z = e^{-p_k t_s}$ location on the z-plane is how we approximate the impulse response of each single-pole analog filter by a single-pole digital filter. (The reader can find the derivation of this $1 - e^{-p_k t_s} z^{-1}$ substitution, illustrated in our Figure 6-25, in references [14] through [16].) So, the kth analog single-pole filter $H_k(s)$ is approximated by a single-pole digital filter whose z-domain transfer function is

$$H_k(z) = \frac{A_k}{1 - e^{-p_k t_s} z^{-1}} . \qquad (6\text{-}50)$$

The final combined discrete filter transfer function $H(z)$ is the sum of the single-poled discrete filters, or

$$H(z) = \sum_{k=1}^{M} H_k(z) = \sum_{k=1}^{M} \frac{A_k}{1 - e^{-p_k t_s} z^{-1}} . \qquad (6\text{-}51)$$

Keep in mind that the above $H(z)$ is not a function of time. The t_s factor in Eq. (6-51) is a constant equal to the discrete-time sample period.

Method 2, Step 5:   Calculate the z-domain transfer function of the sum of the M single-pole digital filters in the form of a ratio of two polynomials in z. Because the $H(z)$ in Eq. (6-51) will be a series of fractions, we'll have to combine those fractions over a common denominator to get a single ratio of polynomials in the familiar form of

$$H(z) = \frac{Y(z)}{X(z)} = \frac{\displaystyle\sum_{k=0}^{N} b(k)z^{-k}}{1 - \displaystyle\sum_{k=1}^{M} a(k)z^{-k}} . \qquad (6\text{-}52)$$

Method 2, Step 6:   Just as in Method 1 Step 6, by inspection, we can express the filter's time-domain equation in the general form of

$$y(n) = b(0)x(n) + b(1)x(n-1) + b(2)x(n-2) + \ldots + b(N)x(n-N)$$
$$+ a(1)y(n-1) + a(2)y(n-2) + \ldots + a(M)y(n-M) . \qquad (6\text{-}53)$$

Again, notice the $a(k)$ coefficient sign changes from Eq. (6-52) to Eq. (6-53). As described in Method 1 Steps 6 and 7, if we choose to make the digital filter's gain equal to the prototype analog filter's gain by multiplying the $b(k)$ coefficients by the sample period t_s, then the IIR filter's time-domain expression will be in the form

$$y(n) = t_s \cdot [b(0)x(n) + b(1)x(n-1) + b(2)x(n-2) + \ldots + b(N)x(n-N)]$$
$$+ a(1)y(n-1) + a(2)y(n-2) + \ldots + a(M)y(n-M) , \qquad (6\text{-}54)$$

yielding a final $H(z)$ z-domain transfer function of

$$H(z) = \frac{Y(z)}{X(z)} = \frac{t_s \cdot \sum\limits_{k=0}^{N} b(k)z^{-k}}{1 - \sum\limits_{k=1}^{M} a(k)z^{-k}} \ . \qquad (6\text{-}54')$$

Finally, we can implement the improved IIR structure shown in Figure 6-22 using the $a(k)$ and $b(k)$ coefficients from Eq. (6-53) or the $a(k)$ and $t_s \cdot b(k)$ coefficients from Eq. (6-54).

To provide a more meaningful comparison between the two impulse invariance design methods, let's dive in and go through an IIR filter design example using both methods.

6.4.1 Impulse Invariance Design Method 1 Example

Assume that we need to design an IIR filter that approximates a second-order Chebyshev prototype analog low-pass filter whose passband ripple is 1 dB. Our f_s sampling rate is 100 Hz ($t_s = 0.01$), and the filter's 1 dB cutoff frequency is 20 Hz. Our prototype analog filter will have a frequency magnitude response like that shown in Figure 6-26.

Given the above filter requirements, assume that the analog prototype filter design effort results in the $H_c(s)$ Laplace transfer function of

$$H_c(s) = \frac{17410.145}{s^2 + 137.94536s + 17410.145} \ . \qquad (6\text{-}55)$$

It's the transfer function in Eq. (6-55) that we intend to approximate with our discrete IIR filter. To find the analog filter's impulse response, we'd

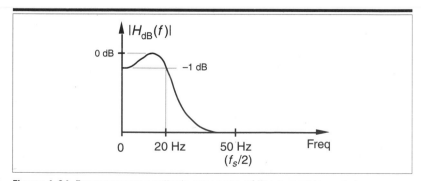

Figure 6-26 Frequency magnitude response of the example prototype analog filter.

like to get $H_c(s)$ into a form that allows us to use Laplace transform tables to find $h_c(t)$. Searching through systems analysis textbooks we find the following Laplace transform pair:

$X(s)$, Laplace transform of $x(t)$:	$x(t)$:	
$$\frac{A\omega}{(s+\alpha)^2+\omega^2}$$	$Ae^{-\alpha t} \cdot \sin(\omega t)$.	(6-56)

Our intent, then, is to modify Eq. (6-55) to get it into the form on the left side of Eq. (6-56). We do this by realizing that the Laplace transform expression in Eq. (6-56) can be rewritten as

$$\frac{A\omega}{(s+\alpha)^2+\omega^2} = \frac{A\omega}{s^2+2\alpha s+\alpha^2+\omega^2} \ . \tag{6-57}$$

If we set Eq. (6-55) equal to the right side of Eq. (6-57), we can solve for A, α, and ω. Doing that,

$$H_c(s) = \frac{17410.145}{s^2+137.94536s+17410.145} = \frac{A\omega}{s^2+2\alpha s+\alpha^2+\omega^2} \ . \tag{6-58}$$

Solving Eq. (6-58) for A, α, and ω, we first find

$$\alpha = \frac{137.94536}{2} = 68.972680 \ ; \tag{6-59}$$

$$\alpha^2+\omega^2 = 17410.145 \ , \tag{6-60}$$

so

$$\omega = \sqrt{17410.145-\alpha^2} = 112.485173; \tag{6-61}$$

and

$$A = \frac{17410.145}{\omega} = 154.77724 \ . \tag{6-62}$$

OK, we can now express $H_c(s)$ in the desired form of the left side of Eq. (6-57) as

$$H_c(s) = \frac{(154.77724)(112.485173)}{(s + 68.972680)^2 + (112.485173)^2} \cdot \qquad (6\text{-}63)$$

Using the Laplace transform pair in Eq. (6-56), the time-domain impulse response of the prototype analog filter becomes

$$h_c(t) = Ae^{-\alpha t} \cdot \sin(\omega t) = 154.77724e^{-68.972680t} \cdot \sin(112.485173t) \ . \qquad (6\text{-}64)$$

OK, we're ready to perform Method 1, Step 4, to determine the discrete IIR filter's z-domain transfer function $H(z)$ by performing the z-transform of $h_c(t)$. Again, scanning through digital signal processing textbooks or a good math reference book, we find the following z-transform pair where the time-domain expression is in the same form as Eq. (6-64)'s $h_c(t)$ impulse response:

$x(t)$:	$X(z), z$ – transform of $x(t)$:	
$Ce^{-\alpha t} \cdot \sin(\omega t)$	$\dfrac{Ce^{-\alpha t} \cdot \sin(\omega t)z^{-1}}{1 - 2[e^{-\alpha t} \cdot \cos(\omega t)]z^{-1} + e^{-2\alpha t}z^{-2}} \cdot$	(6-65)

Remember now, the α and ω in Eq. (6-65) are generic and are not related to the α and ω values in Eq. (6-59) and Eq. (6-61). Substituting the constants from Eq. (6-64) into the right side of Eq. (6-65), we get the z-transform of the IIR filter as

$$H(z) = \frac{154.77724e^{-68.972680t} \cdot \sin(112.485173t)z^{-1}}{1 - 2[e^{-68.972680t} \cdot \cos(112.485173t)]z^{-1} + e^{-2\cdot 68.972680t}z^{-2}} \cdot$$
$$\qquad (6\text{-}66)$$

Performing Method 1, Step 5, we substitute the t_s value of 0.01 for the continuous variable t in Eq. (6-66), yielding the final $H(z)$ transfer function of

$$H(z) = \frac{154.77724e^{-68.972680\cdot 0.01} \cdot \sin(112.485173 \cdot 0.01)z^{-1}}{1 - 2[e^{-68.972680\cdot 0.01} \cdot \cos(112.485173 \cdot 0.01)]z^{-1} + e^{-2\cdot 68.972680\cdot 0.01}z^{-2}}$$

$$= \frac{154.77724e^{-0.68972680} \cdot \sin(1.12485173)z^{-1}}{1 - 2[e^{-0.68972680} \cdot \cos(1.12485173)]z^{-1} + e^{-2\cdot 0.68972680}z^{-2}}$$

$$= \frac{Y(z)}{X(z)} = \frac{70.059517z^{-1}}{1 - 0.43278805z^{-1} + 0.25171605z^{-2}} \cdot \qquad (6\text{-}67)$$

OK, hang in there, we're almost finished. Here are the final steps of Method 1. Because of the transfer function $H(z) = Y(z)/X(z)$, we can cross-multiply the denominators to rewrite the bottom line of Eq. (6-67) as

$$Y(z) \cdot (1 - 0.43278805z^{-1} + 0.25171605z^{-2}) = X(z) \cdot (70.059517z^{-1}) ,$$

or

$$Y(z) = 70.059517 \cdot X(z)z^{-1} + 0.43278805 \cdot Y(z)z^{-1} - 0.25171605 \cdot Y(z)z^{-2}. \quad (6\text{-}68)$$

By inspection of Eq. (6-68), we can now get the time-domain expression for our IIR filter. Performing Method 1, Steps 6 and 7, we multiply the $x(n-1)$ coefficient by the sample period value of $t_s = 0.01$ to allow for proper scaling as

$$y(n) = 0.01 \cdot 70.059517 \cdot x(n-1) + 0.43278805 \cdot y(n-1) - 0.25171605 \cdot y(n-2)$$
$$= 0.70059517 \cdot x(n-1) + 0.43278805 \cdot y(n-1) - 0.25171605 \cdot y(n-2) , \quad (6\text{-}69)$$

and there we (finally) are. The coefficients from Eq. (6-69) are what we use in implementing the improved IIR structure shown in Figure 6-22 to approximate the original second-order Chebyshev analog low-pass filter.

Let's see if we get the same result if we use the impulse invariance design Method 2 to approximate the example prototype analog filter.

6.4.2 Impulse Invariance Design Method 2 Example

Given the original prototype filter's Laplace transfer function as

$$H_c(s) = \frac{17410.145}{s^2 + 137.94536s + 17410.145} , \quad (6\text{-}70)$$

and the value of $t_s = 0.01$ for the sample period, we're ready to proceed with Method 2's Step 3. To express $H_c(s)$ as the sum of single-pole filters, we'll have to factor the denominator of Eq. (6-70) and use partial fraction expansion methods. For convenience, let's start by replacing the constants in Eq. (6-70) with variables in the form of

$$H_c(s) = \frac{c}{s^2 + bs + c} , \quad (6\text{-}71)$$

where $b = 137.94536$, and $c = 17410.145$. Next, using Eq. (6-15) with $a = 1$, we can factor the quadratic denominator of Eq. (6-71) into

$$H_c(s) = \frac{c}{\left(s + \dfrac{b}{2} + \sqrt{\dfrac{b^2 - 4c}{4}}\right) \cdot \left(s + \dfrac{b}{2} - \sqrt{\dfrac{b^2 - 4c}{4}}\right)} . \qquad (6\text{-}72)$$

If we substitute the values for b and c in Eq. (6-72), we'll find that the quantity under the radical sign is negative. This means that the factors in the denominator of Eq. (6-72) are complex. Because we have lots of algebra ahead of us, let's replace the radicals in Eq. (6-72) with the *imaginary* term jR, where $j = \sqrt{-1}$ and $R = |(b^2 - 4c)/4|$, such that

$$H_c(s) = \frac{c}{(s + b/2 + jR)(s + b/2 - jR)} . \qquad (6\text{-}73)$$

OK, partial fraction expansion methods allow us to partition Eq. (6-73) into two separate fractions of the form

$$H_c(s) = \frac{c}{(s + b/2 + jR)(s + b/2 - jR)}$$

$$= \frac{K_1}{(s + b/2 + jR)} + \frac{K_2}{(s + b/2 - jR)} , \qquad (6\text{-}74)$$

where the K_1 constant can be found to be equal to $jc/2R$ and constant K_2 is the complex conjugate of K_1, or $K_2 = -jc/2R$. (To learn the details of partial fraction expansion methods, the interested reader should investigate standard college algebra or engineering mathematics textbooks.) Thus, $H_c(s)$ can be of the form in Eq. (6-48) or

$$H_c(s) = \frac{jc/2R}{(s + b/2 + jR)} + \frac{-jc/2R}{(s + b/2 - jR)} . \qquad (6\text{-}75)$$

We can see from Eq. (6-75) that our second-order prototype filter has two poles, one located at $p_1 = -b/2 - jR$ and the other at $p_2 = -b/2 + jR$. We're now ready to map those two poles from the s-plane to the z-plane as called out in Method 2, Step 4. Making our $1 - e^{-p_k t_s} z^{-1}$ substitution for the $s + p_k$ terms in Eq. (6-75), we have the following expression for the z-domain single-pole digital filters,

$$H(z) = \frac{jc/2R}{1-e^{-(b/2+jR)t_s}z^{-1}} + \frac{-jc/2R}{1-e^{-(b/2-jR)t_s}z^{-1}} \ . \tag{6-76}$$

Our objective in Method 2, Step 5 is to massage Eq. (6-76) into the form of Eq. (6-52), so that we can determine the IIR filter's feed forward and feedback coefficients. Putting both fractions in Eq. (6-76) over a common denominator gives us

$$H(z) = \frac{(jc/2R)(1-e^{-(b/2-jR)t_s}z^{-1}) - (jc/2R)(1-e^{-(b/2+jR)t_s}z^{-1})}{(1-e^{-(b/2+jR)t_s}z^{-1})(1-e^{-(b/2-jR)t_s}z^{-1})}. \tag{6-77}$$

Collecting like terms in the numerator and multiplying out the denominator gives us

$$H(z) = \frac{(jc/2R)(1-e^{-(b/2-jR)t_s}z^{-1} - 1 + e^{-(b/2+jR)t_s}z^{-1})}{1-e^{-(b/2-jR)t_s}z^{-1} - e^{-(b/2+jR)t_s}z^{-1} + e^{-bt_s}z^{-2}} \ . \tag{6-78}$$

Factoring the exponentials and collecting like terms of powers of z in Eq. (6-78),

$$H(z) = \frac{(jc/2R)(e^{-(b/2+jR)t_s} - e^{-(b/2-jR)t_s})z^{-1}}{1-(e^{-(b/2-jR)t_s} + e^{-(b/2+jR)t_s})z^{-1} + e^{-bt_s}z^{-2}} \ . \tag{6-79}$$

Continuing to simplify our $H(z)$ expression by factoring out the real part of the exponentials,

$$H(z) = \frac{(jc/2R)e^{-bt_s/2}(e^{-jRt_s} - e^{jRt_s})z^{-1}}{1-e^{-bt_s/2}(e^{jRt_s} + e^{-jRt_s})z^{-1} + e^{-bt_s}z^{-2}} \ . \tag{6-80}$$

We now have $H(z)$ in a form with all the like powers of z combined into single terms, and Eq. (6-80) looks something like the desired form of Eq. (6-52). Knowing that the final coefficients of our IIR filter must be real numbers, the question is "What do we do with those imaginary j terms in Eq. (6-80)?" Once again, Euler to the rescue.[†] Using Euler's equations for sinusoids, we can eliminate the imaginary exponentials and Eq. (6-80) becomes

---

[†] From Euler, we know that sin(ø) = $(e^{jø} - e^{-jø})/2j$, and cos(ø) = $(e^{jø} + e^{-jø})/2$.

$$H(z) = \frac{(jc/2R)e^{-bt_s/2}[-2j\sin(Rt_s)]z^{-1}}{1 - e^{-bt_s/2}[2\cos(Rt_s)]z^{-1} + e^{-bt_s}z^{-2}}$$

$$= \frac{(c/R)e^{-bt_s/2}[\sin(Rt_s)]z^{-1}}{1 - e^{-bt_s/2}[2\cos(Rt_s)]z^{-1} + e^{-bt_s}z^{-2}} \cdot \qquad (6\text{-}81)$$

If we plug the values c = 17410.145, b = 137.94536, R = 112.48517, and t_s = 0.01 into Eq. (6-81), we get the following IIR filter transfer function:

$$H(z) = \frac{(154.77724)(0.50171312)(0.902203655)z^{-1}}{1 - (0.50171312)(0.86262058)z^{-1} + 0.25171605z^{-2}}$$

$$= \frac{70.059517z^{-1}}{1 - 0.43278805z^{-1} + 0.25171605z^{-2}} \cdot \qquad (6\text{-}82)$$

Because the transfer function $H(z) = Y(z)/X(z)$, we can again cross-multiply the denominators to rewrite Eq. (6-82) as

$$Y(z) \cdot (1 - 0.43278805z^{-1} + 0.25171605z^{-2}) = X(z) \cdot (70.059517z^{-1}) \, ,$$

or

$$Y(z) = 70.059517 \cdot X(z)z^{-1} + 0.43278805 \cdot Y(z)z^{-1} - 0.25171605 \cdot Y(z)z^{-2} . \quad (6\text{-}83)$$

Now we take the inverse z-transform of Eq. (6-83), by inspection, to get the time-domain expression for our IIR filter as

$$y(n) = 70.059517 \cdot x(n-1) + 0.43278805 \cdot y(n-1) - 0.25171605 \cdot y(n-2) . \quad (6\text{-}84)$$

One final step remains. To force the IIR filter gain equal to the prototype analog filter's gain, we multiply the $x(n-1)$ coefficient by the sample period t_s as suggested in Method 2, Step 6. In this case, there's only one $x(n)$ coefficient, giving us

$$y(n) = 0.01 \cdot 70.059517 \cdot x(n-1) + 0.43278805 \cdot y(n-1) - 0.25171605 \cdot y(n-2)$$

$$= 0.70059517 \cdot x(n-1) + 0.43278805 \cdot y(n-1) - 0.25171605 \cdot y(n-2), \quad (6\text{-}85)$$

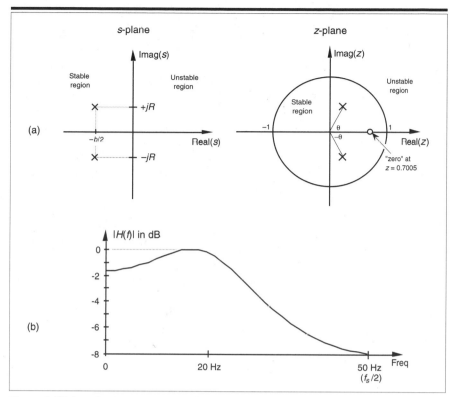

Figure 6-27 Impulse invariance design example filter characteristics: (a) s-plane pole locations of prototype analog filter and z-plane pole locations of discrete IIR filter; (b) frequency magnitude response of the discrete IIR filter.

that compares well with the Method 1 result in Eq. (6-69). (Isn't it comforting to work a problem two different ways and get the same result?)

Figure 6-27 shows, in graphical form, the result of our IIR design example. The s-plane pole locations of the prototype filter and the z-plane poles of the IIR filter are shown in Figure 6-27(a). Because the s-plane poles are to the left of the origin and the z-plane poles are inside the unit circle, both the prototype analog and the discrete IIR filters are stable. We find the prototype filter's s-plane pole locations by evaluating $H_c(s)$ in Eq. (6-75). When $s = -b/2 - jR$, the denominator of the first term in Eq. (6-75) becomes zero and $H_c(s)$ is infinitely large. That $s = -b/2 - jR$ value is the location of the lower s-plane pole in Figure 6-27(a). When $s = -b/2 + jR$ the denominator of the second term in Eq. (6-75) becomes zero and $s = -b/2 + jR$ is the location of the second s-plane pole.

The IIR filter's z-plane pole locations are found from Eq. (6-76). If we multiply the numerators and denominators of Eq. (6-76) by z,

$$H(z) \cdot \frac{z}{z} = \frac{z(jc/2R)}{z(1 - e^{-(b/2+jR)t_s} z^{-1})} + \frac{z(-jc/2R)}{z(1 - e^{-(b/2-jR)t_s} z^{-1})}$$

$$= \frac{(jc/2R)z}{z - e^{-(b/2+jR)t_s}} + \frac{(-jc/2R)z}{z - e^{-(b/2-jR)t_s}} \quad . \tag{6-86}$$

In Eq. (6-86), when z is set equal to $e^{(-b/2+jR)t_s}$, the denominator of the first term in Eq. (6-86) becomes zero and $H(z)$ becomes infinitely large. The value of z of

$$z = e^{-(b/2+jR)t_s} = e^{-bt_s/2} e^{-jRt_s} = e^{-bt_s/2} \angle -Rt_s \text{ radians} \tag{6-87}$$

defines the location of the lower z-plane pole in Figure 6-27(a). Specifically, this lower pole is located at a distance of $e^{-bt_s/2} = 0.5017$ from the origin, at an angle of $\theta = -Rt_s$ radians, or $-64.45°$. Being conjugate poles, the upper z-plane pole is located the same distance from the origin at an angle of $\theta = Rt_s$ radians, or $+64.45°$. Figure 6-27(b) illustrates the frequency magnitude response of the IIR filter in Hz.

Two different implementations of our IIR filter are shown in Figure 6-28. Figure 6-28(a) is an implementation of our second-order IIR filter based on the general IIR structure given in Figure 6-22, and Figure 6-28(b) shows the second-order IIR filter implementation based on the alternate structure from Figure 6-21(b). Knowing that the $b(0)$ coefficient on the left side of Figure 6-28(b) is zero, we arrive at the simplified structure on the right side of Figure 6-28(b). Looking carefully at Figure 6-28(a) and the right side of Figure 6-28(b), we can see that they are equivalent.

Although both impulse invariance design methods are covered in the literature, we might ask, "Which one is preferred?" There's no definite answer to that question because it depends on the $H_c(s)$ of the prototype analog filter. Although our Method 2 example above required more algebra than Method 1, if the prototype filter's s-domain poles were located only on the real axis, Method 2 would have been much simpler because there would be no complex variables to manipulate. In general, Method 2 is more popular for two reasons: (1) the inverse Laplace and z-transformations, although straightforward in our Method 1 example, can be very difficult for higher-order filters, and (2) unlike Method 1, Method 2 can be coded in a software routine or a computer spreadsheet.

Upon examining the frequency magnitude response in Figure 6-27(b), we can see that this second-order IIR filter's roll-off is not particularly

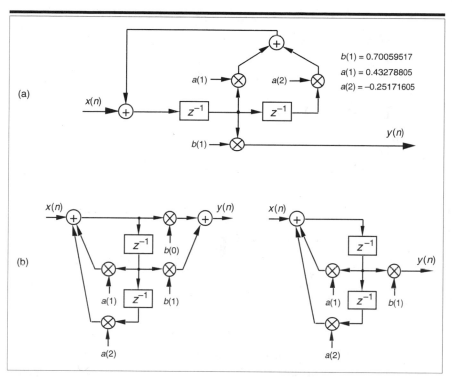

Figure 6-28 Implementations of the impulse invariance design example filter.

steep. This is, admittedly, a simple low-order filter, but it's attenuation slope is so gradual that it doesn't appear to be of much use as a low-pass filter.[†] We can also see that the filter's passband ripple is greater than the desired value of 1 dB in Figure 6-26. What we'll find is that it's not the low order of the filter that contributes to it's poor performance, but the sampling rate used. That second-order IIR filter response is repeated as the shaded curve in Figure 6-29. If we increased the sampling rate to 200 Hz, we'd get the frequency response shown by the dashed curve in Figure 6-29. Increasing the sampling rate to 400 Hz results in the much improved frequency response indicated by the solid line in the figure. Sampling rate changes do not affect our filter order or implementation structure. Remember, if we change the sampling rate, only the sample period t_s changes in our design equations, resulting in a different set of filter

---

[†] A piece of advice. Whenever you encounter any frequency representation (be it a digital filter magnitude response or a signal spectrum) that has nonzero values at $+f_s/2$, be suspicious—be very suspicious—that aliasing is taking place.

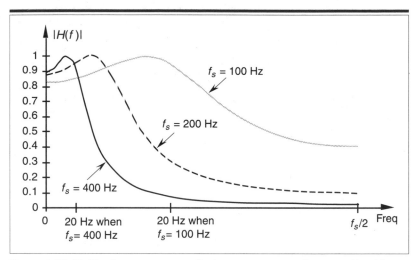

Figure 6-29 IIR filter frequency magnitude response, on a linear scale, at three separate sampling rates. Notice how the filter's absolute cutoff frequency of 20 Hz shifts relative to the different f_s sampling rates.

coefficients for each new sampling rate. So we can see that the smaller we make t_s (larger f_s) the better the resulting filter when either impulse invariance design method is used because the replicated spectral overlap indicated in Figure 6-24(b) is reduced due to the larger f_s sampling rate. The bottom line here is that impulse invariance IIR filter design techniques are most appropriate for narrowband filters; that is, low-pass filters whose cutoff frequencies are much smaller than the sampling rate.

The second analytical technique for analog filter approximation, the bilinear transform method, alleviates the impulse invariance method's aliasing problems at the expense of what's called frequency warping. Specifically, there's a nonlinear distortion between the prototype analog filter's frequency scale and the frequency scale of the approximating IIR filter designed using the bilinear transform. Let's see why.

6.5 Bilinear Transform IIR Filter Design Method

There's a popular analytical IIR filter design technique known as the *bilinear transform* method. Like the impulse invariance method, this design technique approximates a prototype analog filter defined by the continuous Laplace transfer function $H_c(s)$ with a discrete filter whose transfer function is $H(z)$. However, the bilinear transform method has great utility because

- it allows us to simply substitute a function of z for s in $H_c(s)$ to get $H(z)$, thankfully, eliminating the need for Laplace and z-transformations as well as any necessity for partial fraction expansion algebra;

- it maps the entire s-plane to the z-plane, enabling us to completely avoid the frequency-domain aliasing problems we had with the impulse invariance design method; and

- it induces a nonlinear distortion of $H(z)$'s frequency axis, relative to the original prototype analog filter's frequency axis, that sharpens the final roll-off of digital low-pass filters.

Don't worry. We'll explain each one of these characteristics and see exactly what they mean to us as we go about designing an IIR filter.

If the transfer function of a prototype analog filter is $H_c(s)$, then we can obtain the discrete IIR filter z-domain transfer function $H(z)$ by substituting the following for s in $H_c(s)$

$$s = \frac{2}{t_s}\left(\frac{1-z^{-1}}{1+z^{-1}}\right),$$ (6-88)

where, again, t_s is the discrete filter's sampling period $(1/f_s)$. Just as in the impulse invariance design method, when using the bilinear transform method, we're interested in where the analog filter's poles end up on the z-plane after the transformation. This s-plane to z-plane mapping behavior is exactly what makes the bilinear transform such an attractive design technique.[†]

Let's investigate the major characteristics of the bilinear transform's s-plane to z-plane mapping. First we'll show that any pole on the left side of the s-plane will map to the inside of the unit circle in the z-plane. It's easy to show this by solving Eq. (6-88) for z in terms of s. Multiplying Eq. (6-88) by $(t_s/2)(1 + z^{-1})$ and collecting like terms of z leads us to

$$z = \frac{1+(t_s/2)s}{1-(t_s/2)s}.$$ (6-89)

[†] The bilinear transform is a technique in the theory of complex variables for mapping a function on the complex plane of one variable to the complex plane of another variable. It maps circles and straight lines to straight lines and circles, respectively.

If we designate the real and imaginary parts of s as

$$s = \sigma + j\omega_a ,$$ (6-90)

where the subscript in the radian frequency ω_a signifies *analog*, Eq. (6-89) becomes

$$z = \frac{1 + \sigma t_s/2 + j\omega_a t_s/2}{1 - \sigma t_s/2 - j\omega_a t_s/2} = \frac{(1 + \sigma t_s/2) + j\omega_a t_s/2}{(1 - \sigma t_s/2) - j\omega_a t_s/2} .$$ (6-91)

We see in Eq. (6-91) that z is complex, comprising the ratio of two complex expressions. As such, if we denote z as a magnitude at an angle in the form of $z = |z|\angle\theta_z$, we know that the magnitude of z is given by

$$|z| = \frac{\text{Mag}_{\text{numerator}}}{\text{Mag}_{\text{denominator}}} = \sqrt{\frac{(1 + \sigma t_s/2)^2 + (\omega_a t_s/2)^2}{(1 - \sigma t_s/2)^2 + (\omega_a t_s/2)^2}} .$$ (6-92)

OK, if σ is negative ($\sigma < 0$) the numerator of the ratio on the right side of Eq. (6-92) will be less than the denominator, and $|z|$ will be less than 1. On the other hand, if σ is positive ($\sigma > 0$), the numerator will be larger than the denominator, and $|z|$ will be greater than 1. This confirms that, when using the bilinear transform defined by Eq. (6-88), any pole located on the left side of the s-plane ($\sigma < 0$) will map to a z-plane location inside the unit circle. This characteristic ensures that any stable s-plane pole of a prototype analog filter will map to a stable z-plane pole for our discrete IIR filter. Likewise, any analog filter pole located on the right side of the s-plane ($\sigma > 0$) will map to a z-plane location outside the unit circle when using the bilinear transform. This reinforces our notion that, to avoid filter instability, during IIR filter design, we should avoid allowing any z-plane poles to lie outside the unit circle.

Next, let's show that the $j\omega_a$ axis of the s-plane maps to the perimeter of the unit circle in the z-plane. We can do this by setting $\sigma = 0$ in Eq. (6-91) to get

$$z = \frac{1 + j\omega_a t_s/2}{1 - j\omega_a t_s/2} .$$ (6-93)

Here, again, we see in Eq. (6-93) that z is a complex number comprising the ratio of two complex numbers, and we know the magnitude of this z is given by

$$|z|_{\sigma=0}=\frac{Mag_{numerator}}{Mag_{denominator}}=\sqrt{\frac{(1)^2+(\omega_a t_s/2)^2}{(1)^2+(\omega_a t_s/2)^2}}\ . \tag{6-94}$$

The magnitude of z in Eq. (6-94) is *always* 1. So, as we stated, when using the bilinear transform, the $j\omega_a$ axis of the s-plane maps to the perimeter of the unit circle in the z-plane. However, this frequency mapping from the s-plane to the unit circle in the z-plane is not linear. It's important to know why this frequency nonlinearity, or warping, occurs and to understand its effects. So we shall, by showing the relationship between the s-plane frequency and the z-plane frequency that we'll designate as ω_d.

If we define z on the unit circle in polar form as $z = re^{-j\omega d}$ as we did for Figure 6-13, where r is 1 and ω_d is the angle, we can substitute $z = e^{j\omega d}$ in Eq. (6-88) as

$$s=\frac{2}{t_s}\left(\frac{1-e^{-j\omega_d}}{1+e^{-j\omega_d}}\right)\ . \tag{6-95}$$

If we show s in its rectangular form and partition the ratio in brackets into half-angle expressions,

$$s=\sigma+j\omega_a=\frac{2}{t_s}\cdot\frac{e^{-j\omega_d/2}(e^{j\omega_d/2}-e^{-j\omega_d/2})}{e^{-j\omega_d/2}(e^{j\omega_d/2}+e^{-j\omega_d/2})}\ . \tag{6-96}$$

Using Euler's relationships of $\sin(\o) = (e^{j\o} - e^{-j\o})/2j$ and $\cos(\o) = (e^{j\o} + e^{-j\o})/2$, we can convert the right side of Eq. (6-96) to rectangular form as

$$s=\sigma+j\omega_a=\frac{2}{t_s}\cdot\frac{e^{-j\omega_d/2}[2j\sin(\omega_d/2)]}{e^{-j\omega_d/2}[2\cos(\omega_d/2)]}$$

$$=\frac{2}{t_s}\cdot\frac{2e^{-j\omega_d/2}}{2e^{-j\omega_d/2}}\cdot\frac{j\sin(\omega_d/2)}{\cos(\omega_d/2)}$$

$$=\frac{j2}{t_s}\tan(\omega_d/2). \tag{6-97}$$

If we now equate the real and imaginary parts of Eq. (6-97), we see that $\sigma = 0$, and

$$\omega_a = \frac{2}{t_s} \tan\left(\frac{\omega_d}{2}\right) .$$

(6-98)

Let's rearrange Eq. (6-98) to give us the useful expression for the z-domain frequency ω_d, in terms of the s-domain frequency ω_a, of

$$\omega_d = 2 \tan^{-1}\left(\frac{\omega_a t_s}{2}\right) .$$

(6-99)

The important relationship in Eq. (6-99), which accounts for the so-called frequency warping due to the bilinear transform, is illustrated in Figure 6-30. Notice that, because $\tan^{-1}(\omega_a t_s/2)$ approaches $\pi/2$ as ω_a gets large, ω_d must, then, approach twice that value, or π. This means that no matter how large ω_a gets, the z-plane ω_d will never be greater than π.

Remember how we considered Figure 6-14 and stated that only the $-\pi f_s$ to $+\pi f_s$ radians/s frequency range for ω_a can be accounted for on the z-plane? Well, our new mapping from the bilinear transform maps the entire s-plane to the z-plane, not just the primary strip of the s-plane shown in Figure 6-14. Now, just as a walk along the $j\omega_a$ frequency axis on the s-plane takes us to infinity in either direction, a trip halfway around the unit circle in a counterclockwise direction takes us from $\omega_a = 0$ to $\omega_a = +\infty$ radians/s. As such, the bilinear transform maps the s-plane's entire $j\omega_a$ axis onto the unit circle in the z-plane. We illustrate these bilinear transform mapping properties in Figure 6-31.

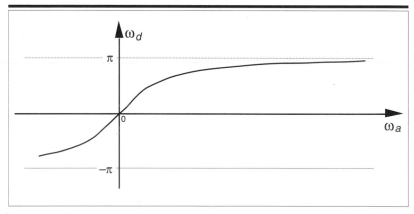

Figure 6-30 Nonlinear relationship between the z-domain frequency ω_d and the s-domain frequency ω_a.

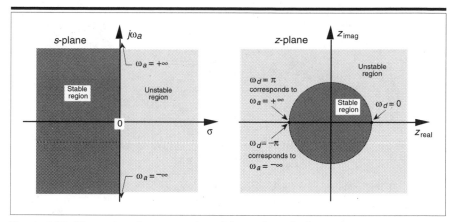

Figure 6-31 Bilinear transform mapping of the s-plane to the z-plane.

To show the practical implications of this frequency warping, let's relate the s-plane and z-plane frequencies to the more practical measure of the f_s sampling frequency. We do this by remembering the fundamental relationship between radians/s and Hz of $\omega = 2\pi f$ and solving for f to get

$$f = \frac{\omega}{2\pi} \ . \tag{6-100}$$

Applying Eq. (6-100) to Eq. (6-99),

$$2\pi f_d = 2\tan^{-1}\left(\frac{2\pi f_a t_s}{2}\right) \ . \tag{6-101}$$

Substituting $1/f_s$ for t_s, we solve Eq. (6-101) for f_d to get

$$f_d = \left(\frac{2}{2\pi}\right)\tan^{-1}\left(\frac{2\pi f_a / f_s}{2}\right) = \frac{\tan^{-1}(\pi f_a / f_s)}{\pi} \ . \tag{6-102}$$

Equation (6-102), the relationship between the s-plane frequency f_a in Hz and the z-plane frequency f_d in Hz, is plotted in Figure 6-32(a) as a function of the IIR filter's sampling rate f_s.

The distortion of the f_a frequency scale, as it translates to the f_d frequency scale, is illustrated in Figure 6-32(b) where an s-plane bandpass frequency magnitude response $|H_a(f_a)|$ is subjected to

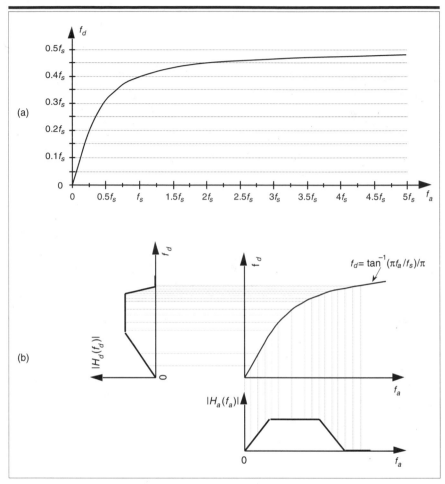

Figure 6-32 Nonlinear relationship between the f_d and f_c frequencies: (a) frequency warping curve scaled in terms of the IIR filter's f_s sampling rate; (b) s-domain frequency response $H_c(f_c)$ transformation to a z-domain frequency response $H_d(f_d)$.

frequency compression as it is transformed to $|H_d(f_d)|$. Notice how the low-frequency portion of the IIR filter's $|H_d(f_d)|$ response is reasonably linear, but the higher frequency portion has been squeezed in toward zero Hz. That's frequency warping. This figure shows why no IIR filter aliasing can occur with the bilinear transform design method. No matter what the shape or bandwidth of the $|H_a(f_a)|$ prototype analog filter, none of its spectral content can extend beyond half the sampling rate of

$f_s/2$ in $|H_d(f_d)|$—and that's what makes the bilinear transform design method as popular as it is.

The steps necessary to perform an IIR filter design using the bilinear transform method are as follows:

Step 1: Obtain the Laplace transfer function $H_c(s)$ for the prototype analog filter in the form of Eq. (6-43).

Step 2: Determine the digital filter's equivalent sampling frequency f_s and establish the sample period $t_s = 1/f_s$.

Step 3: In the Laplace $H_c(s)$ transfer function, substitute the expression

$$\frac{2}{t_s}\left(\frac{1-z^{-1}}{1+z^{-1}}\right) \tag{6-103}$$

for the variable s to get the IIR filter's $H(z)$ transfer function.

Step 4: Multiply the numerator and denominator of $H(z)$ by the appropriate power of $(1 + z^{-1})$ and grind through the algebra to collect terms of like powers of z in the form

$$H(z) = \frac{\displaystyle\sum_{k=0}^{N} b(k)z^{-k}}{1 - \displaystyle\sum_{k=1}^{M} a(k)z^{-k}}. \tag{6-104}$$

Step 5: Just as in the impulse invariance design methods, by inspection, we can express the IIR filter's time-domain equation in the general form of

$$y(n) = b(0)x(n) + b(1)x(n-1) + b(2)x(n-2) + \ldots + b(N)x(n-N)$$

$$+ a(1)y(n-1) + a(2)y(n-2) + \ldots + a(M)y(n-M) \ . \tag{6-105}$$

Although the expression in Eq. (6-105) only applies to the filter structure in Figure 6-18, to complete our design, we can apply the $a(k)$ and $b(k)$ coefficients to the improved IIR structure shown in Figure 6-22.

To show just how straightforward the bilinear transform design method really is, let's use it to solve the IIR filter design problem first presented for the impulse invariance design method.

6.5.1 Bilinear Transform Design Example

Again, our goal is to design an IIR filter that approximates the second-order Chebyshev prototype analog low-pass filter, shown in Figure 6-26, whose passband ripple is 1 dB. The f_s sampling rate is 100 Hz ($t_s = 0.01$), and the filter's 1 dB cutoff frequency is 20 Hz. As before, given the original prototype filter's Laplace transfer function as

$$H_c(s) = \frac{17410.145}{s^2 + 137.94536s + 17410.145} \, , \qquad (6\text{-}106)$$

and the value of $t_s = 0.01$ for the sample period, we're ready to proceed with Step 3. For convenience, let's replace the constants in Eq. (6-106) with variables in the form of

$$H_c(s) = \frac{c}{s^2 + bs + c} \, , \qquad (6\text{-}107)$$

where $b = 137.94536$ and $c = 17410.145$. Performing the substitution of Eq. (6-103) in Eq. (6-107),

$$H(z) = \frac{c}{\left(\dfrac{2}{t_s} \cdot \dfrac{1 - z^{-1}}{1 + z^{-1}}\right)^2 + b\dfrac{2}{t_s}\left(\dfrac{1 - z^{-1}}{1 + z^{-1}}\right) + c} \, . \qquad (6\text{-}108)$$

To simplify our algebra a little, let's substitute the variable a for the fraction $2/t_s$ to give

$$H(z) = \frac{c}{a^2\left(\dfrac{1 - z^{-1}}{1 + z^{-1}}\right)^2 + ab\left(\dfrac{1 - z^{-1}}{1 + z^{-1}}\right) + c} \, . \qquad (6\text{-}109)$$

Proceeding with Step 4, we multiply Eq. (109)'s numerator and denominator by $(1 + z^{-1})^2$ to yield

$$H(z) = \frac{c(1+z^{-1})^2}{a^2(1-z^{-1})^2 + ab(1+z^{-1})(1-z^{-1}) + c(1+z^{-1})^2}. \qquad (6\text{-}110)$$

Multiplying through by the factors in the denominator of Eq. (6-110), and collecting like powers of z,

$$H(z) = \frac{c(1+2z^{-1}+z^{-2})}{(a^2+ab+c)+(2c-2a^2)z^{-1}+(a^2+c-ab)z^{-2}}. \qquad (6\text{-}111)$$

We're almost there. To get Eq. (6-111) into the form of Eq. (6-104) with a constant term of one in the denominator, we divide Eq. (6-111)'s numerator and denominator by $(a^2 + ab + c)$ giving us

$$H(z) = \frac{\dfrac{c}{(a^2+ab+c)}(1+2z^{-1}+z^{-2})}{1+\dfrac{(2c-2a^2)}{(a^2+ab+c)}z^{-1}+\dfrac{(a^2+c-ab)}{(a^2+ab+c)}z^{-2}}. \qquad (6\text{-}112)$$

We now have $H(z)$ in a form with all the like powers of z combined into single terms, and Eq. (6-112) looks something like the desired form of Eq. (6-104). If we plug the values $a = 2/t_s = 200$, $b = 137.94536$, and $c = 17410.145$ into Eq. (6-112) we get the following IIR filter transfer function:

$$H(z) = \frac{0.20482712(1+2z^{-1}+z^{-2})}{1-0.53153089\,z^{-1}+0.35083938\,z^{-2}}$$

$$= \frac{0.20482712+0.40965424\,z^{-1}+0.20482712\,z^{-2}}{1-0.53153089\,z^{-1}+0.35083938\,z^{-2}}, \qquad (6\text{-}113)$$

and there we are. Now, by inspection of Eq. (6-113), we get the time-domain expression for our IIR filter as

$$y(n) = 0.20482712 + 0.40965424 \cdot x(n-1) + 0.20482712 \cdot x(n-2)$$
$$+ 0.53153089 \cdot y(n-1) - 0.35083938 \cdot y(n-2). \qquad (6\text{-}114)$$

The frequency magnitude response of our bilinear transform IIR design example is shown as the dark curve in Figure 6-33(a), where, for

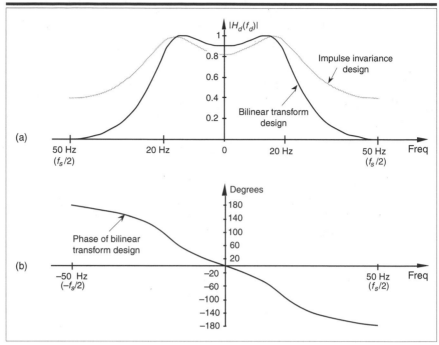

Figure 6-33 Comparison of the bilinear transform and impulse invariance design IIR filters: (a) frequency magnitude responses; (b) phase of the bilinear transform IIR filter.

comparison, we've shown the result of that impulse invariance design example as the shaded curve. Notice how the bilinear transform designed filter's magnitude response approaches zero at the folding frequency of $f_s/2 = 50$ Hz. This is as it should be—that's the whole purpose of the bilinear transform design method. Figure 6-33(b) illustrates the nonlinear phase response of the bilinear transform designed IIR filter.

We might be tempted to think that not only is the bilinear transform design method easier to perform than the impulse invariance design method, but that it gives us a much sharper roll-off for our low-pass filter. Well, the frequency warping of the bilinear transform method does compress (sharpen) the roll-off portion of a low-pass filter, as we saw in Figure 6-32, but an additional reason for the improved response is the price we pay in terms of the additional complexity of the implementation of our IIR filter. We see this by examining the implementation of our IIR filter as shown in Figure 6-34. Notice that our new filter requires five multiplications per filter output sample where the impulse invariance design filter in Figure 6-28(a) required only three multiplications per filter output sample. The additional multiplications are, of course, required by the additional feed forward z terms in the numerator of Eq. (6-113). These added

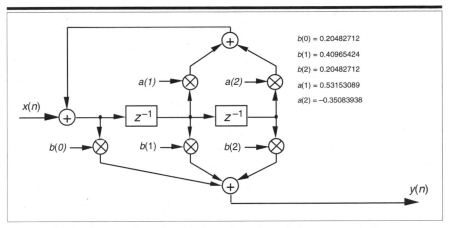

Figure 6-34 Implementation of the bilinear transform design example filter.

$b(k)$ coefficient terms in the $H(z)$ transfer function correspond to zeros in the z-plane created by the bilinear transform that did not occur in the impulse invariance design method.

Because our example prototype analog low-pass filter had a cutoff frequency that was $f_s/5$, we don't see a great deal of frequency warping in the bilinear transform curve in Figure 6-33. (In fact, Kaiser has shown that, when f_s is large, the impulse invariance and bilinear transform design methods result in essentially identical $H(z)$ transfer functions[19].) Had our cutoff frequency been a larger percentage of f_s, bilinear transform warping would have been more serious, and our resultant $|H_d(f_d)|$ cutoff frequency would have been below the desired value. What the pros do to avoid this is to *prewarp* the prototype analog filter's cutoff frequency requirement before the analog $H_c(s)$ transfer function is derived in Step 1.

In that way, they compensate for the bilinear transform's frequency warping before it happens. We can use Eq. (6-98) to determine the prewarped prototype analog filter low-pass cutoff frequency that we want mapped to the desired IIR low-pass cutoff frequency. We plug the desired IIR cutoff frequency ω_d in Eq. (6-98) to calculate the prototype analog ω_a cutoff frequency used to derive the prototype analog filter's $H_c(s)$ transfer function.

Although we explained how the bilinear transform design method avoided the impulse invariance method's inherent frequency response aliasing, it's important to remember that we still have to avoid filter input data aliasing. No matter what kind of digital filter or filter design method is used, the original input signal data must always be obtained using a sampling scheme that avoids the aliasing described in Chapter 2. If the original input data contains errors due to sample rate aliasing, no filter can remove those errors.

   Our introductions to the impulse invariance and bilinear transform design techniques have, by necessity, presented only the essentials of those two design methods. Although rigorous mathematical treatment of the impulse invariance and bilinear transform design methods is inappropriate for an introductory text such as this, more detailed coverage is available to the interested reader[13–16]. References [13] and [15], by the way, have excellent material on the various prototype analog filter types used as a basis for the analytical IIR filter design methods. Although our examples of IIR filter design using the impulse invariance and bilinear transform techniques approximated analog low-pass filters, it's important to remember that these techniques apply equally well to designing bandpass and highpass IIR filters. To design a highpass IIR filter, for example, we'd merely start our design with a Laplace transfer function for the prototype analog highpass filter. Our IIR digital filter design would then proceed to approximate that prototype highpass filter.

   As we have seen, the impulse invariance and bilinear transform design techniques are both powerful and a bit difficult to perform. The mathematics is intricate and the evaluation of the design equations is arduous for all but the simplest filters. As such, we'll introduce a third class of IIR filter design methods based on software routines that take advantage of *iterative optimization* computing techniques. In this case, the designer defines the desired filter frequency response, and the algorithm begins generating successive approximations until the IIR filter coefficients converge (hopefully) to an optimized design.

6.6 Optimized IIR Filter Design Method

The final class of IIR filter design methods we'll introduce are broadly categorized as optimization methods. These IIR filter design techniques were developed for the situation when the desired IIR filter frequency response was not of the standard low-pass, bandpass, or highpass form. When the desired response has an arbitrary shape, closed-form expressions for the filter's z-transform do not exist, and we have no explicit equations to work with to determine the IIR filter's coefficients. For this general IIR filter design problem, algorithms were developed to solve sets of linear, or nonlinear, equations on a computer. These software routines mandate that the designer describe, in some way, the desired IIR filter frequency response. The algorithms, then, assume a filter transfer function $H(z)$ as a ratio of polynomials in z and start to calculate the filter's frequency response. Based on some error criteria, the algorithm begins iteratively adjusting the filter's coefficients to minimize the error between the desired and the actual filter frequency response. The process ends when the error cannot be further minimized, or a predefined number of iterations has occurred, and the final filter coefficients are presented to the filter designer. Although these optimization algorithms are too mathematically complex to cover in any detail here, descriptions of the most popular optimization schemes are readily available in the literature[14,16,20–25].

The reader may ask, "If we're not going to cover optimization methods in any detail, why introduce the subject here at all?" The answer is that if we spend much time at all designing IIR filters, we'll end up using optimization techniques in the form of computer software routines most of the time. The vast majority of commercially available digital signal processing software packages include one or more IIR filter design routines that are based on optimization methods. When a computer-aided design technique is available, filter designers are inclined to use it to design the simpler low-pass, bandpass, or highpass forms even though analytical techniques exist. With all due respect to Laplace, Heaviside, and Kaiser, why plow through all the z-transform design equations when the desired frequency response can be applied to a software routine to yield acceptable filter coefficients in a few seconds?

As it turns out, using commercially available optimized IIR filter design routines is very straightforward. Although they come in several flavors, most optimization routines only require the designer to specify a few key amplitude and frequency values, the desired order of the IIR filter (the number of feedback taps), and the software computes the final feed forward and feedback coefficients. In specifying a low-pass, IIR filter for example, a software design routine might require us to specify the values for δ_p, δ_s, f_1, and f_2 shown in Figure 6-35. Some optimization design routines require the user to specify the order of the IIR

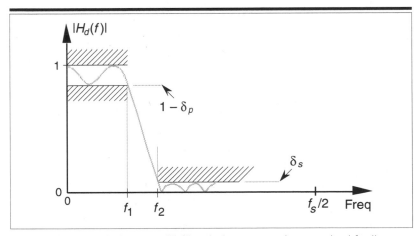

Figure 6-35 Example low-pass IIR filter design parameters required for the optimized IIR filter design method.

filter. Those routines then compute the filter coefficients that best approach the required frequency response. Some software routines, on the other hand, don't require the user to specify the filter order. They compute the minimum order of the filter that actually meets the desired frequency response.

6.7 Pitfalls in Building IIR Digital Filters

There's an old saying in engineering: "It's one thing to design a system on paper, and another thing to actually build one and make it work." (Recall the Tacoma Narrows Bridge episode!) Fabricating a working system based on theoretical designs can be difficult in practice. Let's see why this is often true for IIR digital filters.

Again, the IIR filter structures in Figures 6-18, 6-21(b), and 6-22 are called Direct Form implementations of an IIR filter because they're all equivalent to directly implementing the general time-domain expression for an Mth-order IIR filter given in Eq. (6-21). As it turns out, there can be stability problems and frequency response distortion errors when direct form implementations are used. Such problems arise because we're forced to represent the IIR filter coefficients and the results of intermediate filter calculations with binary numbers having a finite number of bits. These finite word length effects are particularly serious if a direct form IIR filter's $H(z)$ transfer function polynomial in z is of high order (i.e., a filter having a large number of delay elements). There are three major categories of finite word length errors that plague IIR filter implementations: coefficient quantization, overflow errors, and roundoff errors. Although Chapter 9 discusses the nature and effects of using finite word lengths in more detail, we'll briefly discuss those error types as they relate to IIR filters and see what precautions can be taken to minimize their effects.

Remember, all of our IIR filter calculations, thus far, have resulted in very accurate filter coefficient values. In fact, the fractional coefficient values in our previous IIR design examples were accurate to eight decimal digits as indicated in Eq. (6-114). Knowing that it takes four binary bits to store a single decimal digit, we can say that the coefficient values in Eq. (6-114) require the equivalent of 32 bits of accuracy. Consider the situation where we try to actually build the IIR filter in Figure 6-34, but we're constrained to store our $a(k)$ and $b(k)$ coefficients in four-bit hardware registers. Well, we can, indeed, represent the fractional coefficients with only one decimal digit, but they lose some of their precision. For example, we can represent Eq. (6-114)'s original $b(0) = 0.20482712$ coefficient as $b(0) = 0.2$ and lose some accuracy due to coefficient quantization.

Limited–precision coefficients will result in slight filter pole and zero shifting on the z plane, and a frequency magnitude response that may not meet our requirements.

For example, if we used only four-bits (one decimal digit) to represent the coefficients in Figure 6-34, our reduced-accuracy filter magnitude response is shown by the solid curve in Figure 6-36. For contrast, we've also shown the original high-accuracy coefficient response as the shaded curve in Figure 6-36. Sure enough, using four-bit coefficients distorted our original second-order IIR frequency response. Had our filter cutoff frequency (20 Hz) been higher relative to the sample rate (100 Hz), coefficient quantization would result in a frequency response that was even more skewed from the original. As it turns out, higher order filters are very sensitive to coefficient accuracy and, for them, coefficient quantization is more likely to cause unacceptable frequency response distortion. An even bigger problem can occur in high-order polynomials of $H(z)$ when quantized coefficients are rounded up to the nearest quantization level. After factoring a high-order error-containing $H(z)$ polynomial into its single-order factors, this can result in actually moving a pole outside the unit circle, yielding an unstable filter.

Overflow, the second finite word length effect that troubles digital filters, is what happens when the result of an arithmetic operation is too large to be represented in the fixed-length hardware registers designed to contain that result. Because we perform so many additions when we implement IIR filters, overflow is always a potential problem. With no precautions being made to handle overflow, large nonlinearity errors

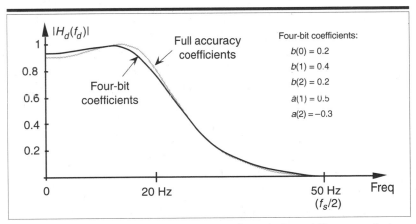

Figure 6-36 Comparison of the bilinear transform designed filter magnitude responses using full-accuracy coefficients (shaded curve) and four-bit coefficients (solid curve).

can result in our filter output samples—often in the form of *overflow oscillations*.

The most common way of dealing with binary overflow errors is called roundoff, or rounding, where a data value is represented by, or rounded off to, the b-bit binary number that's nearest the unrounded data value. It's usually valid to treat roundoff errors as a random process, but conditions occur in IIR filters where rounding can cause the filter output to oscillate forever, even when the filter input sequence is all zeros. This situation, going by the names *limit cycles* and *deadband effects*, has been well analyzed in the literature[26,27]. We can demonstrate limit cycles by considering the second-order IIR filter in Figure 6-37(a) whose time-domain expression is

$$y(n) = x(n) + 1.3y(n-1) - 0.76y(n-2) \, . \tag{6-115}$$

Let's assume that this filter rounds the adder's output to the nearest integral value. If the situation ever arises where $y(-2) = 0$, $y(-1) = 8$, and $x(0)$ and all successive $x(n)$ inputs are zero, the filter output goes into endless oscillation, as shown in Figure 6-37(b). If this filter were to be used in an audio application, when the input signal went silent, the listener could end up hearing an audio tone instead of silence. The shaded line in Figure 6-37(b) shows the filter's stable response to this particular situation if no rounding is used. With rounding, however, this IIR filter certainly lives up to its name.

There are several ways to reduce the ill effects of quantization errors and limit cycles. We can increase the word widths of the hardware registers that contain the results of intermediate calculations. Because roundoff limit cycles affect the least significant bits of an arithmetic result, larger word sizes diminish the impact of limit cycles should they occur. To avoid filter input sequences of all zeros, some practitioners add a *dither sequence* to the filter's input signal sequence. A dither sequence is a sequence of low-amplitude pseudorandom numbers that interferes with an IIR filter's roundoff error generating tendency, allowing the filter output to reach zero should the input signal remain at zero. Dithering, however, decreases the effective signal-to-noise ratio of the filter output[11]. Finally, to avoid limit cycle problems, we can just use an FIR filter. Because FIR filters, by definition, have finite-length impulse responses and have no feedback paths, they cannot support output oscillations of any kind.

As for overflow oscillations, we can eliminate them if we increase the word width of hardware registers, so that overflow never takes place in the IIR filter. Alternatively, it has been shown that the effects of overflow oscillations can be avoided by modifying hardware adders so that their

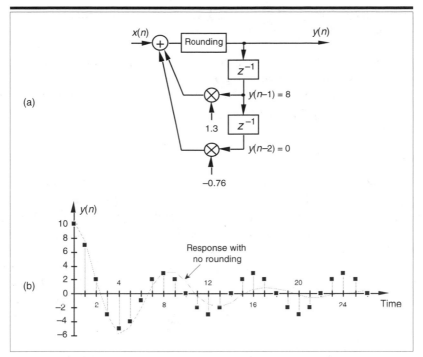

Figure 6-37 Limit cycle oscillations due to rounding in an IIR filter: (a) second-order IIR filter with rounding taking place following the adder; (b) one possible time-domain response of the IIR filter.

sum saturates to a fixed limit when an overflow condition is detected[28, 29]. It may be useful for the reader to keep in mind that, when the signal data is represented in two's complement arithmetic, multiple summations resulting in intermediate overflow errors cause no problems if we can guarantee that the final magnitude of the sum of the numbers is not too large for the final accumulator register. This potential relief from overflow errors, as discussed under the data overflow topic in Section 9.3, is one reason for the popularity of the two's complement format for binary numbers. Some digital filters incorporate additional control circuitry to detect overflow conditions and allow the designer to control (by scaling down) the amplitudes of signals at critical points in the system[14,17]. If your budget can afford floating-point hardware, it has been shown that both standard floating-point and block floating-point data formats can greatly reduce the errors associated with overflow oscillations and limit cycles[30]. (We discuss floating-point number formats in Section 9.4)

Perhaps the most popular technique for minimizing the errors associated with finite word widths is to design filters comprising a cascade string, or parallel combination, of low-order filters. The next section tells us why.

6.8 Cascade and Parallel Combinations of Digital Filters

In practice, we're likely to encounter multiple digital filters connected in cascade or parallel like those in Figure 6-38. As indicated in Figure 6-38(a), the resultant transfer function of two cascaded filter transfer functions is the product of those functions, or

$$H_{cascade}(z) = H_1(z)H_2(z) \, . \tag{6-116}$$

If the filters in Figure 6-38(a) are linear, we can swap their order, having $H_2(z)$ precede $H_1(z)$, with no change in the $Y(z)$ output. By the way, this property of the combined transfer function equaling the product of the two individual transfer functions applies to any combination of linear time invariant IIR and FIR filters. As shown in Figure 6-38(b), the combined transfer function of two parallel filters is the sum of their transfer functions, or

$$H_{parallel}(z) = H_1(z) + H_2(z) \, . \tag{6-117}$$

Hardware filter designers routinely partition high-order digital filters into a string of lower order filters arranged in cascade because the cascaded system usually requires fewer multiplications for a given filter frequency response. It's common practice to subdivide a high-order digital

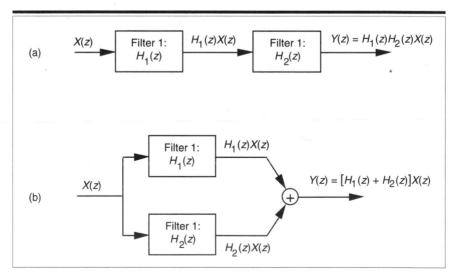

Figure 6-38 Combinations of two digital filters: (a) cascaded filters; (b) parallel filters.

filter into multiple second-order building blocks because second-order filters are easier to design, are less susceptible to coefficient quantization and roundoff errors, and their implementations allow easier data word scaling to reduce the potential overflow effects of data word size growth.

Optimizing the partitioning of a high-order filter into multiple second-order filter sections is a challenging task, however. For example, say we have the sixth-order filter in Figure 6-39(a) that we want to partition into three second-order sections. In factoring the sixth-order filter's $H(z)$ polynomial, we could get up to three separate sets of feedback coefficients in the factored $H(z)$ numerator: $a'(n)$, $a''(n)$, and $a'''(n)$. Likewise, we could have up to three separate sets of feed forward coefficients in the factored

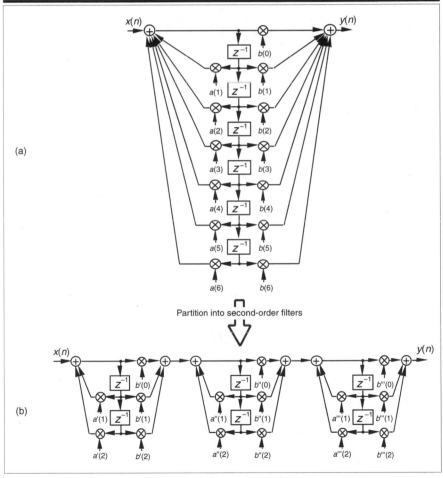

Figure 6-39 IIR filter partitioning: (a) original sixth-order IIR filter; (b) three second-order sections.

denominator: $b'(n)$, $b''(n)$, and $b'''(n)$. Because there are three second-ordered sections, there are three factorial, or six, ways of *pairing* the sets of coefficients. Notice in Figure 6-39(b) how the first section uses the $a'(n)$ and $b'(n)$ coefficients, and the second section uses the $a''(n)$ and $b''(n)$ coefficients. We could just as well have interchanged the sets of coefficients so that the first second-order section uses the a'(n) and b''(n) coefficients, and the second section uses the a''(n) and b'(n) coefficients. So there are six different mathematically equivalent ways of combining the sets of coefficients. Add to this the fact that, for each different combination of low-order sections, there are three factorially distinct ways those three separate second-order sections can be arranged in cascade.

This means that if we want to partition a 2M-order IIR filter into M distinct second-order sections, there are M factorial squared, or $(M!)^2$, ways to do so. As such, there are then $(3!)^2 = 36$ different cascaded filters we could obtain when going from Figure 6-39(a) to 6-39(b).[†] To further complicate this filter sectioning problem, the errors due to quantization will, in general, be different for each possible filter combination. Although this subject is outside the scope of this introductory text, ambitious readers can find further material on optimizing cascaded filter sections in references [19,27], and in part 3 of reference [31].

There's a considerable amount of material in the literature concerning finite word effects as they relate to digital filters. (References [14,16] and [19] discuss quantization noise effects in some detail as well as providing extensive bibliographies on the subject.) Although these books and papers provide useful guidelines in predicting and minimizing quantization errors, prudent designers eventually resort to computer simulations of filter designs. This way they can vary filter hardware characteristics, such as coefficient word widths, accumulator register sizes, sequencing of cascaded sections, and input signal sets. This experimental approach is attractive because it's both reliable and economical.

6.9 A Brief Comparison of IIR and FIR Filters

The question naturally arises as to which filter type, IIR or FIR, is best suited for a given digital filtering application. That's not an easy question to answer, but we can point out a few factors that should be kept in mind. First, we can assume that the differences in the ease of design between the two filter types are unimportant. There are usually more important performance and implementation properties to consider than

---

[†] These combinations can get very large. For example, there are $(5!)^2 = 14,400$ ways to combine five second-order filters to implement a ten-order IIR filter.

design difficulty when choosing between an IIR and an FIR filter. One design consideration that may be significant is the IIR filter's ability to simulate a predefined prototype analog filter. FIR filters do not have this design flexibility.

From a hardware standpoint, with so many fundamental differences between IIR and FIR filters, our choice must be based on those filter characteristics that are most and least important to us. For example, if we needed a filter with exactly linear phase, then an FIR filter is the only way to go. On the other hand, if our design required that a filter accept very high data rates and slight phase nonlinearity is tolerable, we might lean toward IIR filters with their reduced number of necessary multipliers per output sample. Table 6-1 presents a brief comparison between IIR and FIR filters based on several different performance and implementation properties.

Table 6-1 IIR and FIR Filter Characteristics Comparison

Characteristic	IIR	FIR
Number of necessary multiplications	Least	Most
Sensitivity to filter coefficient quantization	Can be high for Direct Form*	Very low
Probability of overflow errors	Can be high for Direct Form*	Very low
Stability	Must be designed in	Guaranteed
Linear phase	No	Guaranteed **
Can simulate prototype analog filters	Yes	No
Required hardware memory	Least	Most
Hardware filter control complexity	Moderate	Simple
Availability of design software	Good	Very good
Ease of design or complexity of design software	Moderately complicated	Simple
Difficulty of quantization noise analysis	Most complicated	Least complicated
Supports adaptive filtering	Yes	Yes

* These problems can be minimized though cascade or parallel implementations.
** Guaranteed so long as the FIR coefficients are symmetrical.

References

[1] Churchill, R. V. *Modern Operational Mathematics in Engineering*, McGraw-Hill, New York, 1944, pp. 307–34.

[2] Aseltine, J. A. *Transform Method in Linear System Analysis*, McGraw-Hill, New York, 1958, pp. 287–92.

[3] Nixon, F. E. *Handbook of Laplace Transformation, Tables and Examples*, Prentice-Hall, Englewood Cliffs, New Jersey, 1960.

[4] Kaiser, J. F. "Digital Filters," in *System Analysis by Digital Computer*. Ed. by F. F. Kuo and J. F. Kaiser, John Wiley and Sons, New York, 1966, pp. 218–77.

[5] Kaiser, J. F. "Design Methods for Sampled Data Filters," Chapter 7, in *S1963 Proc. 1st Allerton Conference*, pp. 221-36.

[6] Ragazzini, J. R. and Franklin, G. F. *Sampled-Data Control Systems*, McGraw-Hill, New York, 1958, pp. 52–83.

[7] Milne-Thomson, L. M. *The Calculus of Finite Differences*, Macmillan, London, 1951, pp. 232–51.

[8] Truxal, J. G. 1955. *Automatic Feedback Control System Synthesis*, McGraw-Hill, New York, 1955, pp. 283.

[9] Blackman, R. B. *Linear Data-Smoothing and Prediction in Theory and Practice*, Addison-Wesley, Reading, Mass., 1965, pp. 81–84.

[10] Gold, B. and Jordan, K. L., Jr. "A Note on Digital Filter Synthesis," *Proceedings of the IEEE*, Vol. 56, October 1968, pp. 1717.

[11] Rabiner, L. R., et al. "Terminology in Digital Signal Processing," *IEEE Trans. on Audio and Electroacoustics*, Vol. AU-20, No. 5, December 1972, pp. 327.

[12] Stearns, S. D. *Digital Signal Analysis*, Hayden Book Co. Inc., Rochelle Park, New Jersey, 1975, pp. 114.

[13] Stanley, W. D., et al., *Digital Signal Processing*, Reston Publishing Co. Inc., Reston, Virginia, 1984, pp. 191.

[14] Oppenheim, A. V., and Schafer, R. W. *Discrete-Time Signal Processing*, Prentice-Hall, Englewood Cliffs, New Jersey, 1989, pp. 406.

[15] Williams, C. S. *Designing Digital Filters*, Prentice-Hall, Englewood Cliffs, New Jersey, 1986, pp. 166–86.

[16] Rabiner, L. R., and Gold, B. *Theory and Application of Digital Signal Processing*, Prentice-Hall, Englewood Cliffs, New Jersey, 1975, pp. 216.

[17] Johnson, M. "Implement Stable IIR Filters Using Minimal Hardware," *EDN*, 14 April 1983.

[18] Oppenheim, A. V., Willsky, A. S., and Young, I. T. *Signals and Systems*, Prentice-Hall, Englewood Cliffs, New Jersey, 1983, pp. 659.

[19] Kaiser, J. F. "Some Practical Considerations in the Realization of Linear Digital Filters," *Proc. Third Annual Allerton Conference on Circuit and System Theory*, 1965, pp. 621-33

[20] Deczky, A. G. "Synthesis of Digital Recursive Filters Using the Minimum P Error Criterion," *IEEE Trans. on Audio and Electroacoustics*, Vol. AU-20, No. 2, October 1972, pp. 257.

[21] Steiglitz, K. "Computer-Aided Design of Recursive Digital Filters," *IEEE Trans. on Audio and Electroacoustics*, Vol. 18, No. 2, 1970, pp. 123.

[22] Richards, M. A. "Application of Deczky's Program for Recursive Filter Design to the Design of Recursive Decimators," *IEEE Trans. on Acoustics, Speech, and Signal Processing*, Vol. ASSP-30, October 1982, pp. 811.

[23] Parks, T. W., and Burrus, C. S. *Digital Filter Design*, John Wiley and Sons, New York, 1987, pp. 244.

[24] Rabiner, L., Graham, Y., and Helms, H. "Linear Programming Design of IIR Digital Filters with Arbitrary Magnitude Functions," *IEEE Trans. on Acoustics, Speech, and Signal Processing.*, Vol. ASSP-22, No. 2, April 1974, pp. 117.

[25] Friedlander, B., and Porat, B. "The Modified Yule-Walker Method of ARMA Spectral Estimation," *IEEE Trans. on Aerospace Electronic Systems*, Vol. AES-20, No. 2, March 1984, pp. 158-73.

[26] Jackson, L. B. "On the Interaction of Roundoff Noise and Dynamic Range and Dynamic Range in Digital Filters," *Bell System Technical Journal*, Vol. 49, February 1970, pp. 159-84.

[27] Jackson, L. B. "Roundoff Noise Analysis for Fixed-Point Digital Filters Realized in Cascade or Parallel Form," *IEEE Trans. Audio Electroacoustics*, Vol. AU-18, June 1970, pp. 107-22.

[28] Sandberg, I. W. "A Theorem Concerning Limit Cycles in Digital Filters," *Proc. Seventh Annual Allerton Conference on Circuit and System Theory*, Monticello, Illinois, October 1969.

[29] Ebert, P. M., et al. "Overflow Oscillations in Digital Filters," *Bell Sys. Tech. Journal*, Vol. 48, November 1969, pp. 2999–3020.

[30] Oppenheim, A. V. "Realization of Digital Filters Using Block Floating Point Arithmetic," *IEEE Trans. Audio Electroacoustics*, Vol. AU-18, June 1970, pp. 130-36.

[31] Rabiner, L. R., and Rader, C. M., Eds., *Digital Signal Processing*, IEEE Press, New York, 1972, pp. 361.

Advanced Sampling Techniques

Beyond the low-pass and bandpass sampling schemes discussed in Chapter 2 there are specialized sampling techniques that we're likely to encounter in practice. In this chapter we introduce several of those advanced sampling methods and show why and when they're used.

7.1 Quadrature Sampling

Many digital signal processing applications use the complex data format with its real (in-phase) and imaginary (quadrature phase) parts. These applications fall in the general category known as *quadrature processing*. For example, the phase preservation characteristic of complex data representation is exploited in high data rate digital communications systems, radar systems, and time difference of arrival (TDOA) processing in radio direction-finding schemes[1,2]. The enhanced phase measurement capabilities of quadrature processing are used in coherent pulse measurement systems, antenna beamforming applications, and single sideband modulators[3–6]. Complex data representation allows us to realize additional processing power through the coherent measurement of the phase of sinusoids comprising an input signal. To obtain the time-domain representation of a continuous signal using complex notation, the signal must be digitized by a technique called *quadrature sampling.*[†]

[†] Quadrature sampling goes by various other names in the literature, such as vector demodulation, complex demodulation, complex down-conversion, quadrature heterodyning, I/Q sampling, and complex sampling[7–9]. Care must be exercised here from a semantic viewpoint. There's a testing scheme referred to as *coherent sampling*, used to characterize the dynamic performance of A/D converters, that is unrelated to quadrature sampling[10,11]. This clever A/D testing technique is discussed in Section 10.9.

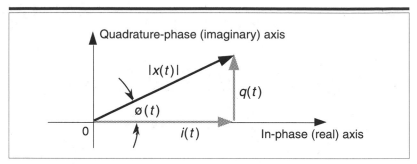

Figure 7-1 In-phase *i(t)* and quadrature-phase *q(t)* components of a
continuous signal *x(t)*.

To illustrate this idea, let's consider a continuous time-domain signal
$x(t)$ that has some instantaneous magnitude at a phase angle of $\varnothing(t)$. This
signal can, equally well, be defined by its in-phase $i(t)$ and quadrature-
phase $q(t)$ components as shown in Figure 7-1. We could use the low-pass,
or bandpass, sampling schemes in Chapter 2 to obtain discrete samples
representing $x(t)$ and then calculate the $i(t)$ and $q(t)$ components.
Quadrature sampling, on the other hand, allows us to obtain the discrete
samples representing $i(t)$ and $q(t)$ directly. Thus, quadrature sampling
produces two separate sampled data sequences, one sequence representing the in-phase and one sequence representing the quadrature-phase
components of the original continuous signal $x(t)$.

Let's keep two things in mind as we begin our discussion of quadrature sampling. First, in studying quadrature processing, we'll encounter
frequency-domain spectra that are not symmetrical about the 0 Hz axis.
Although frequency-domain symmetry is inherent in real signals, the
complex nature of quadrature signals does not have this restriction. It's
the lifting of this symmetry restriction that makes quadrature processing
so powerful. Second, recall from Euler's relationships discussed in
Appendix C that

$$\sin(2\pi ft) = \frac{e^{j2\pi ft} - e^{-j2\pi ft}}{2j} = \frac{-je^{j2\pi ft}}{2} + \frac{je^{-j2\pi ft}}{2} , \qquad (7\text{-}1)$$

and

$$\cos(2\pi ft) = \frac{e^{j2\pi ft} + e^{-j2\pi ft}}{2} = \frac{e^{j2\pi ft}}{2} + \frac{e^{-j2\pi ft}}{2} , \qquad (7\text{-}2)$$

reminding us that real sinusoids or cosinusoids can be represented as complex phasor quantities through the convention of negative frequency, that is, negative exponentials. With these thoughts in mind, we view the hybrid processing known as I/Q demodulation (or Weaver demodulation for those readers with experience in communications theory), shown in Figure 7-2(a).

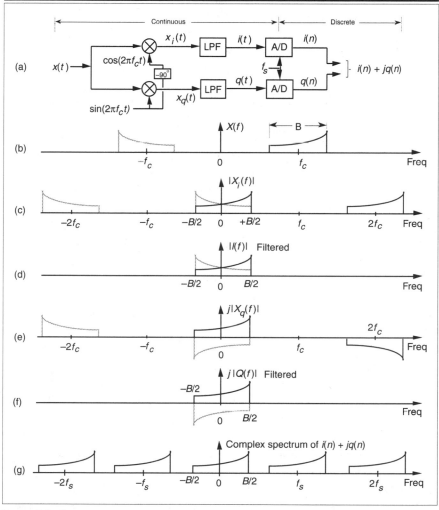

Figure 7-2 Quadrature bandpass sampling: (a) sampling block diagram; (b) continuous spectrum of continuous input $x(t)$; (c) continuous in-phase component of $x(t)$; (d) filtered in-phase component of $x(t)$; (e) continuous quadrature-phase component of $x(t)$; (f) filtered quadrature-phase component of $x(t)$; (g) sampled sum of the filtered in-phase component and the filtered quadrature-phase component of $x(t)$.

Consider a bandpass signal $x(t)$, centered at carrier frequency f_c Hz, that has the real spectrum $X(f)$ shown in Figure 7-2(b). When we mix $x(t)$ with a cosine wave of f_c Hz, the positive phasor portion of the cosine (that is, $e^{j2\pi f_c t}/2$) will translate the spectral components located at $-f_c$ (dashed curve) up to baseband centered at 0 Hz, resulting in what is sometimes called a zero-frequency IF format[4]. Likewise, the negative phasor portion of the cosine ($e^{-j2\pi f_c t}/2$) will translate the spectral components located at $+f_c$ (solid curve) down to baseband also centered at 0 Hz. The resultant continuous signal spectrum is shown as $X_i(f)$ in Figure 7-2(c) where we show the two translated spectra as overlaid, not summed, to see their individuality. $X_i(f)$ is low-pass filtered by an analog filter with a cutoff frequency slightly greater than $B/2$ Hz, called the in-phase component of $x(t)$, and shown as the $I(f)$ magnitude spectrum in Figure 7-2(d). (For the reader who's not at ease with the notion of negative frequencies, it's a good idea to review Appendix C to see why negative frequencies exist when we use complex signal notation.)

The continuous heterodyned product of $x(t)$ and the sinewave of f_c Hz is shown in Figure 7-2(e) where the spectra lying below the frequency axis result from the negative sign of the positive frequency $-je^{j2\pi f_c t}/2$ component in Eq. (7-2). This product is low-pass filtered and represents the quadrature-phase component of $x(t)$, shown as the $jQ(f)$ spectrum in Figure 7-2(f). The original real signal $x(t)$ is now represented in a complex format as $i(t) + jq(t)$. (To recover the real $x(t)$, we need only multiply $i(t)$ by $\cos(2\pi f_c t)$, multiply $q(t)$ by $\sin(2\pi f_c t)$, and sum the two products.)

Figure 7-2 helps remind us that real signals can be represented by a single sequence of sampled values and that their spectra show symmetry about zero Hz. Complex, or quadrature, signals must be represented by two sequences of sampled values, one sequence for the in-phase part and one for the quadrature part. Again, complex signal spectra are not required to be symmetrical about zero Hz. Returning to Figure 7-2(a), the two separate, continuous signals, $i(t)$ and $q(t)$, are digitized by an A/D converter at a sample rate of f_s samples/s. The resultant spectrum for the complex $i(n)+jq(n)$ is shown in Figure 7-2(g). Again, note the spectral replications every f_s Hz. So what's the big deal here? In Figure 7-2(g), notice that the sample rate need only be greater than twice $B/2$ to avoid aliasing.

In quadrature sampling f_s need only be greater than the signal bandwidth, as opposed to twice the signal bandwidth previously described in Chapter 2 for low-pass and bandpass sampling. This means that a given A/D converter can be used to sample signal bandwidths twice as wide with the quadrature sampling technique. Notice, however, that two low-pass filters and two A/D converters are necessary with this scheme.

What we have is two A/D converters operating at half the equivalent first-order bandpass sampling rate. Are we violating the Nyquist criterion? Not really—we still have two discrete sample values being generated for every cycle in the bandwidth of our original signal. So, when we need to digitize very wide bandwidths, the quadrature sampling scheme is attractive, particularly, when our goal is spectral analysis using the FFT. When the two data streams representing the in-phase and quadrature-phase components are applied to a complex radix-2 FFT algorithm, the FFT operates more efficiently than when its inputs are real signals[3,12]. (That particular topic is discussed in Section 10.5.)

7.2 Quadrature Sampling with Digital Mixing

To obtain quadrature samples in practice, the process shown in Figure 7-2(a) is difficult to implement without errors. Over moderately wide bandwidths, it's not easy to configure two analog signal paths that have identical frequency responses (gain and phase)—phase differences of a few degrees are common. Moreover, large DC components are routinely induced by the mixers in the $f_i(t)$ and $f_q(t)$ signals before filtering. Finally, if the low-pass filters are active devices, a kind of continuous noise known as *1/f noise* from the semiconductor components in the mixers/filters induces increased noise energy near 0 Hz.

These disadvantages can be avoided with digital mixers and filters where the sample rate is four times the carrier frequency of the bandpass signal[13–16]. This process is illustrated in Figure 7-3(a). The input signal $x(t)$ from Figure 7-2(a), with the spectrum shown in Figure 7-3(b), is digitized at a sample rate of $f_s = 4f_c$ providing the discrete samples $x(nt_s)$ whose spectrum, $X(m)$, is shown in Figure 7-3(c). The quadrature mixing with $f_c = f_s/4$, to center the input signal's in-phase and quadrature phase components about 0 Hz, is performed digitally with the results of the in-phase replicated components depicted in Figure 7-3(d) and the low-pass filtered in phase spectrum in Figure 7-3(e). Although the A/D converter must sample at four times the carrier frequency, the difficulties in signal-path phase and amplitude matching are eliminated because only one A/D converter is necessary. In addition, the DC bias problems associated with analog signal mixing to zero Hz are avoided.

The real advantage of this technique is how easy the quadrature mixing by f_c can be performed digitally. For example, the input sampled values $x(nt_s)$ can be multiplied by the repetitive four-element sequence 1, 0, –1, 0, etc., to provide $i(n)$. Likewise, $x(nt_s)$ is multiplied by the four-element sequence 0, 1, 0, –1, to generate the $q(n)$ samples. The trick is that

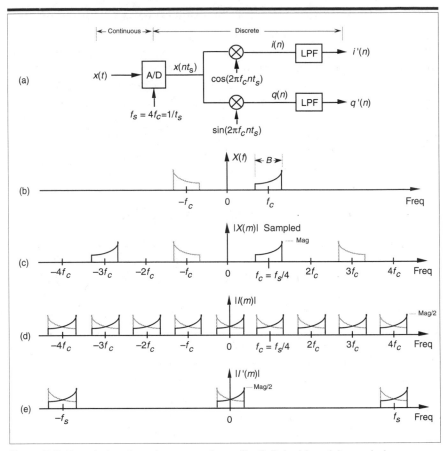

Figure 7-3 Quadrature bandpass sampling with digital mixing: (a) quadrature sampling block diagram; (b) spectrum of continuous bandpass signal $x(t)$; (c) discrete spectrum of sampled bandpass signal $x(nt_s)$; (d) frequency-translated $|I(m)|$ discrete spectrum of in-phase component of $x(nt_s)$; (e) low-pass filtered $|I'(m)|$ spectrum of in-phase component.

these two orthogonal ones sequences translate the input signal's spectrum by exactly $f_s/4$, and that's why our original f_s was made equal to $4f_c$. This clever technique enables complex sampling without multiplications because the mixing is implemented merely by sign changes on $x(nt_s)$.

If we implement this digital mixing process, we'll find that the spectral replication period in Figure 7-3(d) is half what is was in Figure 7-3(c). This comes about from the convolution of the real cosine's positive and negative frequency components with $|X(m)|$ in Figure 7-3(c). To use this digital mixing scheme, we must ensure that the condition $B < f_c = f_s/4$ holds true. In addition to the reduced spectral replication periodicity, we also need to be aware that the DFT magnitude of the frequency-translated $i(n)$, $|I(m)|$, is related to the DFT magnitude of $x(nt_s)$, $|X(m)|$, by

$$|I(m)|_{1,0,-1,0} = \frac{|X(m)|}{2} \; . \tag{7-3}$$

This is indicated by the Mag and Mag/2 factors in Figures 7-3(c) and (d). Likewise, the DFT magnitude of the frequency-translated $q(n)$, $|Q(m)|$, is related to the DFT magnitude $|X(m)|$, by

$$|Q(m)|_{0,1,0,-1} = \frac{|X(m)|}{2} \; . \tag{7-4}$$

So what this all means is that the $i(n)$ and $q(n)$ signals, after translation down to zero Hz, have been reduced in amplitude by a factor of 2. We can find out why Figure 7-3(d)'s replication period and equations (7-3) and (7-4) are true by reviewing Section 10.1, which discusses several different digital quadrature mixing sequences in more detail.

Thus far, we've reviewed the fundamentals of, and an efficient technique for, quadrature bandpass sampling. Let's introduce three other advanced sampling topics, all of which fall in the general category of digital resampling.

7.3 Digital Resampling

The useful and fascinating process of digital resampling is a scheme for changing the effective sampling rate of a discrete signal *after* the signal has already been sampled. As such, resampling has many applications; it's used to minimize computations by reducing data rates when signal bandwidths are narrowed through low-pass filtering. Resampling is mandatory in real-time processing when a hardware digital processor must accept data at a rate that differs from that processor's fundamental clock rate. (A simple example of this is when a digital processor can accept input data at a 1-MHz rate and the incoming data is arriving at a rate of 1.5 MHz. In this case, the incoming data rate must be converted from a 1.5-MHz rate to a 1-MHz sample rate with a minimum of signal distortion.) In satellite and medical image processing, digital resampling is necessary for image enhancement, image scale change, and image rotation. Resampling is also used to reduce the computational complexity of certain narrowband digital filters.

We can define resampling as follows: Consider the process where a continuous signal y_c has been sampled at a rate of $f_{old} = 1/T_{old}$, and the discrete samples are $x_{old}(n) = y_c(nT_{old})$. Resampling is necessary when we need $x_{new}(n) = y_c(nT_{new})$, and direct sampling of the continuous y_c at the rate of $f_{new} = 1/T_{new}$ is not possible. For example, imagine we have an analog-to-digital (A/D) conversion system supplying a sample value every T_{old} seconds. But our processor can accept data only at a rate of one sample every T_{new} seconds. How do we obtain $x_{new}(n)$ directly from $x_{old}(n)$? One possibility is to digital-to-analog (D/A) convert the $x_{old}(n)$ sequence to regenerate the continuous y_c and, then, A/D convert y_c at a sampling rate of f_{new} to obtain $x_{new}(n)$. Due to the spectral distortions induced by D/A followed by A/D conversion, this technique limits our effective dynamic range and is typically avoided in practice. Fortunately, accurate all-digital resampling schemes have been developed, as we shall see.

7.3.1 Resampling by Decimation

Sampling rate changes come in two flavors: rate decreases and rate increases. Decreasing the sampling rate is known as decimation. (The term *decimation* is somewhat of a misnomer because decimation originally meant to reduce by a factor of ten. Currently, *decimation* is the term used for reducing the sample rate by any integral factor.) When the sampling rate is being increased, the process is known as interpolation, i.e., estimating intermediate sample values. Because it's the simplest of the two rate-changing schemes, let's explore decimation first.

We can decimate, or downsample, a sequence of sampled values by a factor of D by retaining every Dth sample and discarding the remaining samples. Relative to the original sample rate, f_{old}, the new sample rate is

$$f_{new} = \frac{f_{old}}{D} . \tag{7-5}$$

For example, to decimate a sequence $x_{old}(n)$ by a factor of $D = 3$, we retain $x_{old}(0)$ and discard $x_{old}(1)$ and $x_{old}(2)$, retain $x_{old}(3)$ and discard $x_{old}(4)$ and $x_{old}(5)$, retain $x_{old}(6)$, and so on. So $x_{new}(n) = x_{old}(3n)$, where $n = 0, 1, 2$, etc. The result of this decimation process is identical to the result of originally sampling at a rate of $f_{new} = f_{old}/3$ to obtain $x_{new}(n)$. The spectral implications of decimation are what we should expect, as shown in Figure 7-4, where the spectrum of an original bandlimited continuous signal is indicated by the solid lines. Figure 7-4(a) shows the discrete replicated spectrum of $x_{old}(n)$, $X_{old}(m)$. With $x_{new}(n) = x_{old}(3n)$, the discrete spectrum

$X_{new}(m)$ is shown in Figure 7-4(b). Two important features are illustrated in Figure 7-4. First, $X_{new}(m)$ could have been obtained directly by sampling the original continuous signal at a rate of f_{new}, as opposed to decimating $x_{old}(n)$ by a factor of 3. And second, there is, of course, a limit to the amount of decimation that can be performed relative to the bandwidth of the original signal B. We must ensure that $f_{new} > 2B$ to prevent aliasing after decimation. When an application requires that f_{new} be less than $2B$, then $x_{old}(n)$ must be low-pass filtered before the decimation process is performed. An example of the need for low-pass filtering before decimation is shown in Figure 7-4(c). If the original signal has a bandwidth B, and we're interested in retaining only the band B', the signal between B' and $f_{new}/2$ must be low-pass filtered before the decimation

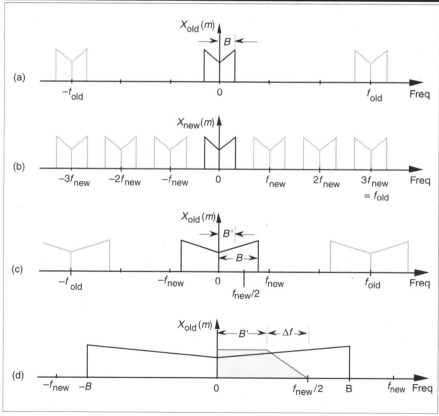

Figure 7-4 Decimation by a factor of three: (a) spectrum and replications of original signal; (b) spectrum of signal decimated by a factor of three; (c) example where only the bandwidth B' from the original bandwidth B is to be retained; (d) relationship of a low-pass filter's cutoff frequency relative to the bandwidth B'.

process is performed. Figure 7-4(d) shows this in more detail where the frequency response of the low-pass filter, shaded, must attenuate the signal amplitude above $f_{new}/2$. In practice, the direct form of FIR filters (the structure shown in Figure 5-13) is the prevailing choice for low-pass *decimation filters* because of the FIR filter's linear phase response[17].

When the desired decimation factor D is large, say $D > 10$, there is an important feature of the filter/decimation process to keep in mind. Significant computational savings may be had by implementing decimation in multiple stages. By way of example, let's assume that we have input data arriving at a sample frequency of 400 kHz, and we must decimate by a factor of $D = 100$ because our digital processor can only accept data at a rate of 4 kHz. Also, let's assume that the baseband frequency range of interest is from 0 to 1.8 kHz. So, with $f_{new} = 4$ kHz, we must filter out any signal above $f_{new}/2$ by having our filter transition band between 1.8 kHz and $f_{new}/2 = 2$ kHz. It's been shown that the number of stages S in the direct form of a FIR low-pass filter is proportional to the ratio of the original sample frequency over the filter transition band, Δf in Figure 7-4(d)[17,18]; that is,

$$S = k \cdot \frac{f_{old}}{(f_{new}/2 - B')} = k \cdot \frac{f_{old}}{\Delta f} , \qquad (7\text{-}6)$$

where $2 < k < 4$ depending on the amount of filter passband and stopband ripple that can be tolerated. So for our case, if k is 3, for example, $S = 6000$. Think of it, a filter with 6000 stages! Fortunately there's a better way. Let's do something that appears to compound our problem and partition our decimation example into two stages: decimation by 50 followed by decimation by 2, as shown in Figure 7-5(a). We'll assume that the original input signal spectrum extends from zero Hz to something greater than 100 kHz, as shown in Figure 7-5(b). If the first low-pass filter LPF_1 has a cutoff frequency of 1.8 kHz and its stopband is defined to begin at 6 kHz, the output of the $D = 50$ decimator will have a spectrum, as shown in Figure 7-5(c), where our 1.8-kHz band of interest is shaded. Notice that there is aliasing of the signal between 2 kHz and 4 kHz—not to worry, LPF_2 will take care of this. When LPF_2 has a cutoff frequency of 1.8 kHz and its stopband is designed to be 2 kHz, the output of the $D = 2$ decimator will have our desired spectrum, shown in Figure 7-5(d). The point is that the total number of stages in the two low-pass filters, S_{Total}, is greatly reduced from 6000. From Eq. (7-6) for the combined LPF_1 and LPF_2 filters,

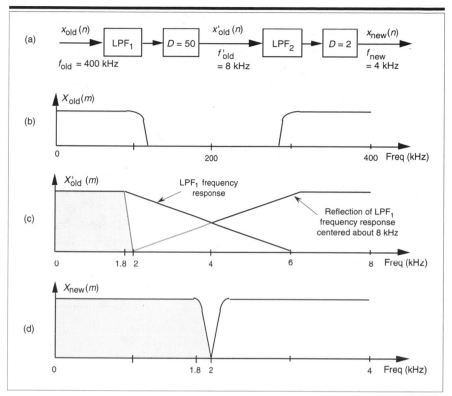

Figure 7-5 Multistage decimation: (a) decimation block diagram; (b) spectrum of original signal; (c) output of the $D = 50$ decimator; (d) output of the $D = 2$ decimator.

$$S_{\text{Total}} = S_{\text{LPF}_1} + S_{\text{LPF}_2} = 3 \cdot \frac{400}{(6 - 1.8)} + 3 \cdot \frac{8}{(2 - 1.8)} = 406 \text{ stages.} \qquad (7\text{-}7)$$

This is an impressive computational reduction. Reviewing Eq. (7-6) for each stage, in the first stage we see that, although $f_{\text{old}} = 400$ kHz remained constant, we increased Δf. In the second stage, both Δf and f_{old} were reduced. The fact to remember in Eq. (7-6) is that the ratio $f_{\text{old}}/\Delta f$ has a much more profound effect than k in determining the number of stages necessary in a low-pass filter. This example, although somewhat exaggerated, shows the kind of computational saving afforded by multistage decimation. Isn't it interesting that adding more processing stages to our original $D = 100$ decimation problem actually decreased the necessary computational requirement?

A sensible way to evaluate FIR decimation filter implementations or any filter implementation, for that matter, is to determine the necessary number of multiplications/second. Going back to our original $D = 3$ decimation example, let's assume $f_{old} = 60$ kHz and our decimation filter need have only $S = 5$ multiplier stages, as shown in Figure 7-6(a), with the filter coefficients being h_0 to h_4. Because the $x_{old}(n)$ samples are arriving at a rate of f_{old}, the low-pass filter requires C_D multiplications/s where

$$C_D = Sf_{old} = 5 \cdot 60,000 = 300,000 \text{ multiplications/s.} \tag{7-8}$$

In Figure 7-6(a), our decimation by three is symbolized by a rotary switch that rotates clockwise 120° each time an $x_{old}(n)$ sample arrives and only every third filter output sample is retained for $x_{new}(n)$. This structure is not so desirable because, for every 15 multiplications, we use only 5 products, an inefficient arrangement even when D is small. Fortunately, the switching (decimation) can be moved ahead of the multipliers, as illustrated in Figure 7-6(b), and we need perform only C_D' multiplications/s where

$$C_D' = \frac{C_D}{D} = \frac{Sf_{old}}{D} = 100,000 \text{ multiplications/s.} \tag{7-9}$$

By moving the decimation process ahead of the multipliers, we give the multipliers $D = 3$ times as long to perform a multiplication, and $x_{new}(n)$ is identical in the two filter structures. (If our odd-order FIR filter coefficients are symmetrical, so that $h(4) = h(0)$, and $h(3) = h(1)$, there's a clever way to further reduce the number of necessary multiplications, as described in Section 10.8.)

Let's step back for a moment to make sure we understand what's going on here. The rotating switches in Figure 7-6 are only conceptual. Again, the switch in Figure 7-6(a) means that we retain only every third output sample of x_{new}. In this case, the filter is operating at the input sample rate of f_{old} and we discard two out of every three filter output samples. In Figure 7-6(b), we don't even bother to calculate the two out of every three samples that are to be discarded; here, the filter is operating at the reduced output sample rate of f_{new} where we calculate only those output samples that we need. By the way, IIR filters cannot take advantage of the reduced computation scheme in Figure 7-6(b) because the various implementations of IIR filters mandate that they must always operate at the input sample rate f_{old}.

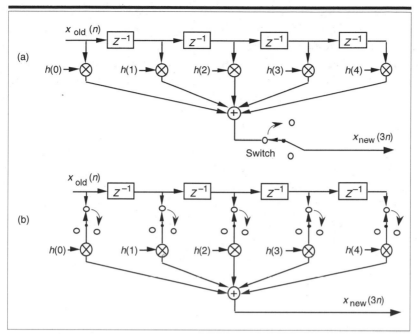

Figure 7-6 Decimation filter implementations: (a) five-stage FIR digital filter; (b) decimation taking place before the filter multiplications.

Before we leave the subject, it's interesting to realize that decimation is one of those rare processes that is not time invariant. From the very nature of its operation, we know that if we delay the input sequence by one sample, a downsampler will generate an entirely different output sequence. For example, if we apply an input sequence $x(n) = x(0), x(1), x(2), x(3), x(4)$, etc., to a downsampler and $D = 3$, the output $y(n)$ will be the sequence $x(0), x(3), x(6)$, etc. Should we delay the input sequence by one sample, our delayed $x'(n)$ input would be $x(1), x(2), x(3), x(4), x(5)$, etc. In this case, the delayed output sequence $y'(n)$ would be $x(1), x(4), x(7)$, etc., which is *not* a delayed version of $y(n)$. Thus, a decimator (downsampler) is not time invariant.

7.3.2 Resampling by Interpolation

As we said before, decimation is only part of the digital resampling story—let's now consider interpolation. Sample rate increase by interpolation is a bit more involved than decimation because, with interpolation, new sample values need to be calculated. Conceptually, interpolation comprises generating a continuous curve that passes through our $x_{old}(n)$ sampled values, followed by resampling that curve at the new sample rate f_{new} to obtain the sequence $x_{new}(n)$. Of course, continuous curves can't exist inside a digital machine, so we'll just have to obtain $x_{new}(n)$ directly from $x_{old}(n)$. To increase a given sample rate or upsample by a factor of U, we have to calculate $U-1$ intermediate values between each sample in $x_{old}(n)$. The process is beautifully straightforward and best understood by way of an example.

Let's assume that we have the sequence $x_{old}(n)$, part of which is shown in Figure 7-7(a), and we want to increase its sample rate by a factor of 4, so, $U = 4$. The sequence's spectrum is provided in Figure 7-7(b) where only the signal spectrum between 0 Hz and $4f_{old}$ is shown. To upsample $x_{old}(n)$, by a factor of four, we must insert three zeros between each sample of $x_{old}(n)$ as shown in Figure 7-7(c), to create the new sequence $x'_{new}(n')$. (This insertion of zeros is called zero padding.) Notice that

$$x'_{new}(n') = x_{old}(n) \text{ , when } n' = 4n \text{ ;} \qquad (7\text{-}10)$$

that is, the old sequence is now embedded in the new sequence. The insertion of the zeros establishes the sample index for the new sequence $x_{new}(n')$ where the interpolated values will be assigned. The spectrum of $x'_{new}(n')$, $X'_{new}(m)$, is shown in Figure 7-7(d) where $f_{new} = 4f_{old}$. In Figure 7-7(d), notice that $x_{old}(n)$'s spectrum is replicated $U = 4$ times in the frequency range of zero Hz to f_{new}. This shouldn't surprise us; $x_{old}(n)$'s spectrum was already replicated 4 times between 0 Hz and f_{new} in Figure 7-7(b). What we've done by adding the zeros is merely increase the effective sample frequency to $f_s = f_{new}$ in Figure 7-7(d). Likewise, we've also increased the new folding frequency by a factor of 4 to $2f_{old}$.

The final step in interpolation is to filter the $x'_{new}(n')$ sequence with a low-pass digital filter whose frequency response about zero Hz and f_{new} Hz is shown as the dashed lines in Figure 7-7(d). This low-pass filter is called an *interpolation filter*, and its output sequence is the desired $x_{new}(n')$, shown in Figure 7-7(e) while its output spectrum is given in Figure 7-7(f). Is that all there is to interpolation? Well, not quite—because we can't implement an ideal low-pass filter, $x_{new}(n')$ will not be an exact interpolation of $x_{old}(n)$. The error manifests itself as the unwanted images indicated by the low-level replications in Figure 7-7(f). With an ideal filter, these images would not exist. We can only approximate an ideal, low-pass filter, and the accuracy of our entire interpolation process depends on the frequency response of our filter approximation! As with decimation,

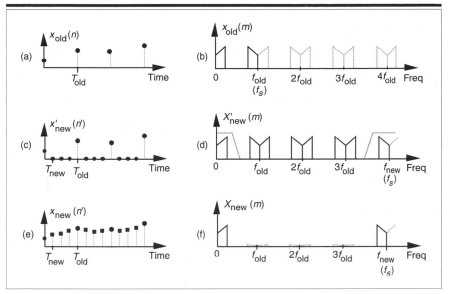

Figure 7-7 Interpolation by a factor of four: (a) original sampled sequence; (b) spectrum of original sequence; (c) zeros inserted in original sequence; (d) spectrum of sequence with inserted zeros; (e) output sequence of interpolation filter; (f) output spectrum of interpolation

interpolation can be thought of as, fundamentally, a low-pass filter design exercise, and the choice of interpolation filter structure deserves attention.

If the FIR interpolation filter design requires S filter stages, we need not perform S multiplications to get each of the $x_{new}(n')$ samples in Figure 7-7(e). By way of an example, returning to our $U = 4$ interpolation case, let's assume we've decided that we need a low-pass filter with $S = 19$ stages. Our example uses an FIR filter with an odd number of stages because this is the optimum structure used for interpolation filters[19]. The job of our low-pass filter is to convolve its impulse response with the $x'_{new}(n')$ sequence. Figure 7-8(a) shows the low-pass filter's impulse response being applied to a portion of the $x'_{new}(n')$ samples to calculate the ninth sample of $x_{new}(n')$, $x_{new}(9)$. The 19 filter coefficients are indicated by the Xs in Figure 7-8(a). With the dots in Figure 7-8(a) representing the $x'_{new}(n')$ sequence, we see that, although there are 19 Xs, only the bold Xs generate nonzero products that contribute to the convolution sum $x_{new}(9)$. The issue here is that we need not perform the multiplications associated with the zero-valued samples in $x'_{new}(n')$. In our example, we need only perform five multiplications to get $x_{new}(9)$. For the general case,

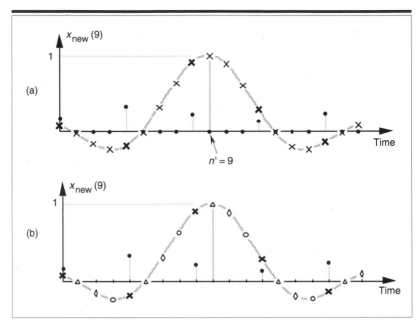

Figure 7-8 Nineteen-stage interpolation filter coefficients: (a) interpolation filter's impulse response shift position to calculate $x_{new}(9)$; (b) Xs mark the only filter coefficients that need be used to calculate $x_{new}(9)$.

the number of multiplications necessitated by an odd-order symmetrical interpolation FIR filter, for each sample of $x_{new}(n')$, cannot be expressed in a closed form for all possible values of S and U, but it can be shown that the minimum number of nonzero multiplications C_{Umin} is

$$C_{Umin} = \lceil \frac{S}{U} \rceil \text{ multiplications,} \qquad (7\text{-}11)$$

where $\lceil \ \rceil$ means "the largest integer contained in." C_{Umin} is the number of $x'_{new}(n')$ values that will be used in the interpolation to determine each $x_{new}(n')$ sample. The number of necessary multiplications/s is $C_{Umin} \cdot f_{new}$.

So the good news is that the number of interpolation filter multiplications necessary is approximately equal to S/U, and that's quite a reduction in computational complexity. The bad news is that we still require that $S = 19$ filter coefficients be stored, and we can't take advantage of any filter coefficient symmetry to reduce the number of multiplications by a further factor of two as described in Section 10.8. Both of these minor problems are the price we must pay to take advantage of Eq. (7-11), and they can be understood by examining Figure 7-8(b). To illustrate our $U = 4$ example, in Figure 7-8(b), we use the filter coefficients indicated by the Xs to calculate $x_{new}(9)$. When we slide the filter's impulse response to the right one sample, we use the coefficients indicated by the circles to calculate $x_{new}(10)$ because the nonzero values of $x'_{new}(n')$ will line up under the circled coefficients. Likewise, when we slide the impulse response to the right one more sample to calculate $x_{new}(11)$, we use the coefficients indicated by the diamonds. Finally, when we slide the impulse response to the right once more, we use the coefficients indicated by the triangles to calculate $x_{new}(12)$. Sliding the filter's impulse response once more to the right, we return to using the coefficients indicated by the Xs to calculate $x_{new}(13)$. You can see the pattern here—there are U, where $U = 4$ in our example, different sets of coefficients that are used to calculate $x_{new}(n')$ from the $x'_{new}(n')$ samples. Each time a new $x_{new}(n')$ sample value is to be calculated, we rotate one step through the sets of coefficients and use

the $\triangle$ coefficient set, when $n' = 0, 4, 8, 12, \ldots, 4n,$

the X coefficient set, when $n' = 1, 5, 9, 13, \ldots, 4n+1,$

the o coefficient set, when $n' = 2, 6, 10, 14, \ldots, 4n+2,$ and

the $\lozenge$ coefficient set, when $n' = 3, 7, 11, 15, \ldots, 4n+3.$

This rotation through U sets of coefficients, where each set contains approximately S/U coefficients, is why interpolation filters are sometimes called *time-varying*, or *polyphase*, filters. At different times, we use different filter coefficients.

So, even though we don't have to perform $S = 19$ multiplications for each $x_{new}(n')$ value, we do have to have $S = 19$ filter coefficients stored and available to calculate each $x_{new}(n')$. (During the calculation of any given $x_{new}(n')$ sample, because the coefficients are not necessarily symmetrical, we can't take advantage of any symmetrical filter structure to reduce the number of multiplications by an additional factor of two as discussed in Section 10.8.) There's one final important feature of the FIR filter's coefficients indicated by the triangles. For our example, when the filter's impulse response peak is aligned above one of the nonzero values of $x'_{new}(n')$, notice that the other coefficients indicated by the remaining triangles have values of zero. This results in the agreeable characteristic that $x_{new}(n') = x'_{new}(n') = x'_{new}(4n)$, so that the output of the filter at the original sampling times is equal to the original samples.

7.3.3 Combining Decimation and Interpolation

While changing sampling rates through decimation or interpolation by integer factors can be useful, what can we do if we need a sample rate change that is not an integer? The good news is that we can implement sample rate changes by any rational fraction U/D with interpolation by a factor of U followed by decimation by a factor of D. Because the fraction U/D can be obtained as accurately as we want with the correct choice of integers U and D, we can change sample rates by almost any factor in practice. For example, a sample rate increase by a factor of 7.125 can be performed by an interpolation of $U = 57$ followed by a decimation of $D = 8$, because $7.125 = 57/8$.

This U/D sample rate change is illustrated by the processes shown in Figure 7-9(a). The really neat part here is that the computational burden of changing the sample rate by the combination of U/D is less than the sum of an individual interpolation followed by an individual decimation because we can combine the interpolation filter LPF_U and the decimation filter LPF_D into a single filter, shown as $\mathrm{LPF}_{U/D}$ in Figure 7-9(b). The process in Figure 7-9(b) is normally called a sample rate converter because, if $U > D$ we have interpolation, and, when $D > U$, we have decimation. (The filter $\mathrm{LPF}_{U/D}$ is sometimes referred to as an interpolation filter, regardless of the ratio U/D, and sometimes it's called a *multirate* filter.) $\mathrm{LPF}_{U/D}$ serves two purposes for us now. First, it must sufficiently attenuate the interpolation images illustrated in Figure 7-7(f), so that, after decimation, they don't contaminate our desired signal beyond acceptable limits. Second, it must have a cutoff frequency that prevents aliasing of our desired signal images resulting from the final decimation.

To illustrate these filter requirements, let's assume we need to resample an input sequence by 4/3 or 1.25, so, $U = 4$, and $D = 3$. The original sequence $x_{\mathrm{old}}(n)$ has the spectrum shown in Figure 7-9(c). The spectrum of the intermediate sequence $x'_{\mathrm{new}}(n')$ after upsampling by 4 and low-pass filtering by $\mathrm{LPF}_{U/D}$ is shown in Figure 7-9(d). The spectral result, after the $D = 3$ decimation to obtain the desired output sequence $x_{\mathrm{new}}(n'')$, is shown in Figure 7-9(e). Notice that the frequency response of $\mathrm{LPF}_{U/D}$ must be designed so that the stopband frequency, f_{stop}, is less than $f_{\mathrm{new}}/2$ to avoid aliasing after the decimation. Of course $\mathrm{LPF}_{U/D}$ is a time-varying filter as described above for Figure 7-8. The attenuation of $\mathrm{LPF}_{U/D}$ beyond f_{stop} must be great enough, so that the images (shown in Figure 7-9(d) at multiples of f_{old}) do not induce intolerable levels of noise when they're aliased by decimation into the final band of 0 to $f_{\mathrm{new}}/2$ Hz. If the number of mul-

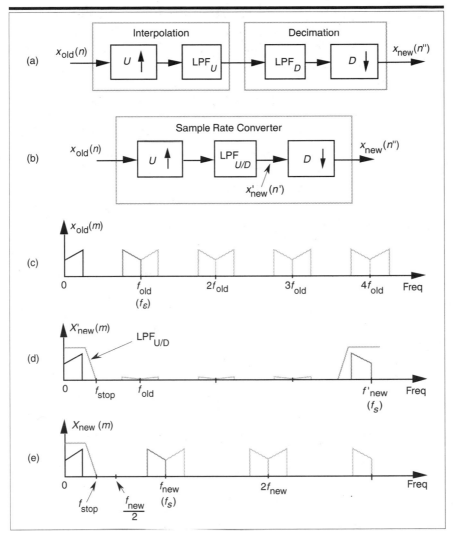

Figure 7-9 Sample rate conversion by a factor of 4/3: (a) combination of interpolation and decimation; (b) sample rate conversion with a single low-pass filter; (c) original discrete input sequence spectrum; (d) spectrum of the low-pass filter output; (e) final discrete spectrum after decimation by a factor of 3.

tiplication stages in $LPF_{U/D}$ is S, the computational complexity of our interpolator/decimator implementation is given by

$$C_{U/D} = \frac{S}{UD} \text{ multiplications/output sample.} \qquad (7\text{-}12)$$

Because the input to $LPF_{U/D}$ is nonzero only for every Uth filter input sample, so, for each filter output sample we need only perform S/U multiplications. Moreover, our decimator discards all but every Dth filter output sample, so the actual number of multiplications necessary is $(S/U)/D$ multiplications per decimator output sample.

Again, our interpolator/decimator problem is an exercise in low-pass filter design, and all the knowledge and tools we have to design low-pass filters can be applied to this task. In software interpolator/decimator design, we want our low-pass filter algorithm to prevent aliasing images and be fast in execution time. For hardware interpolator/decimators, we strive to implement a design that optimizes the conflicting goals of high performance (minimum aliasing), simple architecture, high data throughput speed, and low power and parts count requirements.

This introduction to digital resampling has, by necessity, only touched the surface of this important signal processing technique. Fortunately for us, the excellent work of the first engineers and mathematicians to explore this subject is well documented in the literature. The inquisitive reader can probe further to learn how to choose the number of stages in a multistage process[17,20], the interrelated considerations of designing optimum FIR filters[17,21], the benefits of half-band FIR filters[18,22], when IIR filter structures may be advantageous[21], what special considerations are applicable to resampling in image processing[23–25], guidance in developing the control logic necessary for hardware implementations of resampling algorithms[21], how resampling improves the usefulness of commercial test equipment[26,27], and software development tools for designing multirate filters[28].

References

[1] Floyd, P., and Taylor, J. "Dual-Channel Space Quadrature-Interferometer System," *Microwave System Designer's Handbook*, Fifth Edition, Microwave Systems News, 1987.

[2] Hack, T. "IQ Sampling Yields Flexible Demodulators," *RF Design Magazine*, April 1991.

[3] Mirage Systems, "Automated-Radar-Measurement System for Pulsed I/Q Data Collection," *Microwave System News and Communications Technology*, December 1987.

[4] Chester, D. B., and Phillips, G. "Use DSP Filter Concepts in IF System Design," *Electronic Design*, July 11, 1994.

[5] Steyskal, H. "Digital Beamforming Antennas," *Microwave Journal*, January 1987.

[6] Maruta, R., and Tomozawa, A. "An Improved Method for Digital SSB-FDM Modulation and Demodulation," *IEEE Trans. on Communications*, Vol. COM-26, No. 5, May 1978.

[7] Albaugh, N. "New ADCs for RF Signal Processing," *RF Design*, November 1987.

[8] Oxaal, J. "DSP Hardware Improves Multiband Filters," *EDN*, March 1983.

[9] Sigman, E. "Simplified Digital Down-Converters," *NASA Tech Briefs*, July 1995.

[10] Coleman, B., Meehan, P., Reidy, J., and Weeks, P. "Coherent Sampling Helps When Specifying DSP A/D Converters," *EDN*, October 1987.

[11] Kester, W. "DSP Test Techniques Keep Flash ADCs in Check," *EDN Magazine*, January 18, 1990.

[12] Morris, G., Jr., and Wilck, H. "JPL 2^{20} Channel 330 MHz Bandwidth Digital Spectrum Analyzer," *IEEE Proc., Int. Conference, Acoust., Speech, Signal Processing*, 1978.

[13] Considine, V. "Digital Complex Sampling," *Electronics Letters*, 19, 4 August, 1983.

[14] Rader, C. M. "A Simple Method for Sampling In-Phase and Quadrature Components," *IEEE Trans. on Aerospace and Electronic Systems*, Vol. AES-20, No. 6, November 1984.

[15] Rice, D., and Wu, K. "Quadrature Sampling with High Dynamic Range," *IEEE Trans. on Aerospace and Electronic Systems*, Vol. AES-18, No. 4, November 1982.

[16] Pellon, L. E. "A Double Nyquist Digital Product Detector for Quadrature Sampling," *IEEE Trans. on Signal Processing*, Vol. 40, No. 7, July 1992.

[17] Crochiere, R. E., and Rabiner, L. R. "Optimum FIR Digital Implementations for Decimation, Interpolation, and Narrow-band Filtering," *IEEE Trans. on Acoust. Speech, and Signal Proc.*, Vol. ASSP-23, No. 5, October 1975.

[18] Ballanger, M. G. "Computation Rate and Storage Estimation in Multirate Digital Filtering with Half-Band Filters," *IEEE Trans. on Acoust. Speech, and Signal Proc.*, Vol. ASSP-25, No. 4, August 1977.

[19] Schafer, R. W., and Rabiner, L. R. "A Digital Signal Processing Approach to Interpolation," *Proceedings of the IEEE*, Vol. 61, No. 6, June 1973.

[20] Crochiere, R. E., and Rabiner, L. R. "Decimation and Interpolation of Digital Signals—A Tutorial Review," *Proceedings of the IEEE*, Vol. 69, No. 3, March 1981.

[21] Crochiere, R. E., and Rabiner, L. R. "Further Considerations in the Design of Decimators and Interpolators," *IEEE Trans. on Acoust. Speech, and Signal Proc.*, Vol. ASSP-24, No. 4, August 1976.

[22] Ballanger, M. G., Daguet, J. L., and Lepagnol, G. P. "Interpolation, Extrapolation, and Reduction of Computational Speed in Digital Filters," *IEEE Trans. on Acoust. Speech, and Signal Proc.*, Vol. ASSP-22, No. 4, August 1974.

[23] Hou, H. S., and Andrews, H. C. "Cubic Splines for Image Interpolation and Digital Filtering," *IEEE Trans. on Acoust. Speech, and Signal Proc.*, Vol. ASSP-26, No. 6, August 1978.

[24] Keys, R. G. "Cubic Convolution Interpolation for Digital Image Processing," *IEEE Trans. on Acoust. Speech, and Signal Proc.*, Vol. ASSP-29, No. 6, August 1981.

[25] Parker, J. A., Kenyon, R. V., and Troxel, D. E. "Comparison of Interpolating Methods for Image Resampling," *IEEE Trans. on Medical Imaging*, Vol. MI-2, No. 1, August 1983.

[26] Blue, K. J., et al. "Vector Signal Analyzers for Difficult Measurements on Time-Varying and Complex Modulated Signals," *Hewlett-Packard Journal*, December 1993.

[27] Bartz, M., et al., "Baseband Vector Signal Analyzer Hardware Design," *Hewlett-Packard Journal*, December 1993.

[28] Mitchell, J. A. "Multirate Filters Alter Sampling Rates Even After You've Captured the Data," *EDN*, August 20, 1992.

CHAPTER EIGHT

Signal Averaging

How do we determine the typical amount, a valid estimate, or the true value of some measured parameter? In the physical world, it's not so easy to do because unwanted random disturbances contaminate our measurements. These disturbances are due to both the nature of the variable being measured and the fallibility of our measuring devices. Each time we try to accurately measure some physical quantity, we'll get a slightly different value. Those unwanted fluctuations in a measured value are called *noise*, and digital signal processing practitioners have learned to minimize noise through the process of averaging. In the literature, we can see not only how averaging is used to improve measurement accuracy, but that averaging also shows up in signal detection algorithms as well as in low-pass filter schemes. This chapter introduces the mathematics of averaging and describes how and when this important process is used. Accordingly, as we proceed to quantify the benefits of averaging, we're compelled to make use of the statistical measures known as the mean, variance, and standard deviation.

In digital signal processing, averaging often takes the form of summing a series of time-domain signal samples and then dividing that sum by the number of individual samples. Mathematically, the average of N samples of sequence $x(n)$, denoted x_{ave}, is expressed as

$$x_{\text{ave}} = \frac{1}{N}\sum_{n=1}^{N} x(n) = \frac{x(1) + x(2) + x(3) + \ldots + x(N)}{N} \ .$$

$$(8\text{-}1)$$

(What we call the average, statisticians call the *mean*.) In studying averaging, a key definition that we must keep in mind is the variance of the sequence σ^2 defined as

$$\sigma^2 = \frac{1}{N}\sum_{n=1}^{N}[x(n) - x_{\text{ave}}]^2 \; , \tag{8-2}$$

$$= \frac{[x(1) - x_{\text{ave}}]^2 + [x(2) - x_{\text{ave}}]^2 + [x(3) - x_{\text{ave}}]^2 + \ldots + [x(N) - x_{\text{ave}}]^2}{N} \; . \tag{8-2'}$$

As explained in Appendix D, the σ^2 variance in Eqs. (8-2) and (8-2') gives us a well-defined quantitative measure of how much the values in a sequence fluctuate about the sequence's average. That's because the $x(1) - x_{\text{ave}}$ value in the bracket, for example, is the difference between the $x(1)$ value and the sequence average x_{ave}. The other important quantity that we'll use is the standard deviation defined as the positive square root of the variance, or

$$\sigma = \sqrt{\frac{1}{N}\sum_{n=1}^{N}[x(n) - x_{\text{ave}}]^2} \; . \tag{8-3}$$

To reiterate our thoughts, the average value x_{ave} is the constant level about which the individual sequence values may vary. The variance σ^2 indicates the sum of the magnitudes squared of the noise fluctuations of the individual sequence values about the x_{ave} average value. If the sequence $x(n)$ represents a time series of signal samples, we can say that x_{ave} specifies the constant, or DC, value of the signal, the standard deviation σ reflects the amount of the fluctuating, or AC, component of the signal, and the variance σ^2 is an indication of the power in the fluctuating component. (Appendix D explains and demonstrates the nature of these statistical concepts for those readers who don't use them on a daily basis.)

We're now ready to investigate two kinds of averaging, *coherent* and *incoherent*, to learn how they're different from each other and to see under what conditions they should be used.

8.1 Coherent Averaging

In the coherent averaging process (also known as linear, predetection, or vector averaging), the key feature is the timing used to sample the original signal; that is, we collect multiple sets of signal plus noise samples, and we need the time phase of the signal in each set to be identical. For example, when averaging a sinewave embedded in noise, coherent averaging requires that the phase of the sinewave be the same at the beginning of each measured sample set. When this requirement is met, the sinewave will average to its true sinewave amplitude value. The noise, however, is different in each sample set and will average toward zero.[†] The point is that coherent averaging reduces the variance of the noise, while preserving the amplitude of signals that are synchronous, or coherent, with the beginning of the sampling interval. With coherent averaging, we can actually improve the signal-to-noise ratio of a noisy signal. By way of example, consider the sequence of 128 data points plotted in Figure 8-1(a). Those data points represent the time-domain sampling of a single pulse contaminated with random noise. (For illustrative purposes the pulse, whose peak amplitude is 2.5, is shown in the background of Figure 8-1.) It's very difficult to see a pulse in the bold pulse-plus-noise waveform in the foreground of Figure 8-1(a). Let's say we collect 32 sets of 128 pulse-plus-noise samples of the form

$$\text{Sample Set}_1 = x_1(1),\ x_1(2),\ x_1(3)\ ,\ ...,\ x_1(128)\ ,$$
$$\text{Sample Set}_2 = x_2(1),\ x_2(2),\ x_2(3)\ ,\ ...,\ x_2(128)\ ,$$
$$\text{Sample Set}_3 = x_3(1),\ x_3(2),\ x_3(3)\ ,\ ...,\ x_3(128)\ ,$$

$$...$$
$$...$$

$$\text{Sample Set}_{32} = x_{32}(1),\ x_{32}(2),\ x_{32}(3)\ ,\ ...,\ x_{31}(128)\ . \qquad (8\text{-}4)$$

Here's where the coherent part comes in; the signal measurement times must be synchronized, in some manner, with the beginning of the pulse, so that the pulse is in a constant time relationship with the first sample of each sample set. Coherent averaging of the 32 sets of samples, adding up the columns of Eq. (8-4), takes the form of

$$x_{\text{ave}}(k) = \frac{1}{32}\sum_{n=1}^{32} x_n(k) = [x_1(k) + x_2(k) + x_3(k) + ... + x_{32}(k)] / 32\ ,$$

or

$$x_{\text{ave}}(1) = [x_1(1) + x_2(1) + x_3(1) + ... + x_{32}(1)] / 32$$
$$x_{\text{ave}}(2) = [x_1(2) + x_2(2) + x_3(2) + ... + x_{32}(2)] / 32$$
$$x_{\text{ave}}(3) = [x_1(3) + x_2(3) + x_3(3) + ... + x_{32}(3)] / 32$$

$$...$$
$$...$$

$$x_{\text{ave}}(128) = [x_1(128) + x_2(128) + x_3(128) + ... + x_{32}(128)] / 32\ . \qquad (8\text{-}5)$$

[†] Noise samples are assumed to be uncorrelated with each other and uncorrelated with the sample rate. If some component of the noise is correlated with the sample rate, that noise component *will* be preserved after averaging.

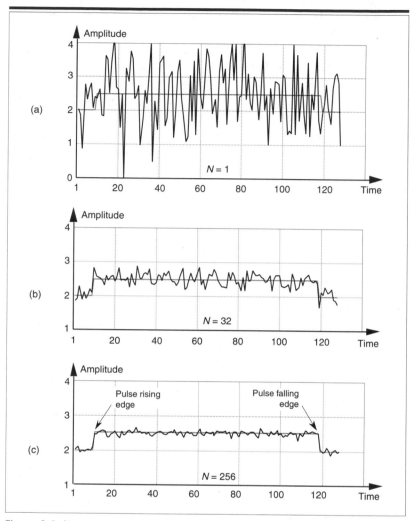

Figure 8-1 Signal pulse plus noise: (a) one sample set; (b) average of 32 sample sets; (c) average of 256 sample sets.

If we perform 32 averages indicated by Eq. (8-5) on a noisy pulse like that in Figure 8-1(a), we'd get the 128-point $x_{ave}(k)$ sequence plotted in Figure 8-1(b). Here, we've reduced the noise fluctuations riding on the pulse, and the pulse shape is beginning to become apparent. The coherent average of 256 sets of pulse measurement sequences results in the plot shown in Figure 8-1(c), where the pulse shape is clearly visible now. We've reduced the noise fluctuations while preserving the pulse amplitude. (An important concept to keep in mind is that summation and averaging both reduce noise variance. Summation is merely implementing Eq. (8-5) without dividing the sum by $N = 32$. If we perform summations and don't divide by N, we merely change the vertical scales for the graphs in Figure 8-1(b) and (c). However, the noise fluctuations will remain unchanged relative to true pulse amplitude on the new scale.)

The mathematics of this averaging process in Eq. (8-5) is both straightforward and important. What we'd like to know is the signal-to-noise improvement gained by coherent averaging as a function of N, the number of sample sets averaged. Let's say that we want to measure some constant time signal with amplitude A, and each time we actually make a measurement we get a slightly different value for A. We realize that our measurements are contaminated with noise such that the nth measurement result $r(n)$ is

$$r(n) = A + \text{noise}(n) \,, \tag{8-6}$$

where noise(n) is the noise contribution. Our goal is to determine A when the $r(n)$ sequence of noisy measurements is all we have to work with. For a more accurate estimate of A, we average N separate $r(n)$ measurement samples and calculate a single average value r_{ave}. To get a feeling for the accuracy of r_{ave}, we decide to take a series of averages $r_{\text{ave}}(k)$, to see how that series fluctuates with each new average; that is,

$$r_{\text{ave}}(1) = [r(1) + r(2) + r(3) + \dots + r(N)]/N \,, \qquad \leftarrow \text{1st } N\text{-point average}$$
$$r_{\text{ave}}(2) = [r(N+1) + r(N+2) + r(N+3) + \dots + r(2N)]/N \,, \qquad \leftarrow \text{2nd } N\text{-point average}$$
$$r_{\text{ave}}(3) = [r(2N+1) + r(2N+2) + r(2N+3) + \dots + r(3N)]/N \,, \qquad \leftarrow \text{3rd } N\text{-point average}$$
$$\dots$$
$$\dots$$
$$r_{\text{ave}}(k) = [r([k{-}1]\cdot N{+}1) + r([k{-}1]\cdot N{+}2) + r([k{-}1]\cdot N{+}3) + \dots + r(k\cdot N)]/N \,. \tag{8-7}$$

or, more concisely,

$$r_{\text{ave}}(k) = \frac{1}{N} \sum_{n=1}^{N} r([k-1]\cdot N + n) \ . \tag{8-8}$$

To see how averaging reduces our measurement uncertainty, we need to compare the standard deviation of our $r_{\text{ave}}(k)$ sequence of averages with the standard deviation of the original $r(n)$ sequence.

If the standard deviation of our original series of measurements $r(n)$ is σ_{in}, it has been shown [1-5] that the standard deviation of our $r_{ave}(k)$ sequence of N-point averages, σ_{ave}, is given by

$$\sigma_{ave} = \frac{\sigma_{in}}{\sqrt{N}} \; . \tag{8-9}$$

Equation (8-9) is significant because it tells us that the $r_{ave}(k)$ series of averages will not fluctuate as much about A as the original $r(n)$ measurement values did; that is, the $r_{ave}(k)$ sequence will be less noisy than any $r(n)$ sequence, and the more we average by increasing N, the more closely an individual $r_{ave}(k)$ estimate will approach the true value of A.[†]

In a different way, we can quantify the noise reduction afforded by averaging. If the quantity A represents the amplitude of a signal and σ_{in} represents the standard deviation of the noise riding on that signal amplitude, we can state that the original signal-amplitude-to-noise ratio is

$$SNR_{in} = \frac{A}{\sigma_{in}} \; . \tag{8-10}$$

Likewise, the signal-amplitude-to-noise ratio at the output of an averaging process, SNR_{ave} is defined as

$$SNR_{ave} = \frac{r_{ave}}{\sigma_{ave}} = \frac{A}{\sigma_{ave}} \; . \tag{8-11}$$

Continuing, the signal-to-noise ratio *gain*, SNR_{coh} gain, that we've realized through coherent averaging is the ratio of SNR_{ave} over SNR_{in}, or

$$SNR_{coh} \text{ gain} = \frac{SNR_{ave}}{SNR_{in}} = \frac{A/\sigma_{ave}}{A/\sigma_{in}} = \frac{\sigma_{in}}{\sigma_{ave}} \; . \tag{8-12}$$

Substituting σ_{ave} from Eq. (8-9) in Eq. (8-12), the SNR gain becomes

$$SNR_{coh} \text{ gain} = \frac{\sigma_{in}}{\sigma_{in}/\sqrt{N}} = \sqrt{N} \; . \tag{8-13}$$

[†] Equation (8-9) is based on the assumptions that the average of the original noise is zero and that neither A nor σ_{in} change during the time we're performing our averages.

Through averaging, we can realize a signal-to-noise ratio improvement proportional to the square root of the number of signal samples averaged. In terms of signal-to-noise ratio measured in decibels, we have a coherent averaging, or *integration*, gain of

$$SNR_{coh} \text{ gain(dB)} = 20 \cdot \log_{10}(SNR_{coh}) = 20 \cdot \log_{10}(\sqrt{N}) = 10 \cdot \log_{10}(N) \,. \quad (8\text{-}14)$$

Again, Eqs. (8-13) and (8-14) are valid if A represents the *amplitude* of a signal and σ_{in} represents the original noise standard deviation.

Another way to view the integration gain afforded by coherent averaging is to consider the standard deviation of the input noise, σ_{in} and the probability of measuring a particular value for the Figure 8-1 pulse amplitude. Assume that we made many individual measurements of the pulse amplitude and created a fine-grained histogram of those measured values to get the dashed curve in Figure 8-2(a). The vertical axis of Figure 8-2(a) represents the probability of measuring a pulse-amplitude value corresponding to the values on the horizontal axis. If the noise fluctuations follow the well-known *normal*, or Gaussian distribution, that dashed probability distribution curve is described by

$$p(x) = \frac{1}{\sigma\sqrt{2\pi}} e^{-(x-\mu)^2 / 2\sigma^2} = K e^{-(x-\mu)^2 / 2\sigma^2} \qquad (8\text{-}15)$$

where $\sigma = \sigma_{in}$ and the true pulse amplitude is represented by $\mu = 2.5$. We see from that dashed curve that any given measured value will most likely (with highest probability) be near the actual pulse-amplitude value of 2.5. Notice, however, that there's a nonzero probability that the measured value could be as low as 1.0 or as high as 4.0. Let's say that the dashed curve represents the probability curve of the pulse-plus-noise signal in Figure 8-1(a). If we averaged a series of 32 pulse-amplitude values and plotted a probability curve of our averaged pulse-amplitude measurements, we'd get the solid curve in Figure 8-2. This curve characterizes the pulse-plus-noise values in Figure 8-1(b). From this solid curve, we see that there's a very low likelihood (probability) that a measured value, after 32-point averaging, will be less than 2.0 or greater than 3.0.

From Eq. (8-9), we know that the standard deviation of the result of averaging 32 signal sample sets

$$\sigma_{ave} = \frac{\sigma_{in}}{\sqrt{32}} = \frac{\sigma_{in}}{5.65} \,. \qquad (8\text{-}16)$$

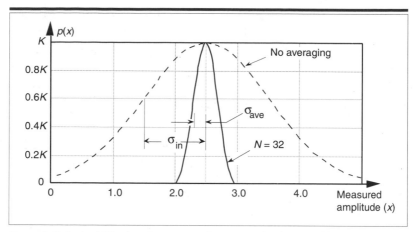

Figure 8-2 Probability density curves of measured pulse amplitudes with no averaging ($N = 1$) and with $N = 32$ averaging.

In Figure 8-2, we can see a statistical view of how an averager's output standard deviation is reduced from the averager's input standard deviation. Taking larger averages by increasing N beyond 32 would squeeze the solid curve in Figure 8-2 even more toward its center value of 2.5, the true pulse amplitude.[†]

Returning to the noisy pulse signal in Figure 8-1, and performing coherent averaging for various numbers of sample sets N, we see in Figure 8-3(a) that as N increases, the averaged pulse amplitude approaches the true amplitude of 2.5. Figure 8-3(b) shows how rapidly the variance of the noise riding on the pulse falls off as N is increased. An alternate way to see how the noise variance decreases with increasing N is the noise power plotted on a logarithmic scale as in Figure 8-3(c). In this plot, the noise variance is normalized to that noise variance when no averaging is performed, i.e., when $N = 1$. Notice that the slope of the curve in Figure 8-3(c) closely approximates that predicted by Eqs. (8-13) and (8-14); that is, as N increases by a factor of 10, we reduce the average noise power by 10 dB. Although the test signal in this discussion was a pulse signal, had the signal been sinusoidal, Eqs. (8-13) and (8-14) would still apply.

[†] The curves in Figure 8-2 are normalized for convenient illustration. From Eq. (8-15) and assuming that $\sigma = 1$ when $N = 1$, then $K = 0.3989$. When $N = 32$, the new standard deviation is $\sigma' = \sigma / \sqrt{N} = 1 / \sqrt{32}$ and $K = 0.3989 \cdot \sqrt{32} = 2.23$.

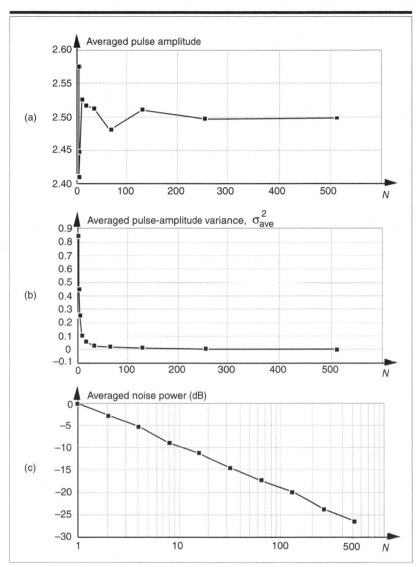

Figure 8-3 Results of averaging signal pulses plus noise: (a) measured pulse amplitude vs. N; (b) measured variance of pulse amplitude vs. N; (c) measured pulse-amplitude noise power vs. N on a logarithmic scale.

8.2 Incoherent Averaging

The process of incoherent averaging (also known as rms, postdetection, scalar, or video averaging) is the averaging of signal samples where no sample timing constraints are used; that is, signal measurement time intervals are not synchronized in any way with the phase of the signal being measured. Think for a moment what the average would be of the noisy pulse signal in Figure 8-1(a) if we didn't in some way synchronize the beginning of the collection of the individual signal sample sets with the beginning of the pulse. The result would be pulses that begin at a different time index in each sample set. The averaging of multiple sample sets would then smear the pulse across the sample set, or just "average the pulse signal away." (For those readers familiar with using oscilloscopes, incoherent averaging would be like trying to view the pulse when the beginning of the scope sweep was not triggered by the signal.) As such, incoherent averaging is not so useful in the time domain.[†] In the frequency domain, however, it's a different story because incoherent averaging can provide increased accuracy in measuring relative signal powers. Indeed, incoherent averaging is used in many test instruments, such as spectrum, network, and signal analyzers.

In some analog test equipment, time-domain signals are represented in the frequency domain using a narrowband sweeping filter followed by a power detector. These devices measure signal power as a function of frequency. The power detector is necessary because the sweeping measurement is not synchronized, in time, with the signal being measured. Thus the frequency-domain data represents power only and contains no signal phase information. Although it's too late to improve the input's signal-amplitude-to-noise ratio, incoherent averaging can improve the accuracy of signal power measurements in the presence of noise; that is, if the signal-power spectrum is very noisy, we can reduce the power estimation fluctuations and improve the accuracy of signal-power and noise-power measurements. Figure 8-4(a) illustrates this idea where we see the power (magnitude squared) output of an FFT of a fundamental tone and several tone harmonics buried in background noise. Notice that the noise-power levels in Figure 8-4(a) fluctuate by almost 20 dB about the true average noise power indicated by the dashed line at −19 dB.

If we take 10 FFTs, average the square of their output magnitudes, and normalize those squared values, we get the power spectrum shown in Figure 8-4(b). Here, we've reduced the variance of the noise in the power spectrum but have not improved the tones' signal-power-to-noise-power ratios; that is, the average noise-power level remains unchanged. Averaging the output magnitudes squared of 100 FFTs results in the spec-

---

[†] The term *incoherent averaging* is a bit of a misnomer. Averaging a set of data is just that, averaging—we add up a set of data values and divide by the number of samples in the set. Incoherent averaging should probably be called *averaging data that's obtained incoherently*.

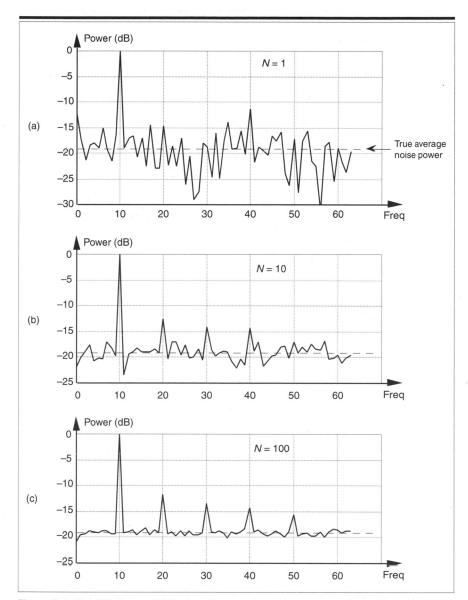

Figure 8-4 Results of averaging signal tones plus noise-power spectra: (a) no averaging, $N = 1$; (b) $N = 10$; (c) $N = 100$.

trum in Figure 8-4(c), which provides a more accurate measure of the relative power levels of the fundamental tone's harmonics.

Just as we arrived at a coherent integration SNR gain expression in Eq. (8-14), we can express an incoherent integration gain, SNR_{incoh} gain, in terms of SNR measured in dB as

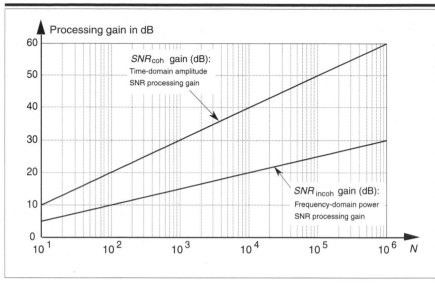

Figure 8-5 Time-domain amplitude SNR processing gain from Eq. (8-14), and the frequency-domain power SNR processing gain from Eq. (8-17), as functions of *N*.

$$SNR_{\text{incoh}}\ \text{gain(dB)} = 10 \cdot \log_{10}(\sqrt{N})\ . \tag{8-17}$$

Equation (8-17) applies when the quantity being averaged represents the *power* of a signal. That's why we used the factor of 10 in Eq. (8-17) as opposed to the factor of 20 used in Eq. (8-14).[†] We can relate the processing gain effects of Eqs. (8-14) and (8-17) by plotting those expressions in Figure 8-5.

8.3 Averaging Multiple Fast Fourier Transforms

We discussed the processing gain associated with a single DFT in Section 3.12 and stated that we can realize further processing gain by increasing the point size of any given *N*-point DFT. Let's discuss this issue when the DFT is implemented using the FFT algorithm. The problem is that large FFTs require a lot of number crunching. Because addition is easier and faster to perform than multiplication, we can average the outputs of multiple FFTs to obtain further FFT signal detection sensitivity; that is, it's eas-

---

[†] Section E.1 of Appendix E explains why the multiplying factor is 10 for signal-power measurements and 20 when dealing with signal-amplitude values.

ier and typically faster to average the outputs of four 128-point FFTs than it is to calculate one 512-point FFT. The increased FFT sensitivity, or noise variance reduction, due to multiple FFT averaging is also called integration gain. So the random noise fluctuations in a FFT's output bins will decrease, while the magnitude of the FFT's signal bin output remains constant when multiple FFT outputs are averaged. (Inherent in this argument is the assumption that the signal is present throughout the observation intervals for all of the FFTs that are being averaged and that the noise sample values are independent both of each other and of the original sample rate.) There are two types of FFT averaging integration gain, incoherent and coherent.

Incoherent integration, relative to FFTs, is averaging the corresponding bin magnitudes of multiple FFTs; that is, to incoherently average k FFTs, the zeroth bin of the incoherent FFT average $F_{incoh}(0)$ is given by

$$F_{incoh}(0) = \frac{|F_1(0)| + |F_2(0)| + |F_3(0)| + ... + |F_k(0)|}{k} , \qquad (8\text{-}18)$$

where $|F_n(0)|$ is the magnitude of the zeroth bin from the nth FFT. Likewise, the first bin of the incoherent FFT average, $F_{incoh}(1)$, is given by

$$F_{incoh}(1) = \frac{|F_1(1)| + |F_2(1)| + |F_3(1)| + ... + |F_k(1)|}{k} , \qquad (8\text{-}18')$$

and, so on, out to the last bin of the FFT average, $F_{incoh}(N-1)$, which is

$$F_{incoh}(N-1) = \frac{|F_1(N-1)| + |F_2(N-1)| + |F_3(N-1)| + ... + |F_k(N-1)|}{k} \qquad (8\text{-}18'')$$

Incoherent integration provides additional reduction in background noise variation to augment a single FFT's inherent processing gain. We can demonstrate this in Figure 8-6(a) where the shaded curve is a single FFT output of random noise added to a tone centered in the 16th bin of a 64-point FFT. The solid curve in Figure 8-6(a) is the incoherent integration of 10 individual 64-point FFT magnitudes. Both curves are normalized to their peak values, so that the vertical scales are referenced to 0 dB. Notice how the variations in the noise power in the solid curve have been reduced by the averaging of the 10 FFTs. The noise power values in the solid curve don't fluctuate as much as the shaded noise-power values. By

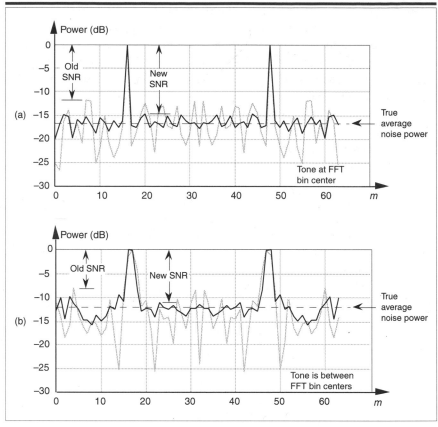

Figure 8-6 Single FFT output magnitudes (shaded) and the average of 10 FFT output magnitudes (solid): (a) tone at bin center; (b) tone between bin centers.

averaging, we haven't raised the power of the tone in the 16th bin, but we have reduced the peaks of the noise-power values. The larger the number of FFTs averaged, the closer the individual noise-power bin values will approach the true average noise power indicated by the dashed horizontal line in Figure 8-6(a).

When the signal tone is not at a bin center, incoherent integration still reduces fluctuations in the FFT's noise-power bins. The shaded curve in Figure 8-6(b) is a single FFT output of random noise added to a tone whose frequency is halfway between the 16th and 17th bins of the 64-point FFT. Likewise, the solid curve in Figure 8-6(b) is the magnitude average of 10 FFTs. The variations in the noise power in the solid curve have again been reduced by the integration of the 10 FFTs. So incoherent integration gain reduces noise-power fluctuations regardless of the frequency location of any signals of interest. As we would expect, the signal peaks are wider, and the true average noise power is larger in Figure 8-6(b) relative to Figure 8-6(a) because leakage raises the average noise-power level and scalloping loss reduces the FFT bin's output power level in Figure 8-6(b). The thing to remember is that incoherent averaging of FFT output magnitudes reduces the *variations* in the background noise power but does not reduce the average background noise power. Equivalent to the incoherent averaging results in Section 8.2, the reduction in the output noise variance[6] of the incoherent average of k FFTs relative to the output noise variance of a single FFT is expressed as

$$\frac{\sigma^2_{\,kFFTs}}{\sigma^2_{\,single\ FFT}} = \frac{1}{k} \ . \tag{8-19}$$

Accordingly, if we average the magnitudes of k separate FFTs, we reduce the noise variance by a factor of k.

In practice, when multiple FFTs are averaged and the FFT inputs are windowed, an overlap in the time-domain sampling process is commonly used. Figure 8-7 illustrates this concept with $5.5Nt_s$ seconds worth of time series data samples, and we wish to average ten separate N-point FFTs where t_s is the sample period $(1/f_s)$. Because the FFTs have a 50 percent overlap in the time domain, some of the input noise in the N time samples for the first FFT will also be contained in the second FFT. The question is

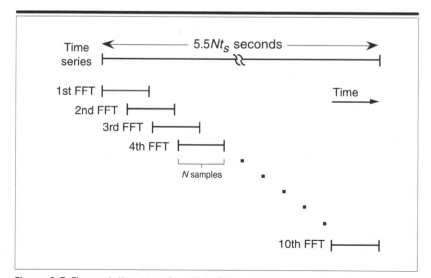

Figure 8-7 Time relationship of multiple FFTs with 50 percent overlap.

"What's the noise variance reduction when some of the noise is common to two FFTs in this averaging scheme?" Well, the answer depends on the window function used on the data before the FFTs are performed. It's been shown that for the most common window functions using an overlap of 50 percent or less, Eq. (8-19) still applies as the level of noise variance reduction[7].

Coherent integration gain is possible when we average the real parts of multiple FFT bin outputs separately from the average of the imaginary parts and then combine the single real average and the single imaginary average into a single complex bin output average value; that is, to coherently average k separate FFTs, the zeroth bin of the complex coherent FFT average, $F_{coh}(0)$, is given by

$$F_{coh}(0) = \frac{F_1(0)_{real} + F_2(0)_{real} + ... + F_k(0)_{real}}{k}$$

$$= j \cdot \frac{F_1(0)_{imag} + F_2(0)_{imag} + ... + F_k(0)_{imag}}{k} . \tag{8-20}$$

The first bin of the complex coherent FFT average, $F_{coh}(1)$, is

$$F_{coh}(1) = \frac{F_1(1)_{real} + F_2(1)_{real} + ... + F_k(1)_{real}}{k}$$

$$+ j \cdot \frac{F_1(1)_{imag} + F_2(1)_{imag} + ... + F_k(1)_{imag}}{k} , \tag{8-20'}$$

and so on out to the last bin of the complex FFT average, $F_{coh}(N{-}1)$, which is

$$F_{coh}(N-1) = \frac{F_1(N-1)_{real} + F_2(N-1)_{real} + ... + F_k(N-1)_{real}}{k}$$

$$+ j \cdot \frac{F_1(N-1)_{imag} + F_2(N-1)_{imag} + ... + F_k(N-1)_{imag}}{k} . \tag{8-20''}$$

Let's consider why the integration gain afforded by coherent averaging is more useful than the integration gain from incoherent averaging. Where incoherent integration does not reduce the background noise-power average level, it does reduce the variations in the averaged FFT's background noise power because we're dealing with magnitudes only, and all FFT noise bin values are positive—they're magnitudes. So their averaged noise bin magnitude values will never be zero. On the other hand, when we average FFT complex bin outputs, those complex noise bin values can be positive *or* negative. Those positive and negative values, in the real or imaginary parts of multiple FFT bin outputs, will tend to cancel each other. This means that noise bin averages can approach zero before we take the magnitude, while a signal bin average will approach its true nonzero magnitude. If we say that the coherently averaged FFT signal-to-noise ratio is expressed by

$$SNR_{coh} = 20 \cdot \log_{10}\left(\frac{\text{signal bin magnitude}}{\text{noise bin magnitude}} \right), \qquad (8\text{-}21)$$

we can see that, should the denominator of Eq. (8-21) approach zero, then SNR_{coh} can be increased. Let's look at an example of coherent integration gain in Figure 8-8(a) where the shaded curve is a single FFT output of random noise added to a tone centered in the 16th bin of a 64-point FFT. The solid curve in Figure 8-8(a) is the coherent integration of ten 64-point FFTs. Notice how the new noise-power average has actually been reduced.[†] That's coherent integration. The larger the number of FFTs averaged, the greater the reduction in the new noise-power average.

When the signal tone is not at bin center, coherent integration still reduces the new noise-power average, but not as much as when the tone is at bin center. The shaded curve in Figure 8-8(b) is a single FFT output of random noise added to a tone whose frequency is halfway between the 16th and 17th bins of the 64-point FFT. Likewise, the solid curve in Figure 8-8(b) is the coherent integration of 10 FFTs. The new noise-power average has, again, been reduced by the integration of the 10 FFTs. Coherent integration provides its maximum gain when the original sample rate f_o is an integral multiple of the tone frequency we're trying to detect. That way the tone will always have the same phase angle at the beginning of the sample interval for each FFT. So, if possible, the f_s sample rate should be

---

[†] The sharp-eyed reader might say, "Wait a minute. Coherent integration seems to reduce the noise-power average, but it doesn't look like it reduces the noise-power variations." Well, the noise-power variations really are reduced in magnitude—remember, we're using a logarithmic scale. The magnitude of a half-division power variation between –20 dB and –30 dB is smaller than the magnitude of a half-division power variation between –10 dB and –20 dB.

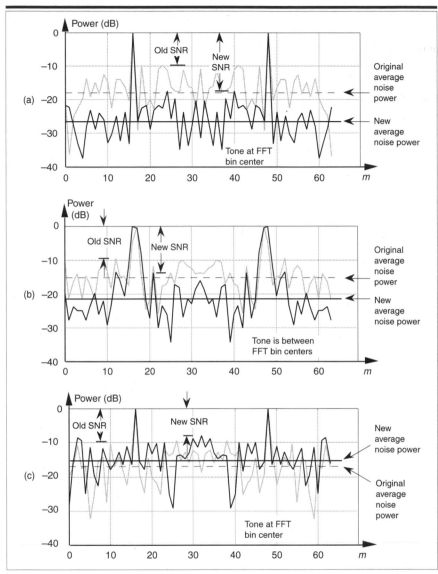

Figure 8-8 Single FFT output magnitudes (shaded) and the coherent integration of 10 FFT outputs (solid): (a) tone at bin center; (b) tone between bin centers; (c) tone at bin center with random phase shift induced prior to the FFT process.

chosen to ensure that the tone of interest, assuming it's not drifting in frequency, will be exactly at a FFT bin center.

Keep two thoughts in mind while using coherent integration of multiple FFTs. First, if the random noise contaminating the time signal does not have an average value of zero, there will be a nonzero DC term in the averaged FFT's $F_{coh}(0)$ bin. Averaging, of course, will not reduce this DC level in the averaged FFT $F_{coh}(0)$ output. Second, coherent integration can actually reduce the averaged FFT's SNR if the tone being detected has a random phase at the beginning of each FFT sample interval. An example of this situation is as follows: the shaded curve in Figure 8-8(c) is a single FFT output of random noise added to a tone centered in the 16th bin of a 64-point FFT. The solid curve in Figure 8-8(c) is the coherent integration of 10 individual 64-point FFTs. However, for each of the 10 FFTs, a random phase shift was induced in the fixed-frequency test tone prior to the FFT. Notice how the averaged FFT's new average noise power is larger than the original average noise power for the single FFT. When the tone has a random phase, we've actually lost rather than gained SNR because the input tone was no longer phase coherent with the analysis frequency of the FFT bin.

There's a good way to understand why coherent integration behaves the way it does. If we look at the phasors representing the successive outputs of a FFT bin, we can understand the behavior of a single phasor representing the sum of those outputs. But, first, let's refresh our memory for a moment concerning the *vector* addition of two phasors.[†] Because an FFT bin output is a complex quantity with a real and an imaginary part, we can depict a single FFT bin output as a phasor, like phasor A shown in Figure 8-9(a). We can depict a subsequent output from the same FFT bin as phasor B in Figure 8-9(b). There are two ways to coherently add the two FFT bin outputs, i.e., add phasors A and B. As shown in Figure 8-9(c), we can add the two real parts and add the two imaginary parts to get the sum phasor C. A graphical method of summing phasors A and B, shown in Figure 8-9(d), is to position the beginning of phasor B at the end of phasor A. Then the sum phasor C is the new phasor from the beginning of phasor A to the end of phasor B. Notice how the two C phasors in Figures 8-9(c) and 8-9(d) are identical. We'll use the graphical phasor summing technique to help us understand coherent integration of FFT bin outputs.

Now we're ready to look at the phasor outputs of an FFT signal bin to see how a single phasor representing the sum of multiple signal bin phasors behaves. Consider the three phasor combinations in Figure 8-10(a). Each phasor is a separate output of the FFT bin containing the signal tone we're trying to detect with coherent integration. The dark arrows are the phasor components due to the tone, and the small shaded arrows are the phasor components due to random noise in the FFT bin containing the tone. In this case, the original signal sample rate is a multiple of the tone

---

[†] Following the arguments put forth in Section A.2 of Appendix A, we'll use the term *phasor*, as opposed to the term *vector*, to describe a single, complex DFT output value.

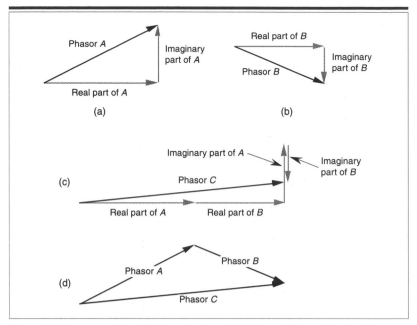

Figure 8-9 Two ways to add phasors A and B, where phasor C = A + B.

frequency, so that the tone phasor phase angle ø is the same at the beginning of the three FFTs, sample intervals. Thus, the three dark tone phasors have a zero phase angle shift relative to one another. The phase angles of the random noise components are random. If we add the three phasor combinations, as a first step in coherent integration, we get the actual phasor sum shown in Figure 8-10(b). The thick shaded vector in Figure 8-10(b) is the ideal phasor sum that would result had there been no noise components in the original three phasor combinations. Because the noise components are reasonably small, the actual phasor sum is not too different from the ideal phasor sum.

Now, had the original signal samples been such that the input tone's phase angles were random at the beginning of the three FFTs, sample intervals, the three phasor combinations in Figure 8-10(c) could result. Summing those three phasor combinations results in the actual phasor sum shown in Figure 8-10(d). Notice that the random tone phases, in Figure 8-10(c), have resulted in an actual phasor sum magnitude (length) that's shorter than the dark lines of the phasor component due to the tone in Figure 8-10(c). What we've done here is degrade, rather than improve,

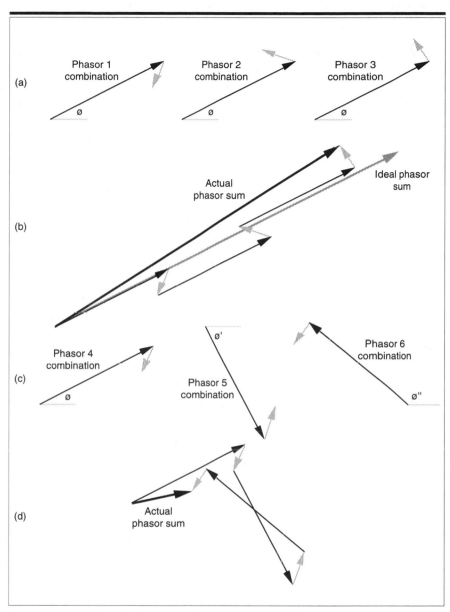

Figure 8-10 FFT bin outputs represented as phasors: (a) three outputs from the same bin when the input tone is at bin center; (b) coherent integration (phasor addition) of the three bin outputs; (c) three outputs from the same bin when input tone has a random phase at the beginning of each FFT; (d) coherent integration of the three bin outputs when input tone has random phase.

the averaged FFT's output SNR when the tone has a random phase at the beginning of each FFT sample interval.

The thought to remember is that, although coherent averaging of FFT outputs is the preferred technique for realizing integration gain, we're rarely lucky enough to work real-world signals with a constant phase at the beginning of every time-domain sample interval. So, in practice, we're usually better off using incoherent averaging. Of course, with either integration technique, the price we pay for improved FFT sensitivity is additional computational complexity and slower system response times because of the additional summation calculations. (A slick way to reduce the computational burden of averaging multiple FFTs is discussed in Section 10.7.)

8.4 Filtering Aspects of Time-Domain Averaging

In Section 5.2 we introduced FIR filters with an averaging example, and that's where we first learned that the process of time-domain averaging performs low-pass filtering. In fact, successive time-domain outputs of an N-point averager are identical to the output of an $(N-1)$-tap FIR filter whose coefficients are all equal to $1/N$, as shown in Figure 8-11.

The question we'll answer here is "What is the frequency magnitude response of a generic N-point averager?" We could evaluate Eq. (6-42), from Section 6.3, that describes the frequency response of a generic N-stage FIR filter. In those expressions, we'd have to set all the $b(0)$ through $b(N)$ coefficient values equal to $1/N$ and calculate $H_{FIR}(\omega)$'s magnitude over the normalized radian frequency range of $0 \le \omega \le \pi$. That range corresponds to an actual frequency range of $0 \le f \le f_s/2$ (where f_s is the equivalent data sample rate in Hz). A simpler approach is to recall, from Section 5.2, that we can calculate the frequency response of an FIR filter by taking the DFT of the filter's coefficients. In doing so, we'd use an M-point FFT

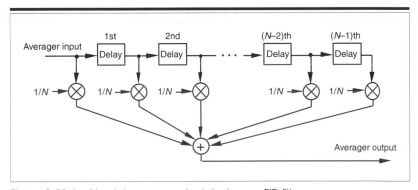

Figure 8-11 An N-point averager depicted as an FIR filter.

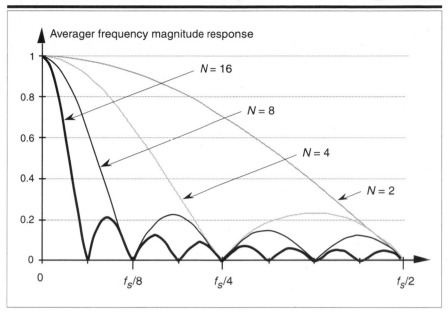

Figure 8-12 *N*-point averager's frequency magnitude response as a function of *N*.

software routine to transform a sequence of N coefficients whose values are all equal to $1/N$. Of course, M should be larger than N so that the $\sin(x)/x$ shape of the frequency response is noticeable. Following through on this by using a 128-point FFT routine, our N-point averager's frequency magnitude responses, for various values of N, are plotted in Figure 8-12. To make these curves more meaningful, the frequency axis is defined in terms of the sample rate f_s in samples/s.

8.5 Exponential Averaging

There is a kind of time-domain averaging used in some power measurement equipment—it's called exponential averaging[8–11]. This technique provides noise reduction by multiplying an input sample by a constant and adding that product to the constant's ones complement multiplied by the most recent averager output. Sounds complicated in words, but the equation for the exponential averager is the simple expression

$$y(n) = \alpha x(n) + (1 - \alpha)y(n - 1), \tag{8-22}$$

where $y(n)$ is the current averager output sample, $y(n–1)$ is the previous averager output sample, and α is the *weighting factor* constant. The

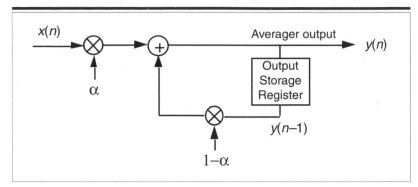

Figure 8-13 Exponential averager.

process described by Eq. (8-22) is implemented as shown in Figure 8-13. The advantage of exponential averaging is that only one storage register is needed to hold the value $y(n–1)$ while waiting for the next input data sample $x(n)$.

The exponential averager's name stems from its time-domain impulse response. Let's assume that the input to the averager is a long string of zeros and we apply a single sample of value 1 at time $t = 0$. Then the input returns again to a string of zero-valued samples. Now, if the weighting factor is $\alpha = 0.4$, the averager's output is shown as the curve in Figure 8-14. When $t = 0$, the input sample is multiplied by α, so the output is 0.4. On the next clock cycle, the input is zero, and the old value of 0.4 is multiplied by $(1 – 0.4)$, or 0.6, to provide an output of 0.24. On the following clock cycle, the input is zero, and the old value of 0.24 is multiplied by 0.6 to provide an output of 0.144. This continues with the averager's output, or impulse response, falling off exponentially because of successive multiplications by 0.6.

A useful feature of the exponential averager is its capability to vary the amount of noise reduction by changing the value of the α weighting factor. If α equals one, input samples are not attenuated, past averager outputs are ignored, and no averaging takes place. In this case, the averager output responds immediately to changes at the input. As α is decreased in value, input samples are attenuated, and past averager outputs begin to affect the present output. These past values represent an exponentially weighted sum of recent inputs, and that summation tends to smooth out noisy signals. The smaller α gets, the more noise reduction is realized. However, with smaller values for α, the slower the averager is in responding to changes in the input. We can demonstrate this behavior by looking

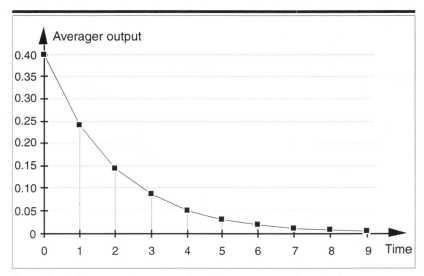

Figure 8-14 Exponential averager impulse response with $\alpha = 0.4$.

at the exponential averager's time-domain step response for various values of α as shown in Figure 8-15.[†]

So we have a trade-off. The more the noise reduction, the more sluggish the averager will be in responding to changes at the input. We can see in Figure 8-15 that, as α gets smaller, affording better noise reduction, the averager's output takes longer to respond and stabilize. Some test instrumentation manufacturers use a clever scheme to resolve this noise reduction versus response time trade-off. They use a large value for α at the beginning of a measurement so that the averager's output responds immediately with a nonzero value. Then, as the measurement proceeds, the value of α is decreased to reduce the noise fluctuations at the input.

The exponential averager's noise variance reduction as a function of the weighting factor α has been shown to be[9,10].

$$\frac{\text{output noise variance}}{\text{input noise variance}} = SNR_{\text{exp}} = \frac{\alpha}{2 - \alpha} . \tag{8-23}$$

[†] The step response is the averager's output when a string of all zeros followed by a string of all ones is applied to the input.

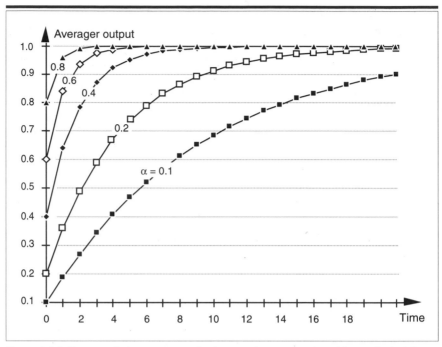

Figure 8-15 Exponential averager output vs. α when a step input is applied at time
t = 0.

Thus, the exponential averager's noise-power reduction in dB is given by

$$SNR_{\exp_{dB}} = 10 \cdot \log_{10}\left(\frac{\alpha}{2-\alpha}\right) . \qquad (8\text{-}24)$$

Equation (8-24) is plotted in Figure 8-16 to illustrate the trade-off between noise reduction and averager response times.

   To demonstrate the exponential averager's noise-power reduction capabilities, Figure 8-17 shows the averager's output with a cosine wave plus noise as an input. The weighting factor α starts out with a value of 1.0 and decreases linearly to a final value of 0.1 at the 180th data input sample. Notice that the noise is reduced as α decreases. However, the cosine wave's peak amplitude also decreases due to the smaller α value.

   The reader may recognize the implementation of the exponential averager in Figure 8-13 as a one-tap infinite impulse response (IIR) digital filter[12]. Indeed it is, and, as such, we can determine its frequency response. We do this by remembering the general expression in Chapter 6 for the frequency response of an IIR filter, or Eq. (6-28) repeated here as

$$H_{\text{exp}}(\omega) = \frac{\displaystyle\sum_{k=0}^{N} b(k) \cdot \cos(k\omega) - j \sum_{k=0}^{N} b(k) \cdot \sin(k\omega)}{1 - \displaystyle\sum_{k=1}^{M} a(k) \cdot \cos(k\omega) + j \sum_{k=1}^{M} a(k) \cdot \sin(k\omega)} \quad . \tag{8-25}$$

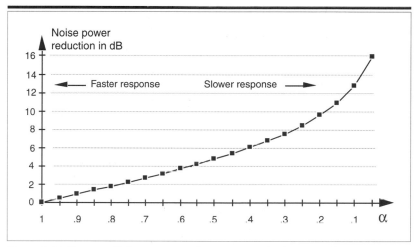

Figure 8-16 Exponential averager noise-power reduction as a function of the weighting factor α.

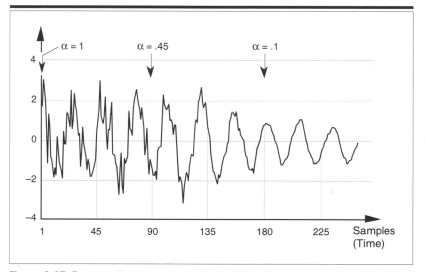

Figure 8-17 Exponential averager output noise reduction as α decreases.

From Figure 6-18, we modify Eq. (8-25) to set $N = 0$, and $M = 1$, so that

$$H_{exp}(\omega) = \frac{b(0) \cdot \cos(0\omega) - jb(0) \cdot \sin(0\omega)}{1 - a(1) \cdot \cos(1\omega) + ja(1) \cdot \sin(1\omega)} \quad . \tag{8-26}$$

Now with $b(0) = \alpha$ and $a(1) = 1 - \alpha$, our exponential averager's frequency response is the complex expression

$$H_{exp}(\omega) = \frac{\alpha}{1 - (1-\alpha) \cdot \cos(\omega) + j(1-\alpha) \cdot \sin(\omega)} \quad . \tag{8-26'}$$

For now, we're interested only in the magnitude response of our filter, so we can express it as

$$
\begin{aligned}
|H_{exp}(\omega)| &= \left| \frac{\alpha}{1 - (1-\alpha) \cdot \cos(\omega) + j(1-\alpha) \cdot \sin(\omega)} \right| \\
&= \frac{\alpha}{\sqrt{[1 - (1-\alpha) \cdot \cos(\omega)]^2 + [(1-\alpha) \cdot \sin(\omega)]^2}} \\
&= \frac{\alpha}{\sqrt{1 - 2 \cdot (1-\alpha) \cdot \cos(\omega) + (1-\alpha)^2}}
\end{aligned}
\tag{8-27}
$$

Evaluating Eq. (8-27) over the normalized angular range of $0 \leq \omega \leq \pi$ (corresponding to a frequency range of $0 \leq f \leq f_s/2$ Hz), the frequency magnitude responses of our exponential averaging filter for various values of α are shown in Figure 8-18(a) using a normalized decibel scale. Notice that, as α decreases, the exponential averager behaves more and more like a low-pass filter.

We can get some practice manipulating complex numbers by deriving the expression for the phase of an exponential averager. We know that the phase angle, as a function of frequency, is the arctangent of the ratio of the imaginary part of $H_{exp}(\omega)$ over the real part of $H_{exp}(\omega)$, or

$$\emptyset_{exp}(\omega) = \tan^{-1} \left(\frac{\text{imaginary part of } H_{exp}(\omega)}{\text{real part of } H_{exp}(\omega)} \right) . \tag{8-27'}$$

To find those real and imaginary parts, knowing that $H_{exp}(\omega)$ is itself a ratio, we determine what the variables are in Eq. (8-26') corresponding to the form of Eq. (A-20), from Appendix A. Doing that, we get

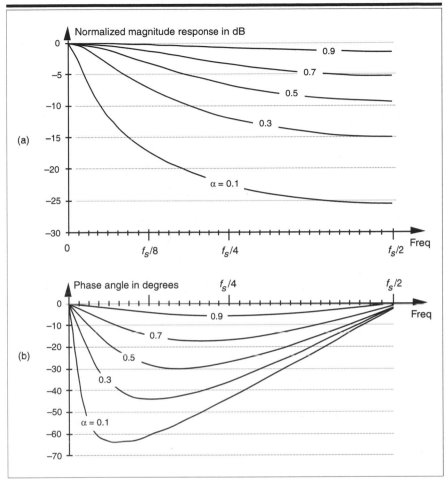

Figure 8-18 Exponential averager frequency response vs. α : (a) normalized magnitude response in dB; (b) phase response in degrees.

$R_1 = \alpha$, $I_1 = 0$, $R_2 = 1 - (1 - \alpha) \cdot \cos(\omega)$, and $I_2 = (1 - \alpha) \cdot \sin(\omega)$.

Substituting those variables in Eq. (A-20), yields

$$H_{exp}(\omega) = \frac{\alpha[1 - (1-\alpha) \cdot \cos(\omega)] - j[\alpha(1-\alpha) \cdot \sin(\omega)]}{[1 - (1-\alpha) \cdot \cos(\omega)]^2 + [(1-\alpha) \cdot \sin(\omega)]^2} \ . \tag{8-28}$$

Representing the denominator of this messy Eq. (8-28) with the term *Den*, we use Eq. (8-27) to express the phase angle of $H_{exp}(\omega)$ as

$$\varnothing_{exp}(\omega) = \tan^{-1}\left(\frac{-\alpha(1-\alpha)\cdot\sin(\omega)/Den}{\alpha[1-(1-\alpha)\cdot\cos(\omega)]/Den} \right)$$

$$= \tan^{-1}\left(\frac{-(1-\alpha)\cdot\sin(\omega)}{1-(1-\alpha)\cdot\cos(\omega)} \right). \qquad (8\text{-}29)$$

The very nonlinear phase of $\varnothing_{exp}(\omega)$ from Eq. (8-29) is calculated over the normalized angular range of $0 \le \omega \le \pi$, corresponding to a frequency range of $0 \le f \le f_s/2$ Hz, and plotted in Figure 8-18(b).

References

[1] Miller, I., and Freund, J. *Probability and Statistics for Engineers*, 2nd Ed., Prentice-Hall, Englewood Cliffs, New Jersey, 1977, pp. 118.

[2] Beller, J., and Pless, W. "A Modular All-Haul Optical Time-Domain Reflectometer for Characterizing Fiber Links," *Hewlett-Packard Journal*, February 1993.

[3] Spiegel, M. R. *Theory and Problems of Statistics*, Shaum's Outline Series, McGraw-Hill Book Co., New York, 1961, pp. 142.

[4] Papoulis, A. *Probability, Random Variables, and Stochastic Processes*, McGraw-Hill Book Co., New York, 1984, pp. 245.

[5] Davenport, W. B., Jr., and Root, W. L. *Random Signals and Noise*, McGraw-Hill Book Co., New York 1958, pp. 81–84.

[6] Welch, P. D. "The Use of Fast Fourier Transform for the Estimation of Power Spectra: A Method Based on Time Averaging over Short, Modified Periodograms," *IEEE Transactions on Audio and Electroacoust.*, Vol. AU-15, No. 2, June 1967

[7] Harris, F. J. "On the Use of Windows for Harmonic Analysis with the Discrete Fourier Transform," *Proceedings of the IEEE*, Vol. 66, No. 1, January 1978.

[8] Booster, D. H., et al. "Design of a Precision Optical Low-Coherence Reflectometer," *Hewlett-Packard Journal*, February 1993.

[9] Witte, R. A. "Averaging Techniques Reduce Test Noise, Improve Accuracy," *Microwaves & RF*, February 1988.

[10] Oxaal, J. "Temporal Averaging Techniques Reduce Image Noise," *EDN*, March 17, 1983.

[11] Lymer, A. "Digital-Modulation Scheme Processes RF Broadcast Signals," *Microwaves & RF*, April 1994.

[12] Hayden, D. "Timer Controls DSP-Filter Frequency Resolution," *EDN*, April 13, 1995.

CHAPTER NINE

Digital Data Formats and Their Effects

In digital signal processing, there are many ways to represent numerical data in computing hardware. These representations, known as *data formats*, have a profound effect on the accuracy and ease of implementation of any given signal processing algorithm. The simpler data formats enable uncomplicated hardware designs to be used at the expense of a restricted range of number representation and susceptibility to arithmetic errors. The more elaborate data formats are somewhat difficult to implement in hardware, but they allow us to manipulate very large and very small numbers while providing immunity to many problems associated with digital arithmetic. The data format chosen for any given application can mean the difference between processing success and failure—it's where our algorithmic rubber meets the road.

In this chapter, we'll introduce the most common types of *fixed-point* digital data formats and show why and when they're used. Next, we'll use analog-to-digital (A/D) converter operation to establish the precision and dynamic range afforded by these fixed-point formats along with the inherent errors encountered with their use. Finally, we'll cover the interesting subject of *floating-point* binary formats.

9.1 Fixed-Point Binary Formats

Within digital hardware, numbers are represented by binary digits known as bits—in fact, the term *bit* originated from the words *Binary digIT*. A single bit can be in only one of two possible states: either a one or a zero.[†] A six-bit binary number could, for example, take the form 101101, with the leftmost bit known as the *most significant bit* (msb), while the rightmost bit is called the *least significant bit* (lsb). The number of bits in a binary number is known as the word length—hence 101101 has a word length of six. Like the decimal number system so familiar to us, the binary number system assumes a weight associated with each digit in the number. That weight is the base of the system (two for binary numbers and ten for decimal numbers) raised to an integral power. To illustrate this with a simple example, the decimal number 4631 is

$$(4 \cdot 10^3) + (6 \cdot 10^2) + (3 \cdot 10^1) + (1 \cdot 10^0)$$

$$= 4000 + 600 + 30 + 1 = 4631 . \tag{9-1}$$

The factors, 10^3, 10^2, 10^1, and 10^0 are the digit weights in Eq. (9-1). Similarly, the six-bit binary number 101101 is equal to decimal 45 as shown by

$$(1 \cdot 2^5) + (0 \cdot 2^4) + (1 \cdot 2^3) + (1 \cdot 2^2) + (0 \cdot 2^1) + (1 \cdot 2^0)$$

$$= 32 + 8 + 4 + 1 = 45 . \tag{9-2}$$

Using subscripts to signify the base of a number, we can write Eq. (9-2) as $101101_2 = 45_{10}$. Equation (9-2) shows us that, like decimal numbers, binary numbers use the *place value* system where the position of a digit signifies its weight. If we use B to denote a number system's base, the place value representation of the four-digit number $a_3a_2a_1a_0$ is

$$(a_3 \cdot B^3) + (a_2 \cdot B^2) + (a_1 \cdot B^1) + (a_0 \cdot B^0) . \tag{9-3}$$

In Eq. (9-3), B^n is the weight multiplier for the digit a_n, where $0 \le a_n \le B-1$. (This place value system of representing numbers is very old—so old, in fact, that it's origin is obscure. However, with its inherent positioning of the decimal or binary point, this number system is so convenient and powerful that its importance has been compared to that of the alphabet[1].)

[†] Binary numbers are used because early electronic computer pioneers quickly realized that it was much more practical and reliable to use electrical devices (relays, vacuum tubes, transistors, etc.) that had only two states, *on* or *off*. Thus, the on/off state of a device could represent a single binary digit.

9.1.1. Octal Numbers

As the use of minicomputers and microprocessors rapidly expanded in the 1960s, people grew tired of manipulating long strings of ones and zeros on paper and began to use more convenient ways to represent binary numbers. One way to express a binary number is an octal format with its base of eight. Converting from binary to octal is as simple as separating the binary number into three-bit groups starting from the right. For example, the binary number 10101001_2 can be converted to octal format as

$$10101001_2 \rightarrow \quad 10 \mid 101 \mid 001 = 251_8 \,.$$

Each of the three groups of bits above are easily converted from their binary formats to a single octal digit because, for three-bit words, the octal and decimal formats are the same. That is, starting with the left group of bits, $10_2 = 2_{10} = 2_8$, $101_2 = 5_{10} = 5_8$, and $001_2 = 1_{10} = 1_8$. The octal format also uses the place value system meaning that $251_8 = (2 \cdot 8^2 + 5 \cdot 8^1 + 1 \cdot 8^0)$. Octal format enables us to represent the eight-digit 10101001_2 with the three-digit 251_8. Of course, the only valid digits in the octal format are 0 to 7—the digits 8 and 9 have no meaning in octal representation.

9.1.2. Hexadecimal Numbers

Another popular binary format is the hexadecimal number format using 16 as its base. Converting from binary to hexadecimal is done, this time, by separating the binary number into four-bit groups starting from the right. The binary number 10101001_2 is converted to hexadecimal format as

$$10101001_2 \rightarrow \quad 1010 \mid 1001 = A9_{16} \,.$$

If you haven't seen the hexadecimal format used before, don't let the A9 digits confuse you. In this format, the characters A, B, C, D, E, and F represent the digits whose decimal values are 10, 11, 12, 13, 14, and 15 respectively. We convert the two groups of bits above to two hexadecimal digits by starting with the left group of bits, $1010_2 = 10_{10} = A_{16}$, and $1001_2 = 9_{10} = 9_{16}$. Hexadecimal format numbers also use the place value system, meaning that $A9_{16} = (A \cdot 16^1 + 9 \cdot 16^0)$. For convenience, then, we can represent the eight-digit 10101001_2 with the two-digit number $A9_{16}$. Table 9-1 lists the permissible digit representations in the number systems discussed thus far.

Table 9-1 Allowable Digit Representations vs. Number System Base

Binary	Octal	Decimal	Hexadecimal	Decimal Equivalent
0	0	0	0	0
1	1	1	1	1
	2	2	2	2
	3	3	3	3
	4	4	4	4
	5	5	5	5
	6	6	6	6
	7	7	7	7
		8	8	8
		9	9	9
			A	10
			B	11
			C	12
			D	13
			E	14
			F	15

9.1.3. Fractional Binary Numbers

Fractions (numbers whose magnitudes are greater than zero and less than one) can also be represented by binary numbers if we use a binary point identical in function to our familiar decimal point. In the binary numbers we've discussed so far, the binary point is assumed to be fixed just to the right of the rightmost digit. Using the symbol ◊ to denote the binary point, the six-bit binary fraction $11_◊0101$ is equal to decimal 3.3125 as shown by

$$(1 \cdot 2^1) + (1 \cdot 2^0) + (0 \cdot 2^{-1}) + (1 \cdot 2^{-2}) + (0 \cdot 2^{-3}) + (1 \cdot 2^{-4})$$

$$= (1 \cdot 2) + (1 \cdot 1) + (0 \cdot \frac{1}{2}) + (1 \cdot \frac{1}{4}) + (0 \cdot \frac{1}{8}) + (1 \cdot \frac{1}{16})$$

$$= 2 + 1 + 0 + 0.25 + 0 + 0.0625 = 3.3125 \ . \tag{9-4}$$

For the example in Eq. (9-4), the binary point is set between the second and third most significant bits. Having a stationary position for the binary point is why binary numbers are often called *fixed-point* binary.

For some binary number formats (like the floating-point formats that we'll cover shortly), the binary point is fixed just to the left of the most significant bit. This forces the number values to be restricted to the range between zero and one. In this format, the largest and smallest values possible for a b-bit fractional word are $1-2^{-b}$ and 2^{-b}, respectively. Taking a six-bit binary fraction, for example, the largest value is $_\diamond 111111_2$, or

$$_\diamond(1 \cdot 2^{-1}) + (1 \cdot 2^{-2}) + (1 \cdot 2^{-3}) + (1 \cdot 2^{-4}) + (1 \cdot 2^{-5}) + (1 \cdot 2^{-6})$$

$$=_\diamond (1 \cdot \frac{1}{2}) + (1 \cdot \frac{1}{4}) + (1 \cdot \frac{1}{8}) + (1 \cdot \frac{1}{16}) + (1 \cdot \frac{1}{32}) + (1 \cdot \frac{1}{64})$$

$$= 0.5 + 0.25 + 0.125 + 0.0625 + 0.03125 + 0.015625 = 0.984375 , \qquad (9\text{-}5)$$

which is $1-2^{-6} = 1-\frac{1}{64}$ in decimal. The smallest nonzero value is $_\diamond 000001_2$, equaling a decimal $\frac{1}{64} = 0.015625_{10}$.

9.1.4. Sign-Magnitude Binary Format

For binary numbers to be at all useful in practice, they must be able to represent negative values. Binary numbers do this by dedicating one of the bits in a binary word to indicate the sign of a number. Let's consider a popular binary format known as sign-magnitude. Here, we assume that a binary word's leftmost bit is a sign bit and the remaining bits represent the magnitude of a number which is always positive. For example, we can say that the four-bit number 0011_2 is $+3_{10}$ and the binary number 1011_2 is equal to -3_{10}, or

magnitude bits   magnitude bits

$\downarrow\downarrow\downarrow$ $\downarrow\downarrow\downarrow$

$0011_2 = 3_{10}$,   and $1011_2 = -3_{10}$.

$\uparrow$ $\uparrow$

sign bit of zero   sign bit of one
signifies positive   signifies negative

Of course, using one of the bits as a sign bit reduces the magnitude of the numbers we can represent. If an unsigned binary number's word length is b bits, the number of different values that can be represented is 2^b. An eight-bit word, for example, can represent $2^8 = 256$ different integral values. With zero being one of the values we have to express, a b-bit unsigned binary word can represent integers from 0 to 2^b-1. The largest value represented by an unsigned eight-bit word is $2^8-1 = 255_{10} = 11111111_2$. In the sign-magnitude binary format a b-bit word can represent only a magnitude of $\pm 2^{b-1}-1$, so the largest positive or negative value we can represent by an eight-bit sign-magnitude word, then, is $\pm 2^{8-1}-1 = \pm 127$.

9.1.5 Two's Complement Format

Another common binary number scheme, known as the *two's complement* format, also uses the leftmost bit as a sign bit. The two's complement format is the most convenient numbering scheme from a hardware design standpoint and has been used for decades. It enables computers to perform both addition and subtraction using the same hardware adder logic. To get the negative version of a positive two's complement number, we merely complement (change a one to a zero and change a zero to a one) each bit and add a one to the complemented word. For example, with 0011_2 representing a decimal 3 in two's complement format, we obtain a negative decimal 3 through the following steps:

$$+3 \text{ in two's complement} \rightarrow \quad 0\,0\,1\,1$$

$$\text{complement of } +3 \rightarrow \quad 1\,1\,0\,0$$
$$\text{add one} \rightarrow \quad {}^+\underline{0\,0\,0\,1}$$
$$-3 \text{ in two's complement} \rightarrow \quad 1\,1\,0\,1\,.$$

In the two's complement format, a b-bit word can represent positive amplitudes as great as $2^{b-1}-1$, and negative amplitudes as large as -2^{b-1}. Table 9-2 shows four-bit word examples of sign-magnitude and two's complement binary formats.

While using two's complement numbers, we have to be careful when adding two numbers of different word lengths. Consider the case where a four-bit number is added to a eight-bit number:

$$+15 \text{ in two's complement} \rightarrow \quad 0\,0\,0\,0\,1\,1\,1\,1$$
$$\text{add } +3 \text{ in two's complement} \rightarrow \quad {}^+\underline{0\,0\,1\,1}$$
$$+18 \text{ in two's complement} \rightarrow \quad 0\,0\,0\,1\,0\,0\,1\,0\,.$$

Table 9-2 Examples of Binary Number Formats

Sign-Magnitude	Two's Complement	Offset Binary	Decimal Equivalent
0111	0111	1111	7
0110	0110	1110	6
0101	0101	1101	5
0100	0100	1100	4
0011	0011	1011	3
0010	0010	1010	2
0001	0001	1001	1
0000	0000	1000	0
1001	1111	0111	−1
1010	1110	0110	−2
1011	1101	0101	−3
1100	1100	0100	−4
1101	1011	0011	−5
1110	1010	0010	−6
1111	1001	0001	−7
–	1000	0000	−8

No problem so far. The trouble occurs when our four-bit number is negative. Instead of adding a +3 to the +15, let's try to add a −3 to the +15:

$$+15 \text{ in two's complement} \rightarrow \quad 0\,0\,0\,0\,1\,1\,1\,1$$
$$\text{add a } -3 \text{ in two's complement} \rightarrow \quad {}^+\underline{1\,1\,0\,1}$$
$$+20 \text{ in two's complement} \rightarrow \quad 0\,0\,0\,1\,0\,1\,0\,0 \quad . \quad \leftarrow \textbf{Wrong answer}$$

The above arithmetic error can be avoided by performing what's called a *sign-extend* operation on the four-bit number. This process, typically performed automatically in hardware, extends the sign bit of the four-bit negative number to the left, making it an eight-bit negative number. If we sign-extend the −3 and, then perform the addition, we'll get the correct answer:

$$+15 \text{ in two's complement} \rightarrow \quad 0\,0\,0\,0\,1\,1\,1\,1$$
$$\text{add a sign-extended } -3 \text{ in two's complement} \rightarrow {}^+\underline{1\,1\,1\,1\,1\,1\,0\,1}$$
$$+12 \text{ in two's complement} \rightarrow 1\,0\,0\,0\,0\,1\,1\,0\,0 \quad . \quad \leftarrow \textbf{That's better}$$
$$\uparrow$$
$$\textbf{overflow bit is ignored}$$

9.1.6. Offset Binary Format

Another useful binary number scheme is known as the *offset binary* format. While this format is not as common as two's complement, it still shows up in some hardware devices. Table 9-2 shows offset binary format examples for four-bit words. Offset binary represents negative numbers by subtracting 2^{b-1} from an unsigned binary value. For example, in the second row of Table 9-2, the offset binary number is 1110_2. When this number is treated as an unsigned binary number it's equivalent to 14_{10}. For four-bit words $b = 4$ and $2^{b-1} = 8$, so $14_{10} - 8_{10} = 6_{10}$ which is the decimal equivalent of 1110_2 in offset binary. The difference between the unsigned binary equivalent and the actual decimal equivalent of the offset binary numbers in Table 9-2 is always -8. This kind of offset is sometimes referred to as a *bias* when the offset binary format is used. (It may interest the reader that we can convert back and forth between the two's complement and offset binary formats merely by complementing a word's most significant bit.)

The history, arithmetic, and utility of the many available number formats is a very broad field of study. A thorough and very readable discussion of the subject is given by Knuth in reference [2].

9.2 Binary Number Precision and Dynamic Range

As we implied before, for any binary number format, the number of bits in a data word is a key consideration. The more bits used in the word, the better the resolution of the number, and the larger the maximum value that can be represented.[†] Assuming that a binary word represents the amplitude of a signal, digital signal processing practitioners find it useful to quantify the dynamic range of various binary number schemes. For a signed integer binary word length of $b+1$ bits (one sign bit and b magnitude bits), the dynamic range in decibels is defined by

$$\text{dynamic range}_{dB} = 20 \cdot \log_{10}\left(\frac{\text{largest possible word value}}{\text{smallest possible word value}}\right)$$

$$= 20 \cdot \log_{10}\left(\frac{2^b - 1}{1}\right) = 20 \cdot \log_{10}(2^b - 1) \ . \qquad (9\text{-}6)$$

[†] Some computers use 64-bit words. Now 2^{64} is approximately equal to $1.8 \cdot 10^{19}$—that's a pretty large number. So large, in fact, that if we started incrementing a 64-bit counter once per second at the beginning of the universe ($\approx$20 billion years ago), the most significant four bits of this counter would *still* be all zeros today.

When 2^b is much larger than 1, we can ignore the -1 in Eq. (9-6) and state that

$$\text{dynamic range}_{dB} = 20 \cdot \log_{10}(2^b)$$

$$= 20 \cdot \log_{10}(2) \cdot b = 6.02 \cdot b \text{ dB}. \qquad (9\text{-}6')$$

Equation (9-6'), dimensioned in dB, tells us that the dynamic range of our number system is directly proportional to the word length. Thus, an eight-bit two's complement word, with seven bits available to represent signal magnitude, has a dynamic range of $6.02 \cdot 7 = 42.14$ dB. Most people simplify Eq. (9-6') by using the rule of thumb that the dynamic range is equal to "six dB per bit."

9.3 Effects of Finite Fixed-Point Binary Word Length

The effects of finite binary word lengths touches all aspects of digital signal processing. Using finite word lengths prevents us from representing values with infinite precision, increases the background noise in our spectral estimation techniques, creates nonideal digital filter responses, induces noise in analog-to-digital (A/D) converter outputs, and can (if we're not careful) lead to wildly inaccurate arithmetic results. The smaller the word lengths, the greater these problems will be. Fortunately, these finite, word-length effects are rather well understood. We can predict their consequences and take steps to minimize any unpleasant surprises. The first finite, word-length effect we'll cover is the errors that occur during the A/D conversion process.

9.3.1 A/D Converter Quantization Errors

Practical A/D converters are constrained to have binary output words of finite length. Commercial A/D converters are categorized by their output word lengths which are normally in the range from 8 to 16 bits. A typical A/D converter input analog voltage range is from -1 to $+1$ volt. If we used such an A/D converter having eight-bit output words, the least significant bit would represent

$$\text{lsb value} = \frac{\text{full voltage range}}{2^{\text{word length}}} = \frac{2 \text{ volts}}{2^8} = 7.81 \text{ millivolts}. \qquad (9\text{-}7)$$

What this means is that we can represent continuous (analog) voltages perfectly as long as they're integral multiples of 7.81 millivolts—any intermediate input voltage will cause the A/D converter to output a *best estimate* digital data value. The inaccuracies in this process are called *quantization errors* because an A/D output least significant bit is an indivisible quantity. We illustrate this situation in Figure 9-1(a) where the continuous waveform is being digitized by an eight-bit A/D converter whose output is in the sign-magnitude format. When we start sampling at time $t = 0$, the continuous waveform happens to have a value of 31.25 millivolts (mv), and our A/D output data word will be exactly correct for sample $x(0)$. At time T when we get the second A/D output word for sample $x(1)$, the continuous voltage is between 0 and –7.81 mv. In this case, the A/D converter outputs a sample value of 10000001 representing –7.81 mv, even though the continuous input was not quite as negative as –7.81mv. The 10000001

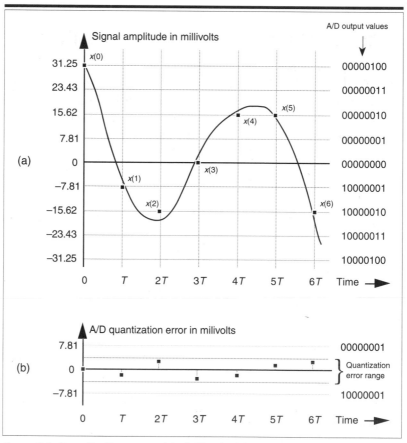

Figure 9-1 Quantization errors: (a) digitized *x(n)* values of a continuous signal; (b) quantization error between the actual analog signal values and the digitized signal values.

A/D output word contains some quantization error. Each successive sample contains quantization error because the A/D's digitized output values must lie on a horizontal dashed line in Figure 9-1(a). The difference between the actual continuous input voltage and the A/D converter's representation of the input is shown as the quantization error in Figure 9-1(b). For an ideal A/D converter, the quantization error, a kind of *roundoff* noise, can never be greater than ±1/2 an lsb, or ±3.905 mv.

While Figure 9-1(b) shows A/D quantization noise in the time domain, we can also illustrate this noise in the frequency domain. Figure 9-2(a) depicts a continuous sinewave of one cycle over the sample interval shown as the dotted line and a quantized version of the time-domain samples of that wave as the dots. Notice how the quantized version of the wave is constrained to have only integral values, giving it a *stair step* effect oscillating above and below the true unquantized sinewave. The quantization here is 4 bits, meaning that we have a sign bit and three bits to represent the magnitude of the wave. With three bits, the maximum peak values for the wave are ±7. Figure 9-2(b) shows the discrete Fourier transform (DFT) of a discrete version of the sinewave whose time-domain sample values are not forced to be integers, but have high precision. Notice in this case that the DFT has a nonzero value only at $m = 1$. On the other hand, Figure 9-2(c) shows the spectrum of the 4-bit quantized samples in Figure 9-2(a) where quantization effects have induced noise components across the entire spectral band. If the quantization noise depictions in Figures 9-1(b) and 9-2(c) look random, that's because they are. As it turns out, even though A/D quantization noise is random, we can still quantify its effects in a useful way.

In the field of communications, people often use the notion of output signal-to-noise ratio, or SNR = (signal power)/(noise power), to judge the usefulness of a process or device. We can do likewise and obtain an important expression for the output SNR of an ideal A/D converter, $SNR_{A/D}$, accounting for finite word-length quantization effects. Because quantization noise is random, we can't explicitly represent its power level, but we can use its statistical equivalent of variance to define $SNR_{A/D}$ measured in decibels as

$$SNR_{A/D} = 10 \cdot \log_{10}\left(\frac{\text{input signal variance}}{\text{A/D quantization noise variance}}\right)$$

$$= 10 \cdot \log_{10}\left(\frac{\sigma^2_{signal}}{\sigma^2_{A/D \, noise}}\right). \tag{9-8}$$

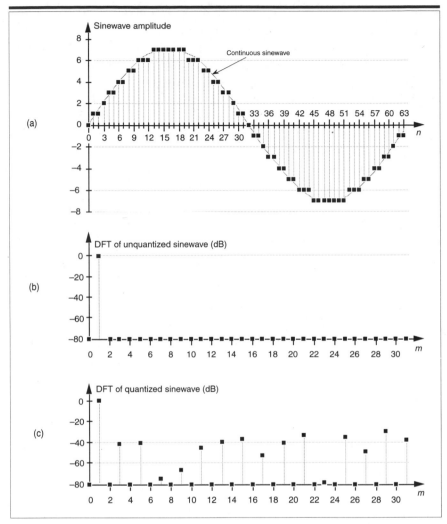

Figure 9-2 Quantization noise effects: (a) input sinewave applied to a 64-point DFT; (b) theoretical DFT magnitude of high-precision sinewave samples; (c) DFT magnitude of a sinewave quantized to 4 bits.

Next, we'll determine an A/D converter's quantization noise variance relative to the converter's maximum input peak voltage V_p. If the full scale ($-V_p$ to $+V_p$ volts) continuous input range of a b-bit A/D converter is $2Vp$, a single quantization level q is that voltage range divided by the number of possible A/D output binary values, or $q = 2V_p/2^b$. (In Figure 9-1, for example, the quantization level q is the lsb value of 7.81 mv.) A depiction of the likelihood of encountering any given quantization error value, called the probability density function $p(e)$ of the quantization error, is shown in Figure 9-3.

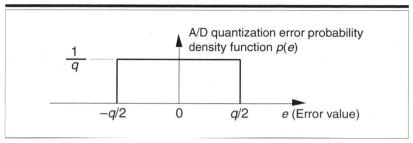

Figure 9-3 Probability density function of A/D conversion roundoff error (noise).

This simple rectangular function has much to tell us. It indicates that there's an equal chance that any error value between $-q/2$ and $+q/2$ can occur. By definition, because probability density functions have an area of unity (i.e., the probability is 100 percent that the error will be somewhere under the curve), the amplitude of the $p(e)$ density function must be the area divided by the width, or $p(e) = 1/q$. From Figure D-4 and Eq. (D-12) in Appendix D, the variance of our uniform $p(e)$ is

$$\sigma^2_{\text{A/D noise}} = \int_{-q/2}^{q/2} e^2 p(e)\,de = \frac{1}{q} \cdot \int_{-q/2}^{q/2} e^2\,de = \frac{q^2}{12} \ . \qquad (9\text{-}9)$$

We can now express the A/D noise error variance in terms of A/D parameters by replacing q in Eq. (9-9) with $q = 2V_p/2^b$ to get

$$\sigma^2_{\text{A/D noise}} = \frac{(2V_p)^2}{12 \cdot (2^b)^2} = \frac{V_p^2}{3 \cdot 2^{2b}} \ . \qquad (9\text{-}10)$$

OK, we're halfway to our goal—with Eq. (9-10) giving us the denominator of Eq. (9-8), we need the numerator. To arrive at a general result, let's express the input signal in terms of its root mean square (rms), the A/D converter's peak voltage, and a loading factor LF defined as

$$LF = \frac{\text{rms of the input signal}}{V_p} = \frac{\sigma_{\text{signal}}}{V_p} \ . \dagger \qquad (9\text{-}11)$$

---

† Recall from Section D.2 that, although the variance σ^2 is associated with the power of a signal, the standard deviation σ is associated with the rms value of a signal.

With the loading factor defined as the input rms voltage over the A/D converter's peak input voltage, we square and rearrange Eq. (9-11) to show the signal variance σ_{signal}^2 as

$$\sigma_{signal}^2 = (LF)^2 V_p^2 \; . \tag{9-12}$$

Substituting Eqs. (9-10) and (9-12) in Eq. (9-8),

$$\mathrm{SNR}_{A/D} = 10 \cdot \log_{10}\left(\frac{(LF)^2 V_p^2}{V_p^2 / (3 \cdot 2^{2b})} \right) = 10 \cdot \log_{10}[(LF)^2 \cdot (3 \cdot 2^{2b})]$$

$$= 6.02 \cdot b + 4.77 + 20 \cdot \log_{10}(LF) \; . \tag{9-13}$$

Eq. (9-13) gives us the $SNR_{A/D}$ of an ideal b-bit A/D converter in terms of the loading factor and the number of bits b. Figure 9-4 plots Eq. (9-13) for various A/D word lengths as a function of the loading factor. Notice that the loading factor in Figure 9-4 is never greater than –3dB, because the maximum continuous A/D input peak value must not be greater V_p volts. Thus, for a sinusoid input, its rms value must not be greater than $V_p / \sqrt{2}$ volts (3 dB below V_p).

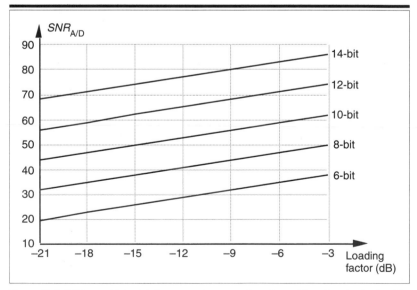

Figure 9-4 $SNR_{A/D}$ of ideal A/D converters as a function of loading factor in dB.

When the input sinewave's peak amplitude is equal to the A/D converter's full-scale voltage V_p, the full-scale *LF* is

$$LF_{\text{full scale}} = \frac{V_p / \sqrt{2}}{V_p} = \frac{1}{\sqrt{2}} . \tag{9-14}$$

Under this condition, the maximum A/D output SNR from Eq. (9-13) is

$$SNR_{A/D\text{-max}} = 6.02 \cdot b + 4.77 + 20 \cdot \log_{10}(1/\sqrt{2})$$

$$= 6.02 \cdot b + 4.77 - 3.01 = 6.02 \cdot b + 1.76 \text{ dB} . \tag{9-15}$$

This discussion of SNR relative to A/D converters means three important things to us:

- An ideal A/D converter will have an $SNR_{A/D}$ defined by Eq. (9-13), so any continuous signal we're trying to digitize with a *b*-bit A/D converter will never have an $SNR_{A/D}$ greater than Eq. (9-13) after A/D conversion. For example, let's say we want to digitize a continuous signal whose SNR is 55 dB. Using an ideal eight-bit A/D converter with its full-scale $SNR_{A/D}$ of 6.02 · 8 + 1.76 = 49.9 dB from Eq. (9-15), the quantization noise will contaminate the digitized values, and the resultant digital signal's SNR can be no better than 49.9 dB. We'll have lost signal SNR through the A/D conversion process. (A ten-bit A/D, with its ideal $SNR_{A/D} \approx 62$ dB, could be used to digitize a 55 dB SNR continuous signal to reduce the SNR degradation caused by quantization noise.) Equations (9-13) and (9-15) apply to ideal A/D converters and don't take into account such additional A/D noise sources as aperture jitter error, missing output bit patterns, and other nonlinearities. So actual A/D converters are likely to have SNRs that are lower than that indicated by theoretical Eq. (9-13). To be safe in practice, it's sensible to assume that $SNR_{A/D\text{-max}}$ is 3 to 6 dB lower than that indicated by Eq. (9-15).

- Equation (9-15) is often expressed in the literature, but it can be a little misleading because it's imprudent to force an A/D converter's input to full scale. It's wise to drive an A/D converter to some level below full scale because inadvertent overdriving will lead to signal clipping and will induce distortion in the A/D's output. So Eq. (9-15) is overly optimistic, and, in practice, A/D converter SNRs will be less than that indicated by Eq. (9-15). The best approximation for an A/D's SNR is to determine the input signal's rms value that will never (or rarely) overdrive the converter input, and plug that value in Eq. (9-11) to get the loading factor value for use in Eq. (9-13).[†] Again, using an A/D converter with a wider word length will alleviate this problem by increasing the available $SNR_{A/D}$.

- Remember now, real-world continuous signals always have their own inherent continuous SNR, so using an A/D converter, whose $SNR_{A/D}$ is a great deal larger than the continuous signal's SNR serves no purpose. In this case, we'd be using the A/D converter's extra bits to digitize the continuous signal's noise to a greater degree of accuracy.

A word of caution is appropriate here concerning our analysis of A/D converter quantization errors. The derivations of Eqs. (9-13) and (9-15) are based upon three assumptions:

- The cause of A/D quantization errors is a stationary random process; that is, the performance of the A/D converter does not change over time. Given the same continuous input voltage, we always expect an A/D converter to provide exactly the same output binary code.

- The probability density function of the A/D quantization error is uniform. We're assuming that the A/D converter is ideal in its operation and all possible errors between $-q/2$ and $+q/2$ are equally likely. An A/D converter having stuck bits or missing output codes would violate this assumption. High-quality A/D converters being driven by continuous signals that cross many quantization levels will result in our desired uniform quantization noise probability density function.

- The A/D quantization errors are uncorrelated with the continuous input signal. If we were to digitize a single continuous sinewave whose frequency was harmonically related to the A/D sample rate, we'd end up sampling the same input voltage repeatedly and the quantization error sequence would not be random. The quantization

---

[†] By the way, some folks use the term *crest factor* to describe how hard an A/D converter's input is being driven. The crest factor is the reciprocal of the loading factor, or $CF = V_p/($rms of the input signal$)$.

error would be predictable and repetitive, and our quantization noise variance derivation would be invalid. In practice, complicated continuous signals such as music or speech, with their rich spectral content, avoid this problem.

To conclude our discussion of A/D converters, let's consider one last topic. In the literature the reader is likely to encounter the expression

$$b_{eff} = \frac{SNR - 1.76}{6.02} \; .$$

$$(9\text{-}16)$$

Equation (9-16) is used by test equipment manufacturers to specify the sensitivity of test instruments using a b_{eff} parameter known as the number of *effective bits*, or effective number of bits (ENOB) [3–8]. Equation (9-16) is merely Eq. (9-15) solved for b. Test equipment manufacturers measure the actual SNR of their product indicating its ability to capture continuous input signals relative to the instrument's inherent noise characteristics. Given this true SNR, they use Eq. (9-16) to determine the b_{eff} value for advertisement in their product literature. The larger b_{eff}, the greater the continuous voltage that can be accurately digitized relative to the equipment's intrinsic quantization noise.

9.3.2 Data Overflow

The next finite, word-length effect we'll consider is called *overflow*. Overflow is what happens when the result of an arithmetic operation has too many bits, or digits, to be represented in the hardware registers designed to contain that result. We can demonstrate this situation to ourselves rather easily using a simple four-function, eight-digit pocket calculator. The sum of a decimal 9.9999999 plus 1.0 is 10.9999999, but on an eight-digit calculator the sum is 10.999999 as

```
    9.9999999
 + 1.0000000
  10.9999999.
          ↑
  this digit gets discarded
```

The hardware registers, which contain the arithmetic result, and drive the calculator's display, can hold only eight decimal digits; so the least significant (of course) digit is discarded. Although the above error is less than one part in ten million, overflow effects can be striking when we work with large numbers. If we use our calculator to add 99,999,999 plus 1, instead of getting the correct result of 100 million, we'll get a result of 1. Now that's an authentic overflow error!

Let's illustrate overflow effects with examples more closely related to our discussion of binary number formats. First, adding two unsigned binary numbers is as straightforward as adding two decimal numbers. The sum of 42 plus 39 is 81, or

$$
\begin{array}{ll}
& 1\ 1\ 1 \quad \leftarrow \text{carry bits}\\
\text{+42 in unsigned binary} \rightarrow & 1\ 0\ 1\ 0\ 1\ 0\\
\text{+39 in unsigned binary} \rightarrow & ^+ 1\ 0\ 0\ 1\ 1\ 1\\
\text{+81 in unsigned binary} \rightarrow & 1\ 0\ 1\ 0\ 0\ 0\ 1\ .
\end{array}
$$

In this case, two 6-bit binary numbers required 7 bits to represent the results. The general rule is *the sum of* m *individual* b*-bit binary numbers can require as many as* $[b + log_2(m)]$ *bits to represent the results*. So, for example, a 24-bit result register (accumulator) is needed to accumulate the sum of sixteen 20-bit binary numbers, or $20 + log_2(16) = 24$. The sum of 256 eight-bit words requires an accumulator whose word length is $[8 + log_2(256)]$, or 16 bits, to ensure that no overflow errors occur.

In the preceding example, if our accumulator word length was 6 bits, an overflow error occurs as

$$
\begin{array}{ll}
& 1\ 1\ 1 \quad \leftarrow \text{carry bits}\\
\text{+42 in unsigned binary} \rightarrow & 1\ 0\ 1\ 0\ 1\ 0\\
\text{+39 in unsigned binary} \rightarrow & ^+ 1\ 0\ 0\ 1\ 1\ 1\\
\text{+17 in unsigned binary} \rightarrow & 1\ 0\ 1\ 0\ 0\ 0\ 1\ . \quad \leftarrow \text{Overflow error}\\
& \uparrow
\end{array}
$$

an overflow out of the sign bit is ignored, causing an overflow error

Here, the most significant bit of the result overflowed the 6-bit accumulator, and an error occurred.

With regard to overflow errors, the two's complement binary format has two interesting characteristics. First, under certain conditions, overflow during the summation of two numbers causes no error. Second, with multiple summations, intermediate overflow errors cause no problems if the final magnitude of the sum of the *b*-bit two's complement numbers is less than 2^{b-1}. Let's illustrate these properties by considering the four-bit two's complement format in Figure 9-5, whose binary values are taken from Table 9-2.

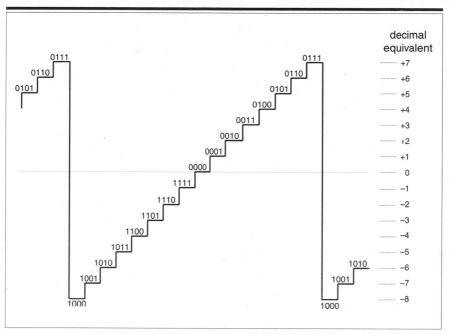

decimal
equivalent

Figure 9-5 Four-bit two's complement binary numbers.

The first property of two's complement overflow, which sometimes causes no errors, can be shown by the following examples:

$$
\begin{array}{rl}
 & 0\ 1\ 0 \quad \leftarrow \text{carry bits} \\
-5 \text{ in two's complement } \rightarrow & 1\ 0\ 1\ 1 \\
+2 \text{ in two's complement } \rightarrow & {}^{+}\underline{0\ 0\ 1\ 0} \\
-3 \text{ in two's complement } \rightarrow & 0\ 1\ 1\ 0\ 1 \quad \leftarrow \text{valid negative result} \\
 & \uparrow
\end{array}
$$

zero overflow out of the sign bit

$$
\begin{array}{rl}
 & 1\ 1\ 0 \quad \leftarrow \text{carry bits} \\
-2 \text{ in two's complement } \rightarrow & 1\ 1\ 1\ 0 \\
+6 \text{ in two's complement } \rightarrow & {}^{+}\underline{0\ 1\ 1\ 0} \\
+4 \text{ in two's complement } \rightarrow & 1\ 0\ 1\ 0\ 0 \quad \leftarrow \text{valid positive result} \\
 & \uparrow
\end{array}
$$

overflow out of the sign bit ignored, no harm done

Then again, the following examples show how two's complement overflow sometimes does cause errors:

```
                              0 0 0        ← carry bits
  –7 in two's complement  →    1 0 0 1
  –6 in two's complement  →  ⁺ 1 0 1 0
  +3 in two's complement  →  1 0 0 1 1    ← invalid positive result
                                 ↑
```

overflow out of the sign bit ignored, causing overflow error

```
                              1 1 1        ← carry bits
  +7 in two's complement  →    0 1 1 1
  +7 in two's complement  →    0 1 1 1
  –2 in two's complement  →  0 1 1 1 0    ← invalid negative result
                               ↑
```

zero overflow out of the sign bit

The rule with two's complement addition is *if the carry bit into the sign bit is the same as the overflow bit out of the sign bit, the overflow bit can be ignored causing no errors; if the carry bit into the sign bit is different from the overflow bit out of the sign bit, the result is invalid.* An even more interesting property of two's complement numbers is that a series of b-bit word summations can be performed where intermediate sums are invalid, but the final sum will be correct if its magnitude is less than 2^{b-1}. We show this by the following example. If we add a +6 to a +7, and then add a –7, we'll encounter an intermediate overflow error but our final sum will be correct as

```
  +7 in two's complement  →    0 1 1 1
  +6 in two's complement  →  ⁺ 0 1 1 0
  –3 in two's complement  →    1 1 0 1    ← overflow error here
  –7 in two's complement  →  ⁺ 1 0 0 1
  +6 in two's complement  →  1 0 1 1 0    ← valid positive result
                               ↑
```

overflow ignored, with no harm done

The magnitude of the sum of the three four-bit numbers was less than 2^{4-1} (<8), so our result was valid. If we add a +6 to a +7, and next add a –5, we'll encounter an intermediate overflow error, and our final sum will also be in error because its magnitude is not less than 8.

```
  +7 in two's complement  →    0 1 1 1
  +6 in two's complement  →  ⁺ 0 1 1 0
  –3 in two's complement  →    1 1 0 1    ← overflow error here
  –5 in two's complement  →  ⁺ 1 0 1 1
  –8 in two's complement  →  1 1 0 0 0    ← invalid negative result
```

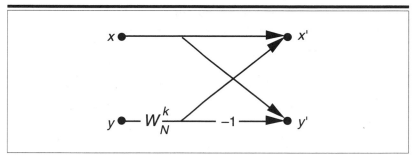

Figure 9-6 Single decimation-in-time FFT butterfly.

Another situation where overflow problems are conspicuous is during the calculation of the fast Fourier transform (FFT). It's difficult at first to imagine that multiplying complex numbers by sines and cosines can lead to excessive data word growth—particularly, because sines and cosines are never greater than unity. Well, we can show how FFT data word growth occurs by considering a decimation-in-time FFT butterfly from Figure 4-14(c) repeated here as Figure 9-6, and grinding through a little algebra. The expression for the x' output of this FFT butterfly, from Eq. (4-26), is

$$x' = x + W_N^k \cdot y \ . \tag{9-17}$$

Breaking up the butterfly's x and y inputs into their real and imaginary parts and remembering that $W_N^k = e^{-j2\pi k/N}$, we can express Eq. (9-17) as

$$x' = x_{\text{real}} + jx_{\text{imag}} + (e^{-j2\pi k/N}) \cdot (y_{\text{real}} + jy_{\text{imag}}) \ . \tag{9-18}$$

If we let α be the twiddle factor angle of $2\pi k/N$, and recall that $e^{-j\alpha} = \cos(\alpha) - j\sin(\alpha)$, we can simplify Eq. (9-18) as

$$x' = x_{\text{real}} + jx_{\text{imag}} + [\cos(\alpha) - j\sin(\alpha) \cdot (y_{\text{real}} + jy_{\text{imag}})$$

$$= x_{\text{real}} + \cos(\alpha)y_{\text{real}} + \sin(\alpha)y_{\text{imag}} + j(x_{\text{imag}} + \cos(\alpha)y_{\text{imag}} - \sin(\alpha)y_{\text{real}}) \ . \tag{9-19}$$

If we look, for example, at just the real part of the x' output, x'_{real}, it comprises the three terms

$$x'_{\text{real}} = x_{\text{real}} + \cos(\alpha)y_{\text{real}} + \sin(\alpha)y_{\text{imag}} \tag{9-20}$$

If x_{real}, y_{real}, and y_{imag} are of unity value when they enter the butterfly and the twiddle factor angle $\alpha = 2\pi k/N$ happens to be $\pi/4 = 45°$, then, x'_{real} can be greater than 2 as

$$x'_{real} = 1 + \cos(45°) \cdot 1 + \sin(45°) \cdot 1$$

$$= 1 + 0.707 + 0.707 = 2.414 \ . \tag{9-21}$$

So we see that the real part of a complex number can more than double in magnitude in a single stage of an FFT. The imaginary part of a complex number is equally likely to more than double in magnitude in a single FFT stage. Without mitigating this word growth problem, overflow errors could render an FFT algorithm useless.

OK, overflow problems are handled in one of two ways: by truncation or rounding—each inducing its own individual kind of quantization errors as we shall see.

9.3.3 Truncation

Truncation is the process where a data value is represented by the largest quantization level that is less than or equal to that data value. If we're quantizing to integral values, for example, the real value 1.2 would be quantized to 1. An example of truncation to integral values is shown in Figure 9-7(a), where all values of x in the range of $0 \leq x < 1$ are set equal to 0, values of x in the range of $1 \leq x < 2$ are set equal to 1, x values in the range of $2 \leq x < 3$ are set equal to 2, and so on.

As we did with A/D converter quantization errors, we can call upon the concept of probability density functions to characterize the errors induced by truncation. The probability density function of truncation errors, in terms of the quantization level, is shown in Figure 9-7(b). In Figure 9-7(a) the quantization level q is 1, so, in this case, we can have truncation errors as great as –1. Drawing upon our results from Eqs. (D-11) and (D-12) in Appendix D, the mean and variance of our uniform truncation probability density function are expressed as

$$\mu_{truncation} = \frac{-q}{2} \tag{9-22}$$

and

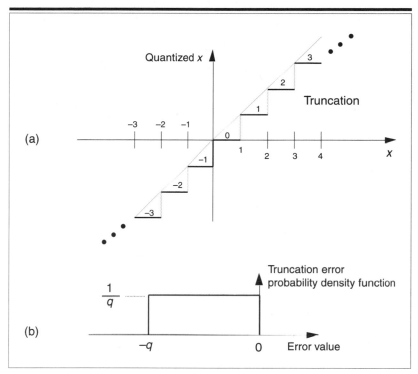

Figure 9-7 Truncation: (a) quantization nonlinearities; (b) error probability density function.

$$\sigma^2_{\text{truncation}} = \frac{q^2}{12} \, . \qquad (9\text{-}23)$$

In a sense, truncation error is the price we pay for the privilege of using integer binary arithmetic. One aspect of this is the error introduced when we use truncation to implement division by some integral power of 2. We often speak of a quick way of dividing a binary value by 2^T is to shift that binary word T bits to the right; that is, we're truncating the data value (not the data word) by lopping off the rightmost T bits after the shift. For example, let's say we have the value 31 represented by the five-bit binary number 11111_2, and we want to divide it by 16 through shifting the bits $T = 4$ places to the right and ignoring (truncating) those shifted bits. After the right shift and truncation, we'd have a binary quotient of $31/16 = 00001_2$. Well, we see the significance of the problem because our quick division gave us an answer of one instead of the correct $31/16 = 1.9375$. Our division-by-truncation error here is almost 50 percent of the correct answer. Had our original dividend been a 63 represented by the six-bit binary number 111111_2, dividing it by 16 through a four-bit shift would give us an answer of binary 000011_2, or decimal three. The correct answer, of course, is $63/16 = 3.9375$. In this case the percentage error is $0.9375/3.9375$, or about 23.8 percent. So, the larger the dividend, the lower the truncation error.

If we study these kinds of errors, we'll find that the resulting truncation error depends on three things: the number of value bits shifted and truncated, the values of the truncated bits (were those dropped bits ones or zeros), and the magnitude of the binary number left over after truncation. Although a complete analysis of these truncation errors is beyond the scope of this book, we can quantify the maximum error that can occur with our division-by-truncation scheme when using binary integer arithmetic. The worst case scenario is when all of the T bits to be truncated are ones. For binary integral numbers the value of T bits, all ones, to the right of the binary point is $1 - 2^T$. If the resulting quotient N is small, the significance of those truncated ones aggravates the error. We can normalize the maximum division error by comparing it to the correct quotient using a percentage. So the maximum percentage error of the binary quotient N after truncation for a T-bit right shift, when the truncated bits are all ones, is

$$\%\text{truncation error}_{max} = 100 \cdot \frac{\text{Correct quotient} - \text{Quotient after truncation}}{\text{Correct quotient}}$$

$$= 100 \cdot \frac{\text{Truncation error}}{\text{Correct quotient}} = 100 \cdot \frac{1 - 2^{-T}}{N + (1 - 2^{-T})} . \qquad (9\text{-}24)$$

Let's plug a few numbers into Eq. (9-24) to understand its significance. In the first example above, $31/16$, where $T = 4$ and the resulting binary quotient was $N = 1$, the percentage error is

$$100 \cdot (1 - 0.0625)/(1 + 1 - 0.0625) = 100 \cdot (0.9375/1.9375) = 48.4\%.$$

Plotting Eq. (9-24) for three different shift values as functions of the quotient, N, after truncation results in Figure 9-8.

So, to minimize this type of division error, we need to keep the resulting binary quotient N, after the division, as large as possible. Remember, Eq. (9-24) was a worst case condition where all of the truncated bits

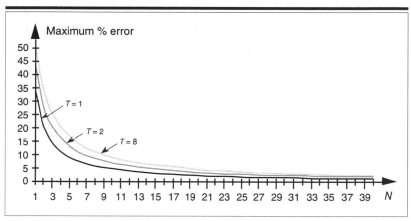

Figure 9-8 Maximum error when data shifting and truncation are used to implement division by 2^T. $T = 1$ is division by 2; $T = 2$ is division by 4; and $T = 8$ is division by 256.

were ones. On the other hand, if all of the truncated bits happened to be zeros, the error will be zero. In practice, the errors will be somewhere between these two extremes. A practical example of how division-by-truncation can cause serious numerical errors is given in reference [9].

9.3.4 Data Rounding

Rounding is an alternate process of dealing with overflow errors where a data value is represented by, or rounded off to, its nearest quantization level. If we're quantizing to integral values, the number 1.2 would be quantized to 1, and the number 1.6 would be quantized to 2. This is shown in Figure 9-9(a) where all values of x in the range of $-0.5 \leq x < 0.5$ are set equal to 0, values of x in the range of $0.5 \leq x < 1.5$ are set equal to 1, x values in the range of $1.5 \leq x < 2.5$ are set equal to 2, etc. The probability density function of the error induced by rounding, in terms of the quantization level, is shown in Figure 9-9(b). In Figure 9-9(a), the quantization level q is 1, so, in this case, we can have truncation error magnitudes no greater than $q/2$, or $1/2$. Again, using our Eqs. (D-11) and (D-12) results from Appendix D, the mean and variance of our uniform roundoff probability density function are expressed as

$$\mu_{roundoff} = 0 \, , \tag{9-25}$$

and

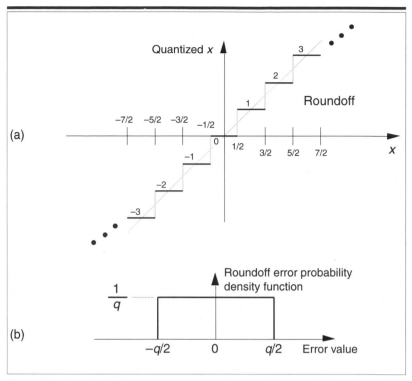

Figure 9-9 Roundoff: (a) quantization nonlinearities; (b) error probability density function.

$$\sigma^2_{\text{roundoff}} = \frac{q^2}{12} \ . \tag{9-26}$$

Because the mean (average) and maximum error possibly induced by roundoff is less than that of truncation, rounding is generally preferred, in practice, to minimize overflow errors.

In digital signal processing, statistical analysis of quantization error effects is exceedingly complicated. Analytical results depend on the types of quantization errors, the magnitude of the data being represented, the numerical format used, and which of the many FFT or digital filter structures we happen to use. Be that as it may, digital signal processing experts have developed simplified error models whose analysis has proved useful. Although discussion of these analysis techniques and their results is beyond the scope of this introductory text, many references are available for the energetic reader[10–18]. (Reference [11] has an extensive reference list of its own on the topic of quantization error analysis.)

Again, the overflow problems using fixed-point binary formats—that we try to alleviate with truncation or roundoff—arise because so many digital signal processing algorithms comprise large numbers of additions or multiplications. This obstacle, particularly in hardware implementations of digital filters and the FFT, is avoided by hardware designers through the use of floating-point binary formats.

9.4 Floating-Point Binary Formats

Floating-point binary formats allow us to overcome most of the limitations of precision and dynamic range mandated by fixed-point binary formats, particularly in reducing the ill effects of overflow [19]. Floating-point formats segment a data word into two parts: a mantissa m and an exponent e. Using these parts, the value of a binary floating-point number n is evaluated as

$$n = m \cdot 2^e , \qquad (9\text{-}27)$$

that is, the number's value is the product of the mantissa and 2 raised to the power of the exponent. (*Mantissa* is a somewhat unfortunate choice of terms because it has a meaning here very different from that in the mathematics of logarithms—*mantissa* originally meant the decimal fraction of a logarithm.[†] However, due to its abundance in the literature we'll continue using the term *mantissa* here.) Of course, both the mantissa and the exponent in Eq. (9-27) can be either positive or negative numbers.

Let's assume that a b-bit floating-point number will use b_e bits for the fixed-point signed exponent and b_m bits for the fixed-point signed mantissa. The greater the number of b_e bits used, the larger the dynamic range of the number. The more bits used for b_m, the better the resolution, or precision, of the number. Early computer simulations conducted by the developers of b-bit floating-point formats indicated that the best trade-off occurred with $b_e \approx b/4$ and $b_m \approx 3b/4$. We'll see that, for typical 32-bit floating-point formats used today, $b_e \approx 8$ bits and $b_m \approx 24$ bits. To take advantage of a mantissa's full dynamic range, most implementations of floating-point numbers treat the mantissa as a fractional fixed-

---

[†] For example, the common logarithm (log to the base 10) of 256 is 2.4082. The 2 to the left of the decimal point is called the characteristic of the logarithm and the 4082 digits are called the mantissa. The 2 in 2.4082 does not mean that we multiply .4082 by 10^2. The 2 means that we take the antilog of .4082 to get 2.56 and multiply that by 10^2 to get 256.

point binary number, shift the mantissa bits to the right or left, so that its most significant bit is a one, and adjust the exponent accordingly. This convention is called *normalization*. When normalized, the mantissa bits are typically called the *fraction* of the floating-point number, instead of the mantissa. For example, the decimal value 3.6875_{10} can be represented by the fractional binary number 11.1011_2. If we use a two-bit exponent with a six-bit fraction floating-point word, we can just as well represent 11.1011_2 by shifting it to the right two places and setting the exponent to two as

$$11.1011_2 \ = \ \boxed{1\,0\ \vert\ _\lozenge 1\,1\,1\,0\,1\,1}\ .$$ (9-28)

exponent   fraction

binary point

The floating-point word above can be evaluated to retrieve our decimal number again as

$$[_\lozenge(1\cdot 2^{-1})+(1\cdot 2^{-2})+(1\cdot 2^{-3})+(0\cdot 2^{-4})+(1\cdot 2^{-5})+(1\cdot 2^{-6})]\cdot 2^2$$

$$=[_\lozenge(1\cdot\frac{1}{2})+(1\cdot\frac{1}{4})+(1\cdot\frac{1}{8})+(0\cdot\frac{1}{16})+(1\cdot\frac{1}{32})+(1\cdot\frac{1}{64})]\cdot 2^2$$

$$=[0.5+0.25+0.125+0.0625+0.03125+0.015625]\cdot 2^2$$

$$=0.921875\cdot 4 = 3.6875\ .$$ (9-29)

After some experience using floating-point normalization, users soon realized that always having a one in the most significant bit of the fraction was wasteful. That redundant one was taking up a single bit position in all data words and serving no purpose. So practical implementations of floating-point formats discard that one, assume its existence, and increase the useful number of fraction bits by one. This is why the term *hidden bit* is used to describe some floating-point formats. While increasing the fraction's precision, this scheme uses less memory because the hidden bit is merely accounted for in the hardware arithmetic logic. Using a hidden bit, the fraction in Eq. (9-28)'s floating point number is shifted to the left one place and would now be

exponent   fraction
↓   ↓

$$11.1011_2 \; = \; \boxed{1\,0 \; | \; _\diamond 1\,1\,0\,1\,1\,0} \; . \qquad\qquad (9\text{-}30)$$

↑
binary point

Recall that the exponent and mantissa bits were fixed-point signed binary numbers, and we've discussed several formats for representing signed binary numbers, i.e., sign magnitude, two's complement, and offset binary. As it turns out, all three signed binary formats are used in industry-standard floating-point formats. The most common floating-point formats, all using 32-bit words, are listed in Table 9-3.

Table 9-3 Floating–Point Number Formats

IEEE Standard P754 Format

Bit	31	30	29	28	27	26	25	24	23	22	21	20	⋯	2	1	0
	S	2^7	2^6	2^5	2^4	2^3	2^2	2^1	2^0	2^{-1}	2^{-2}	2^{-3}	⋯	2^{-21}	2^{-22}	2^{-23}

Sign (s) | ← Exponent (e) → | ← Fraction (f) →

IBM Format

Bit	31	30	29	28	27	26	25	24	23	22	21	20	⋯	2	1	0
	S	2^6	2^5	2^4	2^3	2^2	2^1	2^0	2^{-1}	2^{-2}	2^{-3}	2^{-4}	⋯	2^{-22}	2^{-23}	2^{-24}

Sign (s) | ← Exponent (e) → | ← Fraction (f) →

DEC (Digital Equipment Corp.) Format

Bit	31	30	29	28	27	26	25	24	23	22	21	20	⋯	2	1	0
	S	2^7	2^6	2^5	2^4	2^3	2^2	2^1	2^0	2^{-2}	2^{-3}	2^{-4}	⋯	2^{-22}	2^{-23}	2^{-24}

Sign (s) | ← Exponent (e) → | ← Fraction (f) →

MIL–STD 1750A Format

Bit	31	30	29	⋯	11	10	9	8	7	6	5	4	3	2	1	0
	2^0	2^{-1}	2^{-2}	⋯	2^{-20}	2^{-21}	2^{-22}	2^{-23}	2^7	2^6	2^5	2^4	2^3	2^2	2^1	2^0

← Fraction (f) → | ← Exponent (e) →

The IEEE P754 floating-point format is the most popular because so many manufacturers of floating-point integrated circuits comply with this standard [8, 20–22]. Its exponent e is offset binary (biased exponent), and its fraction is a sign-magnitude binary number with a hidden bit that's assumed to be 2^0. The decimal value of a normalized IEEE P754 floating-point number is evaluated as

$$\text{value}_{\text{IEEE}} = (-1)^s \cdot 1_{\Diamond}f \cdot 2^{e-127}. \tag{9-31}$$

$$\uparrow$$
$$\textbf{hidden bit}$$

The IBM floating point format differs somewhat from the other floating-point formats because it uses a base of 16 rather than 2. Its exponent is offset binary, and its fraction is sign magnitude with no hidden bit. The decimal value of a normalized IBM floating-point number is evaluated as

$$\text{value}_{\text{IBM}} = (-1)^s \cdot 0_{\Diamond}f \cdot 16^{e-64}. \tag{9-32}$$

The DEC floating-point format uses an offset binary exponent, and its fraction is sign magnitude with a hidden bit that's assumed to be 2^{-1}. The decimal value of a normalized DEC floating-point number is evaluated as

$$\text{value}_{\text{DEC}} = (-1)^s \cdot 0_{\Diamond}1f \cdot 2^{e-128}. \tag{9-33}$$

$$\uparrow$$
$$\textbf{hidden bit}$$

MIL-STD 1750A is a United States Military Airborne floating-point standard. Its exponent e is a two's complement binary number residing in the least significant eight bits. MIL-STD 1750A's fraction is also a two's complement number (with no hidden bit), and that's why no sign bit is specifically indicated in Table 9-3. The decimal value of a MIL-STD 1750A floating-point number is evaluated as

$$\text{value}_{1750A} = f \cdot 2^e. \tag{9-34}$$

Notice how the floating-point formats in Table 9-3 all have word lengths of 32 bits—this was not accidental. Using 32-bit words makes these formats easier to handle using 8-, 16-, and 32-bit hardware processors. That fact not withstanding and given the advantages afforded by floating-point number formats, these formats do require a significant amount of logical comparisons and branching to correctly perform arithmetic operations. Reference [23] provides useful flow charts showing what procedural steps must be taken when floating-point numbers are added and multiplied.

9.4.1 Floating-Point Dynamic Range

Attempting to determine the dynamic range of an arbitrary floating-point number format is a challenging exercise. We start by repeating the expression for a number system's dynamic range from Eq. (9-6) as

$$
\text{dynamic range}_{dB} = 20 \cdot \log_{10}\left(\frac{\text{largest possible word value}}{\text{smallest possible word value}}\right). \tag{9-35}
$$

When we attempt to determine the largest and smallest possible values for a floating-point number format, we quickly see that they depend on such factors as

- the position of the binary point

- whether a hidden bit is used or not (If used, its position relative to the binary point is important.)

- the base value of the floating-point number format

- the signed binary format used for the exponent and the fraction (For example, recall from Table 9-2 that the binary two's complement format can represent larger negative numbers than the sign-magnitude format.)

- how unnormalized fractions are handled, if at all. (Unnormalized, also called gradual underflow, means a nonzero number that's less than the minimum normalized format but can still be represented when the exponent and hidden bit are both zero.)

- how exponents are handled when they're either all ones or all zeros. (For example, the IEEE P754 format treats a number having an all ones exponent and a nonzero fraction as an invalid number, whereas the DEC format handles a number having a sign = 1 and a zero exponent as a special instruction instead of a valid number.)

Trying to develop a dynamic range expression that accounts for all the possible combinations of the above factors is impractical. What we can do is derive a rule of thumb expression for dynamic range that's often used in practice[8,22,24].

Let's assume the following for our derivation: the exponent is a b_e-bit offset binary number, the fraction is a normalized sign-magnitude number having a sign bit and b_m magnitude bits, and a hidden bit is used just left of the binary point. Our hypothetical floating-point word takes the following form:

Bit	b_m+b_e-1	b_m+b_e-2	$\cdots$	b_m+2	b_m	b_m-1	b_m-2	$\cdots$	1	0
S	2^{be-1}	2^{be-2}	$\cdots$	2^1	2^0	2^{-1}	2^{-2}	$\cdots$	2^{-bm+1}	2^{-bm}

Sign (s) $\leftarrow$ Exponent (e) $\rightarrow$ $\leftarrow$ Fraction (f) $\rightarrow$

First we'll determine what the largest value can be for our floating-point word. The largest fraction is a one in the hidden bit, and the remaining b_m fraction bits are all ones. This would make fraction $f = [1 + (1 - 2^{-b_m})]$. The first 1 in this expression is the hidden bit to the left of the binary point, and the value in parenthesis is all b_m bits equal to ones to the right of the binary point. The greatest positive value we can have for the b_e-bit offset binary exponent is $2^{(2^{b_e-1}-1)}$. So the largest value that can be represented with the floating-point number is the largest fraction raised to the largest positive exponent or

$$\text{largest possible word value} = [1 + (1 - 2^{-b_m})] \cdot 2^{(2^{b_e-1}-1)} \ . \tag{9-36}$$

The smallest value we can represent with our floating-point word is a one in the hidden bit times two raised to the exponent's most negative value, $2^{-(2^{b_e-1})}$, or

$$\text{smallest possible word value} = 1 \cdot 2^{-(2^{b_e-1})} \ . \tag{9-37}$$

Plugging Eqs. (9-36) and (9-37) into Eq. (9-35),

$$\text{dynamic range}_{dB} = 20 \cdot \log_{10}\left(\frac{[1 + (1 - 2^{-b_m})] \cdot 2^{(2^{b_e-1}-1)}}{1 \cdot 2^{-(2^{b_e-1})}}\right) \ . \tag{9-38}$$

Now here's where the thumb comes in—when b_m is large, say over seven, the 2^{-b_m} value approaches zero; that is, as b_m increases, the all ones fraction

$(1 - 2^{-b_m})$ value in the numerator approaches 1. Assuming this, Eq. (9-38) becomes

$$\text{dynamic range}_{dB} \approx 20 \cdot \log_{10}\left(\frac{[1+1] \cdot 2^{(2^{b_e-1}-1)}}{1 \cdot 2^{-(2^{b_e-1})}}\right)$$

$$= 20 \cdot \log_{10}\left(\frac{2 \cdot 2^{(2^{b_e-1}-1)}}{2^{-(2^{b_e-1})}}\right) = 20 \cdot \log_{10}\left(\frac{2^{(2^{b_e-1})}}{2^{-(2^{b_e-1})}}\right)$$

$$= 20 \cdot \log_{10}(2 \cdot 2^{(2^{b_e-1})}) = 20 \cdot \log_{10}(2^{(2^{b_e})}) = 6.02 \cdot 2^{b_e} . \quad (9\text{-}39)$$

Using Eq. (9-39) we can estimate, for example, the dynamic range of the single-precision IEEE P754 standard floating-point format with its eight-bit exponent:

$$\text{dynamic range}_{\text{IEEE P754}} = 6.02 \cdot 2^8 = 1529 \text{ dB.} \quad (9\text{-}40)$$

Although we've introduced the major features of the most common floating-point formats, there are still more details to learn about floating-point numbers. For the interested reader the references given in this section provide a good place to start.

9.5 Block Floating-Point Binary Format

A marriage of fixed-point and floating-point binary formats is known as *block floating point*. This scheme is used, particularly in dedicated FFT integrated circuits, when large arrays, or blocks, of associated data are to be manipulated mathematically. Block floating-point schemes begin by examining all the words in a block of data, normalizing the largest-valued word's fraction, and establishing the correct exponent. This normalization takes advantage of the fraction's full dynamic range. Next, the fractions of the remaining data words are shifted appropriately, so that they can use the exponent of the largest word. In this way, all of the data words use the same exponent value to conserve hardware memory.

In FFT implementations, the arithmetic is performed treating the block normalized data values as fixed-point binary. However, when an addition causes an overflow condition, all of the data words are shifted one bit to the right (division by two), and the exponent is incremented by one. As the reader may have guessed, block floating-point formats have increased dynamic range and avoid the overflow problems inherent in fixed-point formats but do not reach the performance level of true floating-point formats[8,25,26].

References

[1] Neugebauer, O. "The History of Ancient Astronomy," *Journal of Near Eastern Studies*, Vol. 4, 1945, pp. 12.

[2] Knuth, D. E. *The Art of Computer Programming: Seminumerical Methods*, Vol. 2, Section 4.1, Addison-Wesley Publishing, Reading, Massachusetts, 1981, pp. 179.

[3] Kester, W. "Peripheral Circuits Can Make or Break Sampling-ADC Systems," *EDN Magazine*, October 1, 1992.

[4] Grove, M. "Measuring Frequency Response and Effective Bits Using Digital Signal Processing Techniques," *Hewlett-Packard Journal*, February 1992.

[5] Tektronix. "Effective Bits Testing Evaluates Dynamic Range Performance of Digitizing Instruments," *Tektronix Application Note*, No. 45W-7527, December 1989.

[6] Ushani, R. "Subranging ADCs Operate at High Speed with High Resolution," *EDN Magazine*, April 11, 1991.

[7] Demler, M. "Time-domain Techniques Enhance Testing of High-speed ADCs," *EDN Magazine*, March 30, 1992.

[8] Hilton, H. "A 10-MHz Analog-to-Digital Converter with 110-dB Linearity," *Hewlett-Packard Journal*, October 1993.

[9] Lyons, R. G. "Providing Software Flexibility for Optical Processor Noise Analysis," *Computer Design*, July 1978, pp. 95.

[10] Knuth, D. E. *The Art of Computer Programming: Seminumerical Methods*, Vol. 2, Section 4.2, Addison-Wesley Publishing, Reading, Massachusetts, 1981, pp. 198.

[11] Rabiner, L. R., and Gold, B. *Theory and Application of Digital Signal Processing*, Chapter 5, Prentice-Hall, Englewood Cliffs, New Jersey, 1975, pp. 353.

[12] Jackson, L. B. "An Analysis of Limit Cycles Due to Multiplicative Rounding in Recursive Digital Filters," *Proc. 7th Allerton Conf. Circuit System Theory*, 1969, pp. 69–78.

[13] Kan, E. P. F, and Aggarwal, J. K. "Error Analysis of Digital Filters Employing Floating Point Arithmetic," *IEEE Trans. Circuit Theory*, Vol. CT-18, November 1971, pp. 678–86.

[14] Crochiere, R. E. "Digital Ladder Structures and Coefficient Sensitivity," *IEEE Trans. Audio Electroacoustics*, Vol. AU-20, October 1972, pp. 240–46.

[15] Jackson, L. B. "On the Interaction of Roundoff Noise and Dynamic Range in Digital Filters," *Bell System Technical Journal*, Vol. 49, February 1970, pp. 159–84.

[16] Roberts, R. A., and Mullis, C. T. *Digital Signal Processing*, Addison-Wesley Publishing, Reading, Massachusetts, 1987, pp. 277.

[17] Jackson, L. B. "Roundoff Noise Analysis for Fixed-Point Digital Filters Realized in Cascade or Parallel Form," *IEEE Trans. Audio Electroacoustics*, Vol. AU-18, June 1970, pp. 107–22.

[18] Oppenheim, A. V., and Schafer, R. W. *Discrete-Time Signal Processing*, Sections 6.8 and 9.8, Prentice-Hall, Englewood Cliffs, New Jersey, 1989, pp. 335.

[19] Larimer, J., and D. Chen. "Fixed or Floating? A Pointed Question in DSPs," *EDN Magazine*, August 3, 1995.

[20] Ashton, C. "Floating Point Math Handles Iterative and Recursive Algorithms," *EDN Magazine*, January 9, 1986.

[21] Windsor, B., and Wilson, J. "Arithmetic Duo Excels in Computing Floating Point Products," *Electronic Design*, May 17, 1984.

[22] Windsor, W. A. "IEEE Floating Point Chips Implement DSP Architectures," *Computer Design*, January 1985.

[23] Texas Instruments Inc., *Digital Signal Processing Applications with the TMS320 Family: Theory, Algorithms, and Implementations*, SPRA012A, Texas Instruments, Dallas, TX, 1986.

[24] Strauss, W. I. "Integer or Floating Point? Making the Choice," *Computer Design Magazine*, April 1, 1990, pp. 85.

[25] Oppenheim and Weinstein. "Effects of Finite Register Length in Digital Filtering and the Fast Fourier Transform," *Proc. IEEE*, August 1972, pp. 957–76.

[26] Woods, R. E. "Transform-based Processing: "How Much Precision Is Needed," *ESD: The Electronic System Design Magazine*, February 1987.

CHAPTER TEN

Digital Signal Processing Tricks

As we study the literature of digital signal processing, we'll encounter some creative techniques that *professionals* use to make their algorithms more efficient. These techniques are straightforward examples of the philosophy "don't work hard, work smart" and studying them will give us a deeper understanding of the underlying mathematical subtleties of digital signal processing. In this chapter, we present a collection of these clever tricks of the trade and explore several of them in detail, because doing so reinforces some of the lessons we've learned in previous chapters.

10.1 Frequency Translation without Multiplication

Frequency translation is often called for in digital signal processing algorithms. A filtering scheme called *transmultiplexing* (using the FFT to efficiently implement a bank of bandpass filters) requires spectral shifting by half the sample rate, or $f_s/2$[1]. Inverting bandpass sampled spectra and converting low-pass FIR filters to highpass filters both call for frequency translation by half the sample rate. Conventional quadrature bandpass sampling uses spectral translation by one quarter of the sample rate, or $f_s/4$, to reduce unnecessary computations[2,3]. There are a couple of tricks used to perform discrete frequency translation, or *mixing*, by $f_s/2$ and $f_s/4$ without actually having to perform any multiplications. Let's take a look at these mixing schemes in detail.

First, we'll consider a slick technique for frequency translating an input sequence by $f_s/2$ by merely multiplying that sequence by $(-1)^n$, or $(-1)^0$, $(-1)^1$, $(-1)^2$, $(-1)^3$, etc. Better yet, this requires only changing the sign of every other input sample value because $(-1)^n = 1, -1, 1, -1$, etc. This process may seem a bit mysterious at first, but it can be explained in a straightforward way if we review Figure 10-1. The figure shows us that

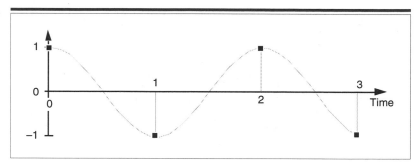

Figure 10-1 Mixing sequence comprising $(-1)^n$; 1, -1, 1, -1, etc.

multiplying a time-domain signal sequence by the $(-1)^n$ mixing sequence is equivalent to multiplying the signal sequence by a sampled cosinusoid where the mixing sequence values are shown as the dots in Figure 10-1. Because the mixing sequence's cosine repeats every two sample values, its frequency is $f_s/2$. Let's look at this situation in detail, not only to understand mixing sequences, but to illustrate the DFT equation's analysis capabilities, to reiterate the nature of complex signals, and to reconfirm the important equivalence of shifting in the time domain and phase shifting in the frequency domain.

We can verify the $(-1)^n$ mixing sequence's frequency translation of $f_s/2$ by taking the DFT of the mixing sequence expressed as $F_{1,-1,1,-1...}(m)$ where

$$F_{1,-1,1,-1,...}(m) = \sum_{n=0}^{N-1}(1,-1,1,-1,...)e^{-j2\pi nm/N} \ . \tag{10-1}$$

Using a 4-point DFT, we expand the sum in Eq. (10-1), with $N = 4$, to

$$F_{1,-1,1,-1}(m) = e^{-j2\pi 0m/4} - e^{-j2\pi 1m/4} + e^{-j2\pi 2m/4} - e^{-j2\pi 3m/4} . \tag{10-2}$$

Notice that the mixing sequence is embedded in the signs of the terms of Eq. (10-2) that we evaluate from $m = 0$ to $m = 3$ to get

$$m = 0: \ F_{1,-1,1,-1}(0) = e^{-j0} - e^{-j0} + e^{-j0} - e^{-j0} = 1 - 1 + 1 - 1 = 0 \ , \tag{10-3}$$

$$m = 1: \ F_{1,-1,1,-1}(1) = e^{-j0} - e^{-j2\pi/4} + e^{-j4\pi/4} - e^{-j6\pi/4} = 1 + j1 - 1 - j1 = 0 \ , \tag{10-4}$$

$$m = 2: \ F_{1,-1,1,-1}(2) = e^{-j0} - e^{-j4\pi/4} + e^{-j8\pi/4} - e^{-j12\pi/4} = 1 + 1 + 1 + 1 = 4 \ , \tag{10-5}$$

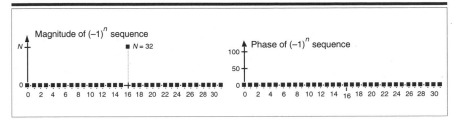

Figure 10-2 Frequency-domain magnitude and phase of an *N*-point $(-1)^n$ sequence.

and

$$m = 3: \quad F_{1,-1,1,-1}(3) = e^{-j0} - e^{-j6\pi/4} + e^{-j12\pi/4} - e^{-j18\pi/4} = 1 - j1 - 1 + j1 = 0 . \quad (10\text{-}6)$$

See how the 1, −1, 1, −1 mixing sequence has a nonzero frequency component only when $m = 2$ corresponding to a frequency of $mf_s/N = 2f_s/4 = f_s/2$. So, in the frequency domain the four-sample 1, −1, 1, −1 mixing sequence is an $f_s/2$ sinusoid with a magnitude of 4 and a phase angle of 0°. Had our mixing sequence contained eight separate values, the results of an 8-point DFT would have been all zeros with the exception of the $m = 4$ frequency sample with its magnitude of 8 and phase angle of 0°. In fact, the DFT of $N (1)^n$ has a magnitude of N at a frequency $f_s/2$. When $N = 32$, for example, the magnitude and phase of a 32-point $(-1)^n$ sequence is shown in Figure 10-2.

Let's demonstrate this $(-1)^n$ mixing with an example. Consider the 32 discrete samples of the sum of 3 sinusoids comprising a real time-domain sequence $x(n)$ shown in Figure 10-3(a) where

$$x(n) = \cos(\frac{2\pi 10 n}{32}) + 0.5 \cdot \cos(\frac{2\pi 11 n}{32} - \frac{\pi}{4}) + 0.25 \cdot \cos(\frac{2\pi 12 n}{32} - \frac{3\pi}{8}) . \quad (10\text{-}7)$$

The frequencies of the three sinusoids are 10/32 Hz, 11/32 Hz, and 12/32 Hz, and they each have an integral number of cycles (10, 11, and 12) over 32 samples of $x(n)$. To show the magnitude and phase shifting effect of using the 1, −1, 1, −1 mixing sequence, we've added arbitrary phase shifts of −π/4 (−45°) and −3π/8 (−67.5°) to the second and third tones. Using a 32-point DFT results in the magnitude and phase of $X(m)$ shown in Figure 10-3(b).

Let's notice something about Figure 10-3(b) before we proceed. The magnitude of the $m = 10$ frequency sample $|X(10)|$ is 16. Remember why this is so? Recall Eq. (3-17) from Chapter 3 where we learned that, if a real input signal contains a sinewave component of peak amplitude A_o with

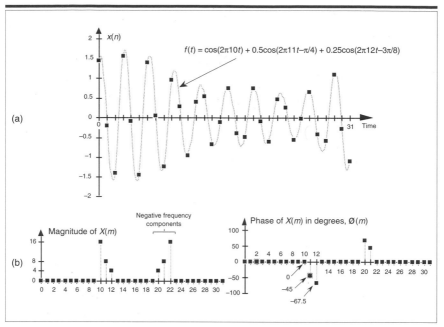

Figure 10-3 Discrete signal sequence $x(n)$: (a) time-domain representation of $x(n)$; (b) frequency-domain magnitude and phase of $X(m)$.

an integral number of cycles over N input samples, the output magnitude of the DFT for that particular sinewave is $A_oN/2$. In our case, then, $|X(10)| = 1 \cdot 32/2 = 16$, $|X(11)| = 8$, and $|X(12)| = 4$.

If we multiply $x(n)$, sample by sample, by the $(-1)^n$ mixing sequence, our new time-domain sequence $x_{1,-1}(n)$ is shown in Figure 10-4(a), and the DFT of the frequency translated $x_{1,-1}$ is provided in Figure 10-4(b). (Remember now, we didn't really perform any explicit multiplications—the whole idea here is to avoid multiplications—we merely changed the sign of alternating $x(n)$ samples to get $x_{1,-1}(n)$.) Notice that the magnitude and phase of $X_{1,-1}(m)$ are the magnitude and phase of $X(m)$ shifted by $f_s/2$, or 16 sample shifts, from Figure 10-3(b) to Figure 10-4(b). The negative frequency components of $X(m)$ are shifted downward in frequency, and the positive frequency components of $X(m)$ are shifted upward in frequency resulting in $X_{1,-1}(m)$. It's a good idea for the reader to be energetic and prove that the magnitude of $X_{1,-1}(m)$ is the convolution of the $(-1)^n$ sequence's spectral magnitude in Figure 10-2 and the magnitude of $X(m)$ in Figure 10-3(b). Another way to look at the $X_{1,-1}(m)$ magnitudes in Figure 10-4(b) is to see that multiplication by the $(-1)^n$ mixing sequence flips the positive frequency band of $X(m)$ from zero to $+f_s/2$ Hz about $f_s/4$ Hz and flips the negative frequency band of $X(m)$ from $-f_s/2$ to zero Hz, about

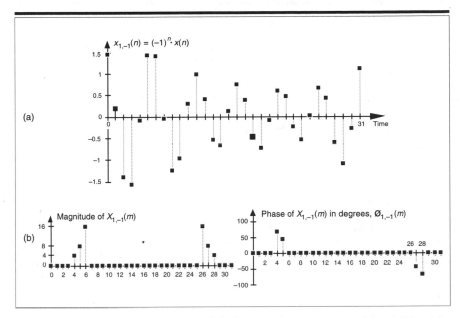

Figure 10-4 Frequency translation by $f_s/2$: (a) mixed sequence $x_{1,-1}(n) = (-1)^n \cdot x(n)$; (b) magnitude and phase of frequency-translated $X_{1,-1}(m)$.

$-f_s/4$ Hz. This process can be used to invert spectra when bandpass sampling is used, as described in Section 2.4.

Another useful mixing sequence is 1,–1,–1,1, etc. It's used to translate spectra by $f_s/4$ in quadrature sampling schemes and is illustrated in Figure 10-5(a). In digital quadrature mixing, we can multiply an input data sequence $x(n)$ by the cosine mixing sequence 1,–1,–1,1 to get the in-phase component of $x(n)$—what we'll call $i(n)$. To get the quadrature-phase product sequence $q(n)$, we multiply the original input data sequence by the sine mixing sequence of 1,1,–1,–1. This sine mixing sequence is out of phase with the cosine mixing sequence by 90°, as shown in Figure 10-5(b).

If we multiply the 1,–1,–1,1 cosine mixing sequence and an input sequence $x(n)$, we'll find that the $i(n)$ product has a DFT magnitude that's related to the input's DFT magnitude $X(n)$ by

$$|I(m)|_{1,-1,-1,1} = \frac{|X(m)|}{\sqrt{2}} . \tag{10-8}$$

To see why, let's explore the cosine mixing sequence 1,–1,–1,1 in the frequency domain. We know that the DFT of the cosine mixing sequence, represented by $F_{1,-1,-1,1}(m)$, is expressed by

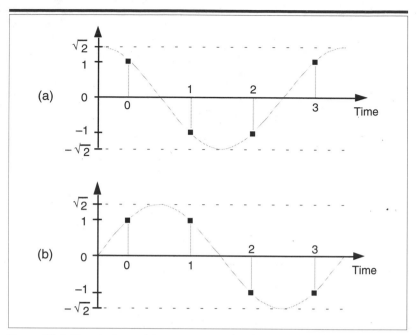

Figure 10-5 Quadrature mixing sequences for downconversion by $f_s/4$:
(a) cosine mixing sequence using 1,–1,–1,1,....; (b) sine mixing
sequence using 1,1,–1,–1,....

$$F_{1,-1,-1,1}(m) = \sum_{n=0}^{N-1}(1,-1,-1,1, \ldots)e^{-j2\pi nm/N} \quad . \tag{10-9}$$

Because a 4-point DFT is sufficient to evaluate Eq. (10-9), with $N = 4$,

$$F_{1,-1,-1,1}(m) = e^{-j2\pi 0m/4} - e^{-j2\pi 1m/4} - e^{-j2\pi 2m/4} + e^{-j2\pi 3m/4} \quad . \tag{10-10}$$

Notice, again, that the cosine mixing sequence is embedded in the signs of the terms of Eq. (10-10). Evaluating Eq. (10-10) for $m = 1$, corresponding to a frequency of $1 \cdot f_s/N$, or $f_s/4$, we find

$$m = 1: \quad F_{1,-1,-1,1}(1) = e^{-j0} - e^{-j\pi/2} - e^{-j\pi} + e^{-j3\pi/2}$$

$$= 1 + j1 + 1 + j1 = 2 + j2 = \frac{4}{\sqrt{2}}\angle 45°. \tag{10-11}$$

So, in the frequency domain, the cosine mixing sequence has an $f_s/4$ magnitude of $4/\sqrt{2}$ at a phase angle of $45°$. Similarly, evaluating Eq. (10-10) for $m = 3$, corresponding to a frequency of $-f_s/4$, we find

$$m = 3: \quad F_{1,-1,-1,1}(3) = e^{-j0} - e^{-j3\pi/2} - e^{-j3\pi} + e^{-j9\pi/2}$$

$$= 1 - j1 + 1 - j1 = 2 - j2 = \frac{4}{\sqrt{2}} - \angle - 45°. \qquad (10\text{-}12)$$

The energetic reader should evaluate Eq. (10-10) for $m = 0$ and $m = 2$, to confirm that the 1,–1,–1,1 sequence's DFT coefficients are zero for the frequencies of 0 and $f_s/2$.

Because the 4-point DFT magnitude of an all positive ones mixing sequence (1, 1, 1, 1) is $4^\dagger$, we see that the frequency-domain scale factor for the 1,–1,–1,1 cosine mixing sequence is expressed as

$$I(m)_{1,-1,-1,1} \text{ scale factor} = \frac{\text{cosine sequence DFT magnitude}}{\text{all ones sequence DFT magnitude}}$$

$$= \frac{4/\sqrt{2}}{4} = \frac{1}{\sqrt{2}}, \qquad (10\text{-}13)$$

which confirms the relationship in Eq. (10-8) and Figure 10-5(a). Likewise, the DFT scale factor for the quadrature-phase mixing sequence (1,1,–1,–1) is

$$Q(m)_{1,1,-1,-1} \text{ scale factor} = \frac{1}{\sqrt{2}},$$

thus

$$|Q(m)|_{1,1,-1,-1} = \frac{|X(m)|}{\sqrt{2}}. \qquad (10\text{-}14)$$

So what this all means is that an input signal's spectral magnitude, after frequency translation, will be reduced by a factor of $\sqrt{2}$. There's really no harm done, however—when the in-phase and quadrature-phase components are combined to find the magnitude of a complex frequency

---

† We can show this by letting $K = N = 4$ in Eq. (3-44) for a four-sample all ones sequence in Chapter 3.

sample $X(m)$, the $\sqrt{2}$ scale factor is eliminated, and there's no overall magnitude loss because

$$|\text{scale factor}| = \sqrt{(I(m) \text{ scale factor})^2 + (Q(m) \text{ scale factor})^2}$$

$$= \sqrt{(1/\sqrt{2})^2 + (1/\sqrt{2})^2} = \sqrt{(1/2) + (1/2)} = 1. \qquad (10\text{-}15)$$

We can demonstrate this quadrature mixing process using the $x(n)$ sequence from Eq. (10-7) whose spectrum is shown in Figure 10-3(b). If we multiply that 32-sample $x(n)$ by 32 samples of the quadrature mixing sequences 1,–1,–1,1 and 1,1,–1,–1, whose DFT magnitudes are shown in Figure 10-6(a) and (b), the new quadrature sequences will have the

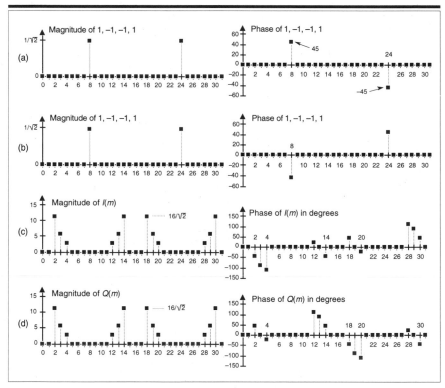

Figure 10-6 Frequency translation by $f_s/4$: (a) normalized magnitude and phase of cosine 1,-1,-1,1 sequence; (b) normalized magnitude and phase of sine 1,1,-1,-1 sequence; (c) magnitude and phase of frequency-translated, in-phase $I(m)$; (d) magnitude and phase of frequency-translated, quadrature-phase $Q(m)$.

frequency-translated $I(m)$ and $Q(m)$ spectra shown in Figure 10-6(c) and (d). (Remember now, we don't actually perform any multiplications; we merely change the sign of appropriate $x(n)$ samples to get the $i(n)$ and $q(n)$ sequences.)

There's a lot to learn from Figure 10-6. First, the positive frequency components of $X(m)$ from Figure 10-3(b) are indeed shifted downward by $f_s/4$ in Figure 10-6(c). Because our total discrete frequency span (f_s Hz) is divided into 32 samples, $f_s/4$ is equal to eight. So, for example, the $X(10)$ component in Figure 10-3(b) corresponds to the $I(10-8) = I(2)$ component in Figure 10-6(c). Likewise, $X(11)$ corresponds to $I(11-8) = I(3)$, and so on. Notice, however, that the positive and negative components of $X(m)$ have each been repeated twice in the frequency span in Figure 10-6(c). This effect is inherent in the process of mixing any discrete time-domain signal with a sinusoid of frequency $f_s/4$. Verifying this gives us a good opportunity to pull convolution out of our toolbox and use it to see why the $I(m)$ spectral replication period is reduced by a factor of 2 from that of $X(m)$. Recall, from the convolution theorem, that the DFT of the time-domain product of $x(n)$ and the $1,-1,-1,1$ mixing sequence $I(m)$ is the convolution of their individual DFTs, or $I(m)$ is equal to the convolution of $X(m)$ and the $1,-1,-1,1$ mixing sequence's magnitude spectrum in Figure 10-6(a). If, for convenience, we denote the $1,-1,-1,1$ cosine mixing sequence's magnitude spectrum as $S_c(m)$, we can say that $I(m) = X(m)*S_c(m)$ where the "$*$" symbol denotes convolution.

Let's look at that particular convolution to make sure we get the $I(m)$ spectrum in Figure 10-6(c). Redrawing $X(m)$ from Figure 10-3(b) to show its positive and negative frequency replications gives us Figure 10-7(a). We also redraw $S_c(m)$ from Figure 10-6(a) showing its positive and negative frequency components in Figure 10-7(b). Before we perform the convolution's shift and multiply, we realize that we don't have to flip $S_c(m)$ about the zero frequency axis because, due to its symmetry, that would have no effect. So now our convolution comprises the shifting of $S_c(m)$ in Figure 10-7(b), relative to the stationary $X(m)$, and taking the product of that shifted sequence and the $X(m)$ spectrum in Figure 10-7(a) to arrive at $I(m)$. No shift of $S_c(m)$ corresponds to the $m = 0$ sample of $I(m)$. The sums of the products for this zero shift is zero, so $I(0) = 0$. If we shift $S_c(m)$ to the right by two samples, we'd have an overlap of $S_c(8)$ and $X(10)$, and that product gives us $I(2)$. One more $S_c(m)$ shift to the right results in an overlap of $S_c(8)$ and $X(11)$, and that product gives us $I(3)$, and so on. So shifting $S_c(m)$ to the right and summing the products of $S_c(m)$ and $X(m)$ results in $I(1)$ to $I(14)$. If we return $S_c(m)$ to its unshifted position in Figure 10-7(b), and then shift it to the left two samples in the negative frequency

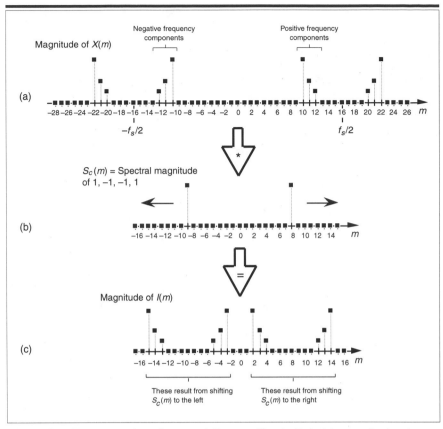

Figure 10-7 Frequency-domain convolution resulting in *I*(m): (a) magnitude of
X(m); (b) spectral magnitude of the cosine's 1,-1,-1,1 time-domain
sequence, S_c(m); this is the sequence we'll shift to the left and right to
perform the convolution; (c) convolution result: the magnitude of
frequency-translated, in-phase *I*(m).

direction, we'd have an overlap of S_c(–8) and *X*(–10), and that product
gives us *I*(–2). One more S_c(m) shift to the left results in an overlap of
S_c(–8) and *X*(–11), and that product gives us *I*(–3), and so on. Continuing
to shift S_c(m) to the left determines the remaining negative frequency
components *I*(–4) to *I*(–14). Figure 10-7(c) shows which *I*(m) samples
resulted from the left and right shifting of S_c(m). By using the convolution theorem, we can see, now, that the magnitudes in Figure 10-7(c) and
Figure 10-6(c) really are the spectral magnitudes of the in-phase component *I*(m) with its reduced spectral replication period.

The upshot of all of this is that we can change the signs of appropriate
x(n) samples to shift *x*(n)'s spectrum by one quarter of the sample rate

without having to perform any explicit multiplications. Moreover, if we change the signs of appropriate $x(n)$ samples in accordance with the mixing sequences in Figure 10-5, we can get the in-phase $i(n)$ and quadrature-phase $q(n)$ components of the original $x(n)$. One important effect of this digital mixing by $f_s/4$ is that the spectral replication periods of $I(m)$ and $Q(m)$ are half the replication period of the original $X(m)$.[†] So we must be aware of the potential frequency aliasing problems that may occur with this frequency-translation method if the signal bandwidth is too wide relative to the sample rate, as discussed in Section 7.3.

Before we leave this particular frequency-translation scheme, let's review two more issues, magnitude and phase. Notice that the untranslated $X(10)$ magnitude is equal to 16 in Figure 10-3(b), and that the translated $I(2)$ and $Q(2)$ magnitudes are $16/\sqrt{2} = 11.314$ in Figure 10-6. This validates Eq. (10-8) and Eq. (10-14). If we use those quadrature components $I(2)$ and $Q(2)$ to determine the magnitude of the corresponding frequency-translated, complex spectral component from the square root of the sum of the squares relationship, we'd find that the magnitude of the peak spectral component is

$$\text{peak component magnitude} = \sqrt{(16/\sqrt{2})^2 + (16/\sqrt{2})^2} - \sqrt{256} = 16, \quad (10\text{-}16)$$

verifying Eq. (10-15). So combining the quadrature components $I(m)$ and $Q(m)$ does not result in any loss in spectral amplitude due to the frequency translation. Finally, in performing the above convolution process, the phase angle samples of $X(m)$ in Figure 10-3(b) and the phase samples of the 1,–1,–1,1 sequence in Figure 10-6(a) add algebraically. So the resultant $I(m)$ phase angle samples in Figure 10-6(c) result from either adding or subtracting 45° from the phase samples of $X(m)$ in Figure 10-3(b).

Another easily implemented mixing sequence used for $f_s/4$ frequency translations to obtain $I(m)$ is the 1, 0, –1, 0, etc., cosine sequence shown in Figure 10-8(a). This mixing sequence's quadrature companion 0, 1, 0, –1, Figure 10-8(b), is used to produce $Q(m)$. To determine the spectra of these sequences, let's, again, use a 4-point DFT to state that

$$F_{1,0,-1,0}(m) = \sum_{n=0}^{N-1} (1,0,-1,0,...)e^{-j2\pi nm/N} \quad . \quad (10\text{-}17)$$

[†] Recall that we saw this reduction in spectral replication period in the quadrature sampling results shown in Figures 7-2(g) and 7-3(d).

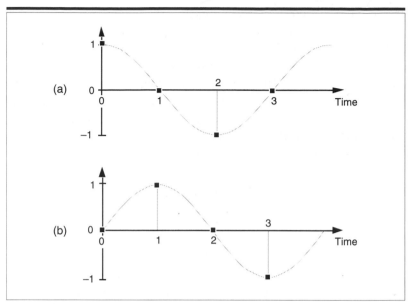

Figure 10-8 Quadrature mixing sequences for downconversion by $f_s/4$: (a) cosine mixing sequence using 1, 0, –1, 0, . . .; (b) sine mixing sequence using 0, 1, 0, –1, . . .

When $N = 4$,

$$F_{1,0,-1,0}(m) = e^{-j2\pi 0m/4} - e^{-j2\pi 2m/4} \quad . \tag{10-18}$$

Again, the cosine mixing sequence is embedded in the signs of the terms of Eq. (10-18), and there are only two terms for our 4-point DFT. We evaluate Eq. (10-18) for $m = 1$, corresponding to a frequency of $f_s/4$, to find that

$$F_{1,0,-1,0}(1) = e^{-j0} - e^{-j\pi} = 1 + 1 = 2\angle 0° \quad . \tag{10-19}$$

Evaluating Eq. (10-18) for $m = 3$, corresponding to a frequency of $-f_s/4$, shows that

$$F_{1,0,-1,0}(3) = e^{-j0} - e^{-j3\pi} = 1 + 1 = 2\angle 0° \quad . \tag{10-20}$$

Using the approach in Eq. (10-13), we can show that the scaling factor for the 1, 0, –1, 0 cosine mixing sequence is given as

$$I(m)_{1,0,-1,0} \text{ scale factor} = \frac{2}{4} = \frac{1}{2} .$$

So

$$|I(m)|_{1,0,-1,0} = \frac{|X(m)|}{2} . \tag{10-21}$$

Likewise, if we went through the same exercise as above, we'd find that the scaling factor for the 0, 1, 0, −1 sine mixing sequence is given by

$$Q(m)_{0,1,0,-1} \text{ scale factor} = \frac{1}{2} .$$

So

$$|Q(m)|_{0,1,0,-1} = \frac{|X(m)|}{2} . \tag{10-22}$$

So these mixing sequences induce a loss in the frequency-translated signal amplitude by a factor of 2.

By way of example, let's show this scale factor loss again by frequency translating the $x(n)$ sequence from Eq. (10-7), whose spectrum is shown in Figure 10-3(b). If we multiply that 32-sample $x(n)$ by 32 samples of the quadrature mixing sequences 1, 0, −1, 0 and 0, 1, 0, −1, whose DFT magnitudes are shown in Figure 10-9(a) and (b), the resulting quadrature sequences will have the frequency-translated $I(m)$ and $Q(m)$ spectra shown in Figure 10-9(c) and (d).

Notice that the untranslated $X(10)$ magnitude is equal to 16 in Figure 10-3(b) and that the translated $I(2)$ and $Q(2)$ magnitudes are $16/2 = 8$ in Figure 10-6. This validates Eq. (10-21) and Eq. (10-22). If we use those quadrature components $I(2)$ and $Q(2)$ to determine the magnitude of the corresponding frequency-translated, complex spectral component from the square root of the sum of the squares relationship, we'd find that the magnitude of the peak spectral component is

$$\text{peak component magnitude} = \sqrt{(16/2)^2 + (16/2)^2} = \sqrt{(16)^2/2} = \frac{16}{\sqrt{2}} . \tag{10-23}$$

When the in-phase and quadrature-phase components are combined to get the magnitude of a complex value, a resultant $\sqrt{2}$ scale factor, for the 1, 0, −1, 0 and 0, 1, 0, −1 sequences, is not eliminated. An overall 3 dB loss remains because we eliminated some of the signal power when we multiplied half of the data samples by zero.

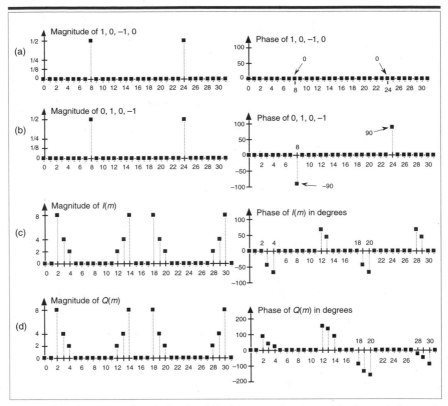

Figure 10-9 Frequency translation by $f_s/4$: (a) normalized magnitude and phase of cosine 1, 0, –1, 0 sequence; (b) normalized magnitude and phase of sine 0, 1, 0, –1 sequence; (c) magnitude and phase of frequency-translated in-phase $I(m)$; (d) magnitude and phase of frequency-translated quadrature-phase $Q(m)$.

The question is "Why would the sequences 1, 0, –1, 0 and 0, 1, 0, –1 ever be used if they induce a signal amplitude loss in $i(n)$ and $q(n)$?" The answer is that the alternating zero-valued samples reduce the amount of follow-on processing performed on $i(n)$ and $q(n)$. Let's say, for example, that an application requires both $i(n)$ and $q(n)$ to be low-pass filtered. When alternating samples of $i(n)$ and $q(n)$ are zeros, the digital filters have only half as many multiplications to perform because multiplications by zero are unnecessary.

Another way to look at this situation is that $i(n)$ and $q(n)$, in a sense, have been decimated by a factor of 2, and the necessary follow-on processing rates (operations/second) are also being reduced by a factor of 2. If $i(n)$ and $q(n)$ are, indeed, applied to two separate FIR digital filters, we can be clever and embed the mixing sequences' plus and minus ones and zeros into the

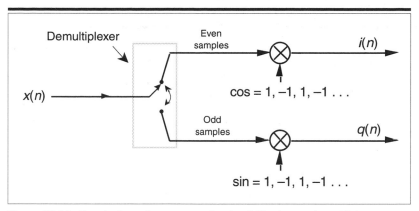

Figure 10-10 Quadrature downconversion by $f_s/4$ using a demultiplexer
(demux) and the sequence 1, -1, 1, -1, ...

filters' coefficient values and avoid actually performing any multiplications. Because some coefficients are zero, they need not be used at all, and
the number of actual multipliers used can be reduced. In that way, we'll
have performed quadrature mixing and FIR filtering in the same process
with a simpler filter. This technique also forces us to be aware of the potential frequency aliasing problems that may occur if the input signal is not
sufficiently bandwidth limited relative to the original sample rate.

Figure 10-10 illustrates an interesting hybrid technique using the $f_s/2$
mixing sequence (1, –1, 1, –1) to perform quadrature mixing and downconversion by $f_s/4$. This scheme uses a demultiplexing process of routing
alternate input samples to one of the two mixer paths[3–6]. Although both
digital mixers use the same mixing sequence, this process is equivalent to
multiplying the input by the two quadrature mixing sequences shown in
Figure 10-8(a) and 10-8(b) with their frequency-domain magnitudes indicated in Figure 10-9(a) and 10-9(b). That's because alternate samples are
routed to the two mixers. Although this scheme can be used for the quadrature sampling and demodulation described in Section 7.2, interpolation
filters must be used to remove the inherent half sample time delay
between $i(n)$ and $q(n)$ caused by using the single mixing sequence of
1,–1,1,–1.

Table 10-1 summarizes the effect of multiplying time-domain signal samples by various digital mixing sequences of ones, zeros, and minus ones.

A "yes" in the last column in Table 10-1 indicates that alternating samples of $i(n)$ and $q(n)$ can be discarded with no adverse impact on their
spectra, allowing a follow-on processing data rate that's half the original
$x(n)$ data rate.

Table 10-1 Digital Mixing Sequences

In-phase sequence	Quadrature sequence	Frequency translation by	Scale factor	Final signal power loss	Decimation can occur
1, –1, 1, –1, . . .	–	$f_s/2$	1	0 dB	no
1, –1, –1, 1, . . .	1, 1, -1, –1, . . .	$f_s/4$	$1/\sqrt{2}$	0 dB	yes
1, 0, –1, 0, . . .	0, 1, 0, –1, . . .	$f_s/4$	1/2	3 dB	yes
1, –1, 1, –1, . . . (with demux)	–1, 1, –1, 1, . . . (with demux)	$f_s/4$	1/2	3 dB	no

10.2 High-Speed Vector-Magnitude Approximation

The quadrature processing techniques employed in spectrum analysis, computer graphics, and digital communications routinely require high-speed determination of the magnitude of a complex vector V given its real and imaginary parts; i.e., the in-phase part I and the quadrature-phase part Q[4]. This magnitude calculation requires a square root operation because the magnitude of V is

$$|V| = \sqrt{I^2 + Q^2} \ . \tag{10-24}$$

Assuming that the sum $I^2 + Q^2$ is available, the problem is to efficiently perform the square root operation.

There are several ways to obtain square roots, but the optimum technique depends on the capabilities of the available hardware and software. For example, when performing a square root using a high-level software language, we employ whatever software square root function is available. Although accurate, software routines can be very slow. In contrast, if a system must accomplish a square root operation in 50 nanoseconds, high-speed magnitude approximations are required[7,8]. Let's look at a neat magnitude approximation scheme that's particularly efficient.

10.2.1 αMax+βMin Algorithm

There is a technique called the αMax+βMin (read as "alpha max plus beta min") algorithm for calculating the magnitude of a complex vector.[†] It's a

---

[†] A "Max+βMin" algorithm had been in use, but in 1988 this author suggested expanding it to the αMax+βMin form, where α could be a value other than unity[9].

linear approximation to the vector-magnitude problem that requires determining which orthogonal vector, I or Q, has the greater absolute value. If the maximum absolute value of I or Q is designated by Max and the minimum absolute value of either I or Q is Min, an approximation of $|V|$, using the αMax+βMin algorithm, is expressed as

$$|V| \approx \alpha\text{Max} + \beta\text{Min} . \qquad (10\text{-}25)$$

There are several pairs for the α and β constants that provide varying degrees of vector-magnitude approximation accuracy to within 0.1dB[7,10]. The αMax+βMin algorithms in reference [10] determine a vector magnitude at whatever speed it takes a system to perform a magnitude comparison, two multiplications, and one addition. But, as a minimum, those algorithms require a 16-bit multiplier to achieve reasonably accurate results. However, if hardware multipliers are not available, all is not lost. By restricting the α and β constants to reciprocals of integral powers of 2, Eq. (10-25) lends itself well to implementation in binary integral arithmetic. A prevailing application of the αMax+βMin algorithm uses α= 1.0 and β = 0.5[11,12]. The 0.5 multiplication operation is performed by shifting the minimum quadrature vector magnitude, Min, to the right by 1 bit. We can gauge the accuracy of any vector magnitude estimation by plotting its error as a function of vector phase angle. Let's do that. The αMax+βMin estimate for a complex vector of unity magnitude, using

$$|V| \approx \text{Max} + \frac{\text{Min}}{2} , \qquad (10\text{-}26)$$

over the vector angular range of 0 to 90°, is shown as the solid curve in Figure 10-11. (The curves in Figure 10-11, of course, repeat every 90°.)

An ideal estimation curve for a unity magnitude vector would have an average value of one and an error standard deviation (σ_e) of zero; that is, having $\sigma_e = 0$ means that the ideal curve is flat—because the curve's value is one for all vector angles and its average error is zero. We'll use this ideal estimation curve as a yardstick to measure the merit of various αMax+βMin algorithms. Let's make sure we know what the solid curve in Figure 10-11 is telling us. It indicates that a unity magnitude vector oriented at an angle of approximately 26° will be estimated by Eq. (10-26) to have a magnitude of 1.118 instead of the correct magnitude of one. The error then, at 26°, is 11.8 percent, or 0.97 dB. Analyzing the entire solid curve in Figure 10-11 results in $\sigma_e = 0.032$ and an average error, over the 0 to 90° range, of 8.6 percent (0.71 dB).

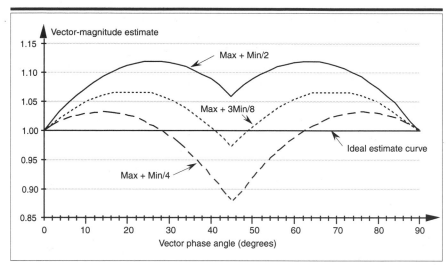

Figure 10-11 Normalized αMax+βMin estimates for α = 1, β = 1/2, and β = 1/4.

To reduce the average error introduced by Eq. (10-26), it is equally convenient to use a β value of 0.25, such as

$$|V| \approx \text{Max} + \frac{\text{Min}}{4} \, . \tag{10-27}$$

Equation (10-27), whose β multiplication is realized by shifting the digital value Min 2 bits to the right, results in the normalized magnitude approximation shown as the dashed curve in Figure 10-11. Although the largest error of 11.6 percent at 45° is similar in magnitude to that realized from Eq. (10-26), Eq. (10-27) has reduced the average error to −0.64 percent (−0.06 dB) and produced a slightly larger standard deviation of $\sigma_e = 0.041$. Though not as convenient to implement as Eqs. (10-26) and (10-27), a β value of 3/8 has been used to provide even more accurate vector magnitude estimates[13]. Using

$$|V| \approx \text{Max} + \frac{3 \cdot \text{Min}}{8} \tag{10-27'}$$

provides the normalized magnitude approximation shown as the dotted curve in Figure 10-11. Equation (10-27') results in magnitude estimates, whose largest error is only 6.8 percent, and a reduced standard deviation of $\sigma_e = 0.026$.

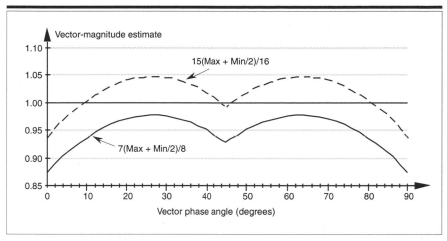

Figure 10-12 αMax+ßMin estimates for α = 7/16, ß = 7/16 and α = 15/16, ß = 15/32.

Although the values for α and β in Figure 10-11 yield rather accurate vector-magnitude estimates, there are other values for α and β that deserve our attention because they result in smaller error standard deviations. Consider α = 7/8 and β = 7/16 where

$$|V| \approx \frac{7}{8}\mathrm{Max} + \frac{7}{16}\mathrm{Min} = \frac{7}{8}\left(\mathrm{Max} + \frac{\mathrm{Min}}{2}\right). \tag{10-28}$$

Equation (10-28), whose normalized results are shown as the solid curve in Figure 10-12, provides an average error of –5.01 percent and $\sigma_e = 0.028$. The 7/8ths factor applied to Eq. (10-26) produces both a smaller σ_e and a reduced average error—it lowers and flattens out the error curve from Eq. (10-26).

A further improvement can be obtained with α = 15/16 and β = 15/32 where

$$|V| \approx \frac{15}{16}\mathrm{Max} + \frac{15}{32}\mathrm{Min} = \frac{15}{16}\left(\mathrm{Max} + \frac{\mathrm{Min}}{2}\right). \tag{10-29}$$

Equation (10-29), whose normalized results are shown as the dashed curve in Figure 10-12, provides an average error of 1.79 percent and $\sigma_e = 0.030$. At the expense of a slightly larger σ_e, Eq. (10-29) provides an average error that is reduced below that provided by Eq. (10-28).

Although Eq. (10-29) appears to require two multiplications and one addition, its digital hardware implementation can be straightforward, as

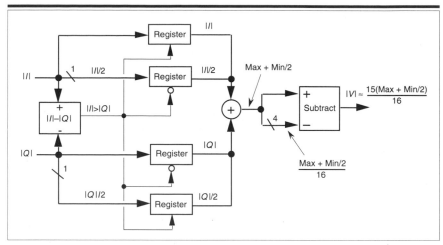

Figure 10-13 Hardware implementation of Eq. (10-29).

shown in Figure 10-13. The diagonal lines, \1 for example, denote a hard-wired shift of 1 bit to the right to implement a divide-by-two operation by truncation. Likewise, the \4 symbol indicates a right shift by 4 bits to realize a divide-by-16 operation. The $|I| > |Q|$ control line is TRUE when the magnitude of I is greater than the magnitude of Q, so that Max = $|I|$ and Min = $|Q|$. This condition enables the registers to apply the values $|I|$ and $|Q|/2$ to the adder. When $|I| > |Q|$ is FALSE, the registers apply the values $|Q|$ and $|I|/2$ to the adder. Notice that the output of the adder, Max + Min/2, is the result of Eq. (10-26). Equation (10-29) is implemented via the subtraction of (Max + Min/2)/16 from Max + Min/2.

In Figure 10-13, all implied multiplications from Eq. (10-29) are performed by hardwired bit shifting, and the total execution time is limited only by the delay times associated with the hardware components.

10.2.2 Overflow Errors

In Figures 10-11 and 10-12, notice that we have a potential overflow problem with the results of Eqs. (10-26), (10-27), and (10-29) because the estimates can exceed the correct normalized vector-magnitude values; i.e., some magnitude estimates are greater than one. This means that, although the correct magnitude value may be within the system's full-scale word width, the algorithm result may exceed the work width of the system and cause overflow errors. With αMax+βMin algorithms, the user must be certain that no true vector magnitude exceeds the value that will produce an estimated magnitude greater than the maximum allowable word width. For example, when using Eq. (10-26), we must ensure that no true vector magnitude exceeds 89.4 percent (1/1.118) of the maximum allowable work width.

10.2.3 Truncation Errors

The penalty we pay for the convenience of having α and β as powers of two is the error induced by the division-by-truncation process; and, thus far, we haven't taken that error into account. The error curves in Figure 10-11 and Figure 10-12 were obtained using a software simulation with its floating-point accuracy and are useful in evaluating different α and β values. However, the true error introduced by the αMax+βMin algorithm will be somewhat different from that shown in Figures 10-11 and 10-12 due to division errors when truncation is used with finite word widths.[†] For αMax+βMin schemes, the truncation errors are a function of the data's word width, the algorithm used, the values of both $|I|$ and $|Q|$, and the vector's phase angle. (These errors due to truncation compound the errors already inherent in our αMax+βMin algorithms.) Thus, a complete analysis of the truncation errors is beyond the scope of this book. What we can do, however, is illustrate a few truncation error examples.

Figure 10-14 shows the percent truncation errors using Eq. (10-29) for vector magnitudes of 4 to 512. Two vector phase angles were chosen to

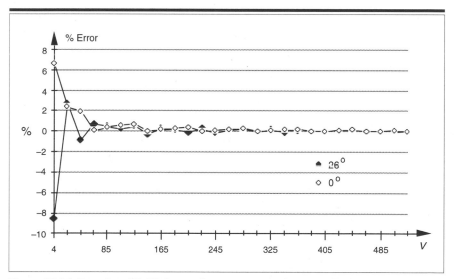

Figure 10-14 Equation (10-29) truncation error vs. vector magnitude $|V|$.

[†] Errors associated with division-by-truncation are covered in more detail in Section 9.3.

Table 10-2 α Max+βMin Algorithm Comparisons

Algorithm $\|V\| \approx$	Largest error (%)	Largest error (dB)	Average error (%)	Average error (dB)	Standard deviation σ_e	Max $\|V\|$ (% F.S.)
Max + Min/2	11.8%	0.97 dB	8.6%	0.71 dB	0.032	89.4%
Max + Min/4	−11.6%	−1.07 dB	−0.64%	−0.06 dB	0.041	97.0%
Max + 3Min/8	6.8%	0.57 dB	3.97%	0.34 dB	0.026	93.6%
7(Max + Min/2)/8	−12.5%	−1.16 dB	−4.99%	−0.45 dB	0.028	100%
15(Max + Min/2)/16	−6.25%	−0.56 dB	1.79%	0.15 dB	0.030	95.4%

illustrate these truncation errors. The first is 26° because this is the phase angle where the most positive algorithm error occurs, and the second is 0° because this is the phase angle that introduces the greatest negative algorithm error. Notice that, at small vector magnitudes, the truncation errors are as great as 9 percent, but for an eight-bit system (maximum vector magnitude = 255) the truncation error is less than 1 percent. As the system word width increases, the truncation errors approach 0 percent. This means that truncation errors add very little to the inherent αMax+βMin algorithm errors.

The relative performance of the various algorithms is summarized in Table 10-2. The last column in Table 10-2 illustrates the maximum allowable true vector magnitude as a function of the system's full-scale (F.S.) word width to avoid overflow errors.

So, the αMax+βMin algorithm enables high-speed, vector-magnitude computation without the need for math coprocessor or hardware multiplier chips. Of course with the recent availability of high-speed, floating-point multiplier integrated circuits—with their ability to multiply or divide by nonintegral numbers in one or two clock cycles—α and β may not always need to be restricted to reciprocals of integral powers of two. It's also worth mentioning that this algorithm can be nicely implemented in a single hardware integrated circuit (for example, a field programmable gate array) affording high-speed operation.

10.3 Data Windowing Tricks

There are two useful schemes associated with using window functions on input data applied to a DFT or an FFT. The first technique is an efficient implementation of the Hanning (raised cosine) and Hamming windows to reduce leakage in the FFT. The second scheme is related to minimizing the amplitude loss associated with using windows.

10.3.1 Windowing in the Frequency Domain

There's a clever technique for minimizing the calculations necessary to implement FFT input data windowing to reduce spectral leakage. There are times when we need the FFT of unwindowed time-domain data, and at the same time, we also want the FFT of that same time data with a window function applied. In this situation, we don't have to perform two separate FFTs. We can perform the FFT of the unwindowed data, and then we can perform frequency-domain windowing on that FFT result to reduce leakage. Let's see how.

Recall from Section 3.9 that the expressions for the Hanning and the Hamming windows were $w_{Han}(n)$ = $0.5-0.5\cos(2\pi n/N)$, and $w_{Ham}(n) = 0.54 -0.46\cos(2\pi n/N)$, respectively. They both have the general cosine function form of

$$w(n) = \alpha - \beta\cos(2\pi n/N) , \qquad (10\text{-}30)$$

for $n = 0, 1, 2, \ldots, N-1$. Looking at the frequency response of the general cosine window function, using the definition of the DFT, the transform of Eq. (10-30) is expressed by

$$W(m) = \sum_{n=0}^{N-1}[\alpha - \beta\cos(2\pi n / N)]e^{-j2\pi nm/N} . \qquad (10\text{-}31)$$

Because $\cos(2\pi n / N) = \dfrac{e^{j2\pi n/N}}{2} + \dfrac{e^{-j2\pi n/N}}{2}$, Eq. (10-31) can be rewritten as

$$W(m) = \sum_{n=0}^{N-1}\alpha e^{-j2\pi nm/N} - \frac{\beta}{2}\sum_{n=0}^{N-1}e^{j2\pi n/N}e^{-j2\pi nm/N} - \frac{\beta}{2}\sum_{n=0}^{N-1}e^{-j2\pi n/N}e^{-j2\pi nm/N}$$

$$= \sum_{n=0}^{N-1}\alpha e^{-j2\pi nm/N} - \frac{\beta}{2}\sum_{n=0}^{N-1}e^{-j2\pi n(m-1)/N} - \frac{\beta}{2}\sum_{n=0}^{N-1}e^{-j2\pi n(m+1)/N} . \quad (10\text{-}32)$$

Equation (10-32) looks pretty complicated, but, using the derivation from Section 3.13 for expressions like those summations, we find that Eq. (10-32) merely results in the superposition of three $\sin(x)/x$ functions in the frequency domain. Their amplitudes are shown in Figure 10-15.

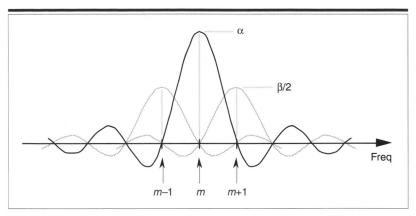

Figure 10-15 General cosine window frequency-response amplitude.

Notice that the two translated $\sin(x)/x$ functions have sidelobes with phase opposite from that of the center $\sin(x)/x$ function. This means that α times the mth bin output, minus $\beta/2$ times the $(m-1)$th bin output, minus $\beta/2$ times the $(m+1)$th bin output will minimize the sidelobes of the mth bin. This frequency-domain convolution process is equivalent to multiplying the input time data sequence by the N-valued window function $w(n)$ in Eq. (10-30)[14,15].

For example, let's say the output of the mth FFT bin is $X(m) = a_m + jb_m$, and the outputs of its two neighboring bins are $X(m-1) = a_{-1} + jb_{-1}$ and $X(m+1) = a_{+1} + jb_{+1}$. Then frequency-domain windowing for the mth bin of the unwindowed $X(m)$ is as follows:

$$X_{\text{windowed}}(m) = \alpha X(m) - \frac{\beta}{2}X(m-1) - \frac{\beta}{2}X(m+1)$$

$$= \alpha(a_m + jb_m) - \frac{\beta}{2}(a_{-1} + jb_{-1}) - \frac{\beta}{2}(a_{+1} + jb_{+1})$$

$$= \alpha a_m - \frac{\beta}{2}(a_{-1} + a_{+1}) + j[\alpha b_m - \frac{\beta}{2}(b_{-1} + b_{+1})] \,.$$

(10-33)

To get a windowed N-point FFT, then, we can apply Eq. (10-33), requiring $4N$ additions and $3N$ multiplications, to the unwindowed FFT result and avoid having to perform the N multiplications of time-domain windowing and a second FFT with its $N\log_2 N$ additions and $2N\log_2 N$ multiplications.

The neat situation here is the α and β values for the Hanning window. They're both 0.5, and the products in Eq. (10-33) can be obtained in hardware with binary shifts by a single bit for α and two shifts for $\beta/2$. Thus, no multiplications are necessary to implement the Hanning frequency-domain windowing scheme. The issues that we need to consider are the window function best for the application and the efficiency of available hardware in performing the frequency-domain multiplications.

Along with the Hanning and Hamming windows, reference [15] describes a family of windows known as Blackman and Blackman-Harris windows that are also very useful for performing frequency-domain windowing. (Be aware that reference [15] has two typographical errors in the 4-Term (–74 dB) window coefficients column on its page 65. Reference [16] specifies that those coefficients should be 0.40217, 0.49703, 0.09892, and 0.00188.) Let's finish our discussion of frequency-domain windowing by saying that this scheme can be efficient because we don't have to window the entire set of FFT data. Frequency-domain windowing need be performed only on those FFT bins that are of interest to us.

10.3.2 Minimizing Window-Processing Loss

In Section 3.9, we stated that nonrectangular window functions reduce the overall signal levels applied to the FFT. Recalling Figure 3-16(a), we see that the peak response of the Hanning window function, for example, is half that obtained with the rectangular window because the input signal is attenuated at the beginning and end edges of the window sample interval, as shown in Figure 10-16(a). In terms of signal power, this attenuation results in a 6 dB loss. Going beyond the signal-power loss, window edge effects can be a problem when we're trying to detect short-duration signals that may occur right when the window function is at its edges. Well, some early digital signal processing practitioners tried to get around this problem by using dual window functions.

The first step in the dual window process is windowing the input data with a Hanning window function and taking the FFT of the windowed data. Then the same input data sequence is windowed against the inverse of the Hanning, and another FFT is performed. (The inverse of the Hanning window is depicted in Figure 10-16(b).) The two FFT results are then averaged. Using the dual window functions shown in Figure 10-16 enables signal energy attenuated by one window to be multiplied by the full gain of the other window. This technique seemed like a reasonable idea at the time, but, depending on the original signal, there could be

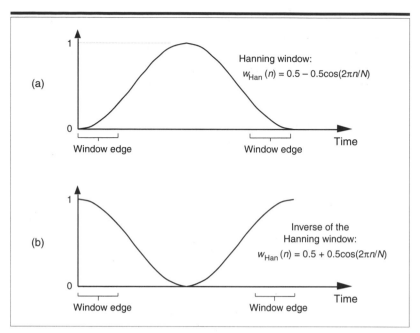

Figure 10-16 in image. Content:

(a) Hanning window:
$$w_{Han}(n) = 0.5 - 0.5\cos(2\pi n/N)$$

(b) Inverse of the Hanning window:
$$w_{Han}(n) = 0.5 + 0.5\cos(2\pi n/N)$$

Figure 10-16 Dual windows used to reduce windowed-signal loss.

excessive leakage from the inverse window in Figure 10-16(b). Remember, the purpose of windowing was to ensure that the first and last data sequence samples, applied to an FFT, had the same value. The Hanning window guaranteed this, but the inverse window could not. Although this dual window technique made its way into the literature, it quickly fell out of favor. The most common technique used today to minimize signal loss due to window edge effects is known as *overlapped windows*.

The use of overlapped windows is depicted in Figure 10-17. It's a straightforward technique where a single *good* window function is applied multiple times to an input data sequence. Figure 10-17 shows an N-point window function applied to the input time series data four times resulting in four separate N-point data sequences. Next, four separate N-point FFTs are performed, and their outputs averaged. Notice that any input sample value that's fully attenuated by one window will be multiplied by the full gain of the following window. Thus, all input samples will contribute to the final averaged FFT results, and the window function keeps leakage to a minimum. (Of course, the user has to decide which particular window function is best for the application.) Figure 10-17 shows a window overlap of 50 percent where each input data sample contributes to the results of two FFTs. It's not uncommon to see an overlap of

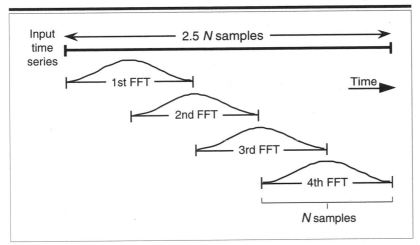

Figure 10-17 Windows overlapped by 50 percent to reduce windowed-signal loss.

75 percent being used where each input data sample would contribute to the results of the three individual FFTs. Of course the 50 percent and 75 percent overlap techniques increase the amount of total signal processing required, but, depending on the application, the improved signal sensitivity may justify the extra number crunching.

10.4 Fast Multiplication of Complex Numbers

The multiplication of two complex numbers is one of the most common functions performed in digital signal processing. It's mandatory in all discrete and fast Fourier transformation algorithms, necessary for graphics transformations, and used in processing digital communications signals. Be it in hardware or software, it's always to our benefit to streamline the processing necessary to perform a complex multiplication whenever we can. If the available hardware can perform three additions faster than a single multiplication, there's a way to speed up a complex multiplication operation[17].

The multiplication of two complex numbers $a + jb$ and $c + jd$, results in the complex product

$$R + jI = (a + jb) \cdot (c + jd) = (ac - bd) + j(bc + ad) . \qquad (10\text{-}34)$$

We can see that Eq. (10-34) required four multiplications and two additions. (From a computational standpoint, we'll assume that a subtraction is equivalent to an addition.) Instead of using Eq. (10-34), we can calculate the following intermediate values:

$$k_1 = a(c + d),$$
$$k_2 = d(a + b),$$

and

$$k_3 = c(b - a). \tag{10-35}$$

Then we perform the following operations to get the final R and I:

$$R = k_1 - k_2,$$

and

$$I = k_1 + k_3. \tag{10-36}$$

The reader is invited to plug the k values from Eq. (10-35) into Eq. (10-36) to verify that the expressions in Eq. (10-36) are equivalent to Eq. (10-34). The intermediate values in Eq. (10-35) required three additions and three multiplications, whereas the results in Eq. (10-36) required two more additions. So we traded one of the multiplications required in Eq. (10-34) for three addition operations needed by Eq. (10-35) and Eq. (10-36). If our hardware uses fewer clock cycles to perform three additions than a single multiplication, we may well gain overall processing speed by using Eq. (10-35) and Eq. (10-36) for complex multiplication, instead of Eq. (10-34).

10.5 Efficiently Performing the FFT of Real Sequences

Upon recognizing its linearity property and understanding the odd and even symmetries of the transform's output, the early investigators of the fast Fourier transform (FFT) realized that two separate, real N-point input data sequences could be transformed using a single N-point complex FFT. They also developed a technique using a single N-point complex FFT to transform a $2N$-point real input sequence. Let's see how these two techniques work.

10.5.1 Performing Two *N*-Point Real FFTs

The standard FFT algorithms were developed to accept complex inputs; that is, the FFT's normal input $x(n)$ sequence is assumed to comprise real and imaginary parts, such as

$$x(0) = x_r(0) + jx_i(0) \ ,$$
$$x(1) = x_r(1) + jx_i(1) \ ,$$
$$x(2) = x_r(2) + jx_i(2) \ ,$$
$$\cdots$$
$$\cdots$$
$$x(N-1) = x_r(N-1) + jx_i(N-1) \ . \tag{10-37}$$

In typical signal processing schemes, FFT input data sequences are usually real. The most common example of this is the FFT input samples coming from an A/D converter that provides real integer values of some continuous (analog) signal. In this case the FFT's imaginary $x_i(n)$'s inputs are all zero. So initial FFT computations performed on the $x_i(n)$ inputs represent wasted operations. Early FFT pioneers recognized this inefficiency, studied the problem, and developed a technique where two independent N-point, *real* input data sequences could be transformed by a single N-point complex FFT. We call this scheme the Two N-Point Real FFTs algorithm. The derivation of this technique is straightforward and described in the literature[18–20]. If two N-point, real input sequences are $a(n)$ and $b(n)$, they'll have discrete Fourier transforms represented by $X_a(m)$ and $X_b(m)$. If we treat the $a(n)$ sequence as the real part of an FFT input and the $b(n)$ sequence as the imaginary part of the FFT input, then

$$x(0) = a(0) + jb(0) \ ,$$
$$x(1) = a(1) + jb(1) \ ,$$
$$x(2) = a(2) + jb(2) \ ,$$
$$\cdots$$
$$\cdots$$
$$x(N-1) = a(N-1) + jb(N-1) \ . \tag{10-38}$$

Applying the $x(n)$ values from Eq. (10-38) to the standard DFT,

$$X(m) = \sum_{n=0}^{N-1} x(n)e^{-j2\pi nm/N} \ , \tag{10-39}$$

we'll get an DFT output $X(m)$ where m goes from 0 to N–1. (We're assuming, of course, that the DFT is implemented by way of an FFT algorithm.) Using the superscript * symbol to represent the complex conjugate, we can extract the two desired FFT outputs $X_a(m)$ and $X_b(m)$ from $X(m)$ by using the following:

$$X_a(m) = \frac{X^*(N-m) + X(m)}{2}, \tag{10-40}$$

and

$$X_b(m) = \frac{j[X^*(N-m) - X(m)]}{2}. \tag{10-41}$$

Let's break Eqs. (10-40) and (10-41) into their real and imaginary parts to get expressions for $X_a(m)$ and $X_b(m)$ that are easier to understand and implement. Using the notation showing $X(m)$'s real and imaginary parts, where $X(m) = X_r(m) + jX_i(m)$, we can rewrite Eq. (10-40) as

$$X_a(m) = \frac{X_r(N-m) + X_r(m) + j[X_i(m) - X_i(N-m)]}{2} \tag{10-42}$$

where $m = 1, 2, 3, \ldots, N-1$. What about the first $X_a(m)$, when $m = 0$? Well, this is where we run into a bind if we actually try to implement Eq. (10-40) directly. Letting $m = 0$ in Eq. (10-40), we quickly realize that the first term in the numerator, $X^*(N-0) = X^*(N)$, isn't available because the $X(N)$ sample does not exist in the output of an N-point FFT! We resolve this problem by remembering that $X(m)$ is periodic with a period N, so $X(N) = X(0)$.[†] When $m = 0$, Eq. (10-40) becomes

$$X_a(0) = \frac{X_r(0) - jX_i(0) + X_r(0) + jX_i(0)}{2} = X_r(0). \tag{10-43}$$

Next, simplifying Eq. (10-41),

$$X_b(m) = \frac{j[X_r(N-m) - jX_i(N-m) - X_r(m) - jX_i(m)]}{2}$$

$$= \frac{X_i(N-m) + X_i(m) + j[X_r(N-m) - X_r(m)]}{2} \tag{10-44}$$

where, again, $m = 1, 2, 3, \ldots, N-1$. By the same argument used for Eq. (10-43), when $m = 0$, $X_b(0)$ in Eq. (10-44) becomes

---

[†] This fact is illustrated in Section 3.8 during the discussion of spectral leakage in DFTs.

$$X_b(0) = \frac{X_i(0) + X_i(0) + j[X_r(0) - X_r(0)]}{2} = X_i(0) \ . \qquad (10\text{-}45)$$

This discussion brings up a good point for beginners to keep in mind. In the literature Eqs. (10-40) and (10-41) are often presented without any discussion of the $m = 0$ problem. So, whenever you're grinding through an algebraic derivation or have some equations tossed out at you, be a little skeptical. Try the equations out on an example—see if they're true. (After all, both authors and book typesetters are human and sometimes make mistakes. We had an old saying in Ohio for this situation: "Trust everybody, but cut the cards.") Following this advice, let's prove that this Two N-Point Real FFTs algorithm really does work by applying the 8-point data sequences from Chapter 3's DFT Examples to Eqs. (10-42) through (10-45). Taking the 8-point input data sequence from Section 3.1's DFT Example 1 and denoting it $a(n)$,

$$
\begin{array}{ll}
a(0) = 0.3535, & a(1) = 0.3535, \\
a(2) = 0.6464, & a(3) = 1.0607, \\
a(4) = 0.3535, & a(5) = -1.0607, \\
a(6) = -1.3535, & a(7) = -0.3535 \ . \qquad (10\text{-}46)
\end{array}
$$

Taking the 8-point input data sequence from Section 3.6's DFT Example 2 and calling it $b(n)$,

$$
\begin{array}{ll}
b(0) = 1.0607, & b(1) = 0.3535, \\
b(2) = -1.0607, & b(3) = -1.3535, \\
b(4) = -0.3535, & b(5) = 0.3535, \\
b(6) = 0.3535, & b(7) = 0.6464 \ . \qquad (10\text{-}47)
\end{array}
$$

Combining the sequences in Eqs. (10-46) and (10-47) into a single complex sequence $x(n)$,

$$
\begin{array}{lll}
 & a(n) & b(n) \\
 & \downarrow & \downarrow \\
x(n) = & 0.3535 & + j\,1.0607 \\
 & + 0.3535 & + j\,0.3535 \\
 & + 0.6464 & - j\,1.0607 \\
 & + 1.0607 & - j\,1.3535 \\
 & + 0.3535 & - j\,0.3535 \\
 & - 1.0607 & + j\,0.3535 \\
 & - 1.3535 & + j\,0.3535 \\
 & - 0.3535 & + j\,0.6464 \ . \qquad (10\text{-}48)
\end{array}
$$

Now, taking the 8-point FFT of the complex sequence in Eq. (10-48) we get

	$X_r(m)$	$X_i(m)$	
	↓	↓	
$X(m) =$	0.0000	$+ j \, 0.0000$	← m = 0 term
	$- 2.8283$	$- j \, 1.1717$	← m = 1 term
	$+ 2.8282$	$+ j \, 2.8282$	← m = 2 term
	$+ 0.0000$	$+ j \, 0.0000$	← m = 3 term
	$+ 0.0000$	$+ j \, 0.0000$	← m = 4 term
	$+ 0.0000$	$+ j \, 0.0000$	← m = 5 term
	$+ 0.0000$	$+ j \, 0.0000$	← m = 6 term
	$+ 2.8283$	$+ j \, 6.8282$	← m = 7 term . (10-49)

So from Eq. (10-43),

$$X_a(0) = X_r(0) = 0 .$$

To get the rest of $X_a(m)$, we have to plug the FFT output's $X(m)$ and $X(N–m)$ values into Eq. (10-42).[†] Doing so,

$$X_a(1) = \frac{X_r(7) + X_r(1) + j[X_i(1) - X_i(7)]}{2} = \frac{2.8283 - 2.8283 + j[-1.1717 - 6.8282]}{2}$$

$$= \frac{0 - j7.9999}{2} = 0 - j4.0 = 4\angle -90^\circ ,$$

$$X_a(2) = \frac{X_r(6) + X_r(2) + j[X_i(2) - X_i(6)]}{2} = \frac{0.0 + 2.8282 + j[2.8282 - 0.0]}{2}$$

$$= \frac{2.8282 + j2.8282}{2} = 1.414 + j1.414 = 2\angle 45^\circ ,$$

$$X_a(3) = \frac{X_r(5) + X_r(3) + j[X_i(3) - X_i(5)]}{2} = \frac{0.0 + 0.0 + j[0.0 - 0.0]}{2} = 0\angle 0^\circ ,$$

$$X_a(4) = \frac{X_r(4) + X_r(4) + j[X_i(4) - X_i(4)]}{2} = \frac{0.0 + 0.0 + j[0.0 - 0.0]}{2} = 0\angle 0^\circ ,$$

[†] Remember, when the FFT's input is complex, the FFT outputs may not be conjugate symmetric; that is, we can't assume that $F(m)$ is equal to $F^*(N–m)$ when the FFT input sequence's real and imaginary parts are both nonzero.

$$X_a(5) = \frac{X_r(3) + X_r(5) + j[X_i(5) - X_i(3)]}{2} = \frac{0.0 + 0.0 + j[0.0 - 0.0]}{2} = 0\angle 0° ,$$

$$X_a(6) = \frac{X_r(2) + X_r(6) + j[X_i(6) - X_i(2)]}{2} = \frac{2.8282 + 0.0 + j[0.0 - 2.8282]}{2}$$

$$= \frac{2.8282 - j2.8282}{2} = 1.414 - j1.414 = 2\angle -45° , \text{ and}$$

$$X_a(7) = \frac{X_r(1) + X_r(7) + j[X_i(7) - X_i(1)]}{2} = \frac{-2.8282 + 2.8282 + j[6.8282 + 1.1717]}{2}$$

$$= \frac{0.0 + j7.9999}{2} = 0 + j4.0 = 4\angle 90° .$$

So Eq. (10-42) really does extract $X_a(m)$ from the $X(m)$ sequence in Eq. (10-49). We can see that we need not solve Eq. (10-42) when m is greater than 4 (or $N/2$) because $X_a(m)$ will always be conjugate symmetric. Because $X_a(7) = X_a(1)$, $X_a(6) = X_a(2)$, etc., only the first $N/2$ elements in $X_a(m)$ are independent and need be calculated.

OK, let's keep going and use Eqs. (10-44) and (10-45) to extract $X_b(m)$ from the FFT output. From Eq. (10-45)

$$X_b(0) = X_i(0) = 0 .$$

Plugging the FFT's output values into Eq. (10-44) to get the next four $X_b(m)$s, we have

$$X_b(1) = \frac{X_i(7) + X_i(1) + j[X_r(7) - X_r(1)]}{2} = \frac{6.8282 - 1.1717 + j[2.8283 + 2.8283]}{2}$$

$$= \frac{5.656 + j5.656}{2} = 2.828 + j2.828 = 4\angle 45° ,$$

$$X_b(2) = \frac{X_i(6) + X_i(2) + j[X_r(6) - X_r(2)]}{2} = \frac{0.0 + 2.8282 + j[0.0 - 2.8282]}{2}$$

$$= \frac{2.8282 - j2.8282}{2} = 1.414 - j1.414 = 2\angle -45° , \text{ and}$$

$$X_b(3) = \frac{X_i(5) + X_i(3) + j[X_r(5) - X_r(3)]}{2} = \frac{0.0 + 0.0 + j[0.0 - 0.0]}{2} = 0\angle 0° \text{ , and}$$

$$X_b(4) = \frac{X_i(4) + X_i(4) + j[X_r(4) - X_r(4)]}{2} = \frac{0.0 + 0.0 + j[0.0 - 0.0]}{2} = 0\angle 0° \text{ .}$$

The question arises "With the additional processing required by Eqs. (10-42) and (10-44) after the initial FFT, how much computational saving (or loss) is to be had by this Two N-Point Real FFTs algorithm?" We can estimate the efficiency of this algorithm by considering the number of arithmetic operations required relative to two separate N-point radix-2 FFTs. First, we estimate the number of arithmetic operations in two separate N-point complex FFTs.

From Section 4.2, we know that a standard radix-2 N-point complex FFT comprises $(N/2) \cdot \log_2 N$ butterfly operations. If we use the optimized butterfly structure, each butterfly requires one complex multiplication and two complex additions. Now, one complex multiplication requires two real additions and four real multiplications, and one complex addition requires two real additions.[†] So a single FFT butterfly operation comprises four real multiplications and six real additions. This means that a single N-point complex FFT requires $(4N/2) \cdot \log_2 N$ real multiplications, and $(6N/2) \cdot \log_2 N$ real additions. Finally, we can say that two separate N-point complex radix-2 FFTs require

two N-point complex FFTs → $\quad 4N \cdot \log_2 N$ real multiplications, and $\quad$ (10-50)

$$6N \cdot \log_2 N \text{ real additions.} \quad (10\text{-}50')$$

Next, we need to determine the computational workload of the Two N-Point Real FFTs algorithm. If we add up the number of real multiplications and real additions required by the algorithm's N-point complex FFT, plus those required by Eq. (10-42) to get $X_a(m)$, and those required by Eq. (10-44) to get $X_b(m)$, the Two N-Point Real FFTs algorithm requires

two N-Point Real FFTs algorithm → $\quad 2N \cdot \log_2 N + N$ real multiplications, and (10-51)

$$3N \cdot \log_2 N + 2N \text{ real additions.} \quad (10\text{-}51')$$

[†] The complex addition $(a+jb) + (c+jd) = (a+c) + j(b+d)$ requires two real additions. A complex multiplication $(a+jb) \cdot (c+jd) = ac-bd + j(ad+bc)$ requires two real additions and four real multiplications.

Equations (10-51) and (10-51') assume that we're calculating only the first $N/2$ independent elements of $X_a(m)$ and $X_b(m)$. The single N term in Eq. (10-51) accounts for the $N/2$ divide by 2 operations in Eq. (10-42) and the $N/2$ divide by 2 operations in Eq. (10-44).

OK, now we can find out how efficient the Two N-Point Real FFTs algorithm is compared to two separate complex N-point radix-2 FFTs. This comparison, however, depends on the hardware used for the calculations. If our arithmetic hardware takes many more clock cycles to perform a multiplication than an addition, then the difference between multiplications in Eqs. (10-50) and (10-51) is the most important comparison. In this case, the percentage gain in computational saving of the Two N-Point Real FFTs algorithm relative to two separate N-point complex FFTs is the difference in their necessary multiplications over the number of multiplications needed for two separate N-point complex FFTs, or

$$\frac{4N \cdot \log_2 N - (2N \cdot \log_2 N + N)}{4N \cdot \log_2 N} \cdot 100\% = \frac{2 \cdot \log_2 N - 1}{4 \cdot \log_2 N} \cdot 100\% . \quad (10\text{-}52)$$

The computational (multiplications only) saving from Eq. (10-52) is plotted as the top curve of Figure 10-18. In terms of multiplications, for $N \geq 32$,

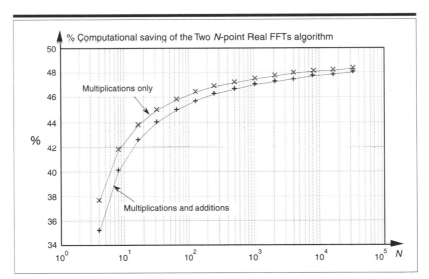

Figure 10-18 Computational saving of the Two N-Point Real FFTs algorithm over that of two separate N-point complex FFTs. The top curve indicates the saving when only multiplications are considered. The bottom curve is the saving when both additions and multiplications are used in the comparison.

the Two N-Point Real FFTs algorithm saves us over 45 percent in computational workload compared to two separate N-point complex FFTs.

For hardware using high-speed multiplier integrated circuits, multiplication and addition can take roughly equivalent clock cycles. This makes addition operations just as important and time consuming as multiplications. Thus the difference between those combined arithmetic operations in Eqs. (10-50) plus (10-50') and Eqs. (10-51) plus (10-51') is the appropriate comparison. In this case, the percentage gain in computational saving of our algorithm over two FFTs is their total arithmetic operational difference over the total arithmetic operations in two separate N-point complex FFTs, or

$$\frac{(4N \cdot \log_2 N + 6N \cdot \log_2 N) - (2N \cdot \log_2 N + N + 3N \cdot \log_2 N + 2N)}{4N \cdot \log_2 N + 6N \cdot \log_2 N} \cdot 100\%$$

$$= \frac{5 \cdot \log_2 N - 3}{10 \cdot \log_2 N} \cdot 100\% \ . \tag{10-53}$$

The full computational (multiplications and additions) saving from Eq. (10-53) is plotted as the bottom curve of Figure 10-18. OK, that concludes our discussion and illustration of how a single N-point complex FFT can be used to transform two separate N-point real input data sequences.

10.5.2 Performing a 2N-Point Real FFT

Similar to the scheme above where two separate N-point real data sequences are transformed using a single N-point FFT, a technique exists where a $2N$-point real sequence can be transformed with a single complex N-point FFT. This 2N-Point Real FFT algorithm, whose derivation is also described in the literature, requires that the $2N$-sample real input sequence be separated into two parts[20,21]. Not broken in two, but unzipped—separating the even and odd sequence samples. The N even-indexed input samples are loaded into the real part of a complex N-point input sequence $x(n)$. Likewise, the input's N odd-indexed samples are loaded into $x(n)$'s imaginary parts. To illustrate this process, let's say we have a $2N$-sample real input data sequence $a(n)$ where $0 \le n \le 2N-1$. We want $a(n)$'s $2N$-point transform $X_a(m)$. Loading $a(n)$'s odd/even sequence values appropriately into an N-point complex FFT's input sequence, $x(n)$,

$$x(0) = a(0) + ja(1) \,,$$
$$x(1) = a(2) + ja(3) \,,$$
$$x(2) = a(4) + ja(5) \,,$$
$$\cdots$$
$$\cdots$$
$$x(N-1) = a(2N-2) + ja(2N-1) \,. \qquad (10\text{-}54)$$

Applying the N complex values in Eq. (10-54) to an N-point complex FFT, we'll get an FFT output $X(m) = X_r(m) + jX_i(m)$, where m goes from 0 to $N-1$. To extract the desired 2N-Point Real FFT algorithm output $X_a(m) = X_{a,real}(m) + jX_{a,imag}(m)$ from $X(m)$, let's define the following relationships

$$X_r^+(m) = \frac{X_r(m) + X_r(N - m)}{2} \,, \qquad (10\text{-}55)$$

$$X_r^-(m) = \frac{X_r(m) - X_r(N - m)}{2} \,, \qquad (10\text{-}56)$$

$$X_i^+(m) = \frac{X_i(m) + X_i(N - m)}{2} \,, \text{ and} \qquad (10\text{-}57)$$

$$X_i^-(m) = \frac{X_i(m) - X_i(N - m)}{2} \,. \qquad (10\text{-}58)$$

The values resulting from Eqs. (10-55) through (10-58) are, then, used as factors in the following expressions to obtain the real and imaginary parts of our final $X_a(m)$:

$$X_{a,real}(m) = X_r^+(m) + \cos(\frac{\pi m}{N}) \cdot X_i^+(m) - \sin(\frac{\pi m}{N}) \cdot X_r^-(m), \qquad (10\text{-}59)$$

and

$$X_{a,imag}(m) = X_i^-(m) - \sin(\frac{\pi m}{N}) \cdot X_i^+(m) - \cos(\frac{\pi m}{N}) \cdot X_r^-(m) \,. \qquad (10\text{-}60)$$

Remember now, the original $a(n)$ input index n goes from 0 to $2N-1$, and our N-point FFT output index m goes from 0 to $N-1$. We apply $2N$ real input time-domain samples to this algorithm and get back N complex frequency-domain samples representing the first half of the equivalent $2N$-point complex FFT, $X_a(0)$ through $X_a(N-1)$. Because this algorithm's

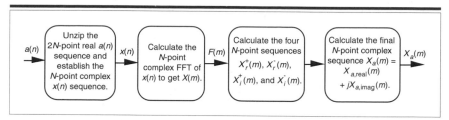

Figure 10-19 Computational flow of the 2N-Point Real FFT algorithm.

$a(n)$ input is constrained to be real, $X_a(N)$ through $X_a(2N-1)$ are merely the complex conjugates of their $X_a(0)$ through $X_a(N-1)$ counterparts and need not be calculated. To help us keep all of this straight, Figure 10-19 depicts the computational steps of the 2N-Point Real FFT algorithm.

To demonstrate this process by way of example, let's apply the 8-point data sequence from Eq. (10-46) to the 2N-Point Real FFT algorithm. Partitioning those Eq. (10-46) samples as dictated by Eq. (10-54), we have our new FFT input sequence:

$$x(0) = 0.3535 + j\,0.3535,$$
$$x(1) = 0.6464 + j\,1.0607,$$
$$x(2) = 0.3535 - j\,1.0607,$$
$$x(3) = -1.3535 - j\,0.3535 . \qquad (10\text{-}61)$$

With $N = 4$ in this example, taking the 4-point FFT of the complex sequence in Eq. (10-61) we get

$$
\begin{array}{ccl}
X_r(m) & X_i(m) & \\
\downarrow & \downarrow & \\
X(m) = 0.0000 & +j\,0.0000 & \leftarrow m = 0 \text{ term} \\
+ 1.4142 & -j\,0.5857 & \leftarrow m = 1 \text{ term} \\
+ 1.4141 & -j\,1.4141 & \leftarrow m = 2 \text{ term} \\
- 1.4142 & +j\,3.4141 & \leftarrow m = 3 \text{ term} . \qquad (10\text{-}62)
\end{array}
$$

Using these values, we now get the intermediate factors from Eqs. (10-55) through (10-58). Calculating our first $X_r^+(0)$ value, again we're reminded that $X(m)$ is periodic with a period N, so $X(4) = X(0)$, and $X_r^+(0) = [X_r(0) + X_r(0)]/2 = 0$. Continuing to use Eqs. (10-55) through (10-58),

$$X_r^+(0) = 0, \quad X_r^-(0) = 0, \quad X_i^+(0) = 0, \quad X_i^-(0) = 0,$$
$$X_r^+(1) = 0, \quad X_r^-(1) = 1.4142, \quad X_i^+(1) = 1.4142, \quad X_i^-(1) = -1.9999,$$
$$X_r^+(2) = 1.4141, \quad X_r^-(2) = 0, \quad X_i^+(2) = -1.4144, \quad X_i^-(2) = 0,$$
$$X_r^+(3) = 0, \quad X_r^-(3) = -1.4142, \quad X_i^+(3) = 1.4142, \quad X_i^-(3) = 1.9999. \quad (10\text{-}63)$$

Using the intermediate values from Eq. (10-63) in Eqs. (10-59) and (10-60),

$$X_{a,real}(0) = (0) + \cos\left(\frac{\pi \cdot 0}{4}\right) \cdot (0) - \sin\left(\frac{\pi \cdot 0}{4}\right) \cdot (0)$$

$$X_{a,imag}(0) = (0) - \sin\left(\frac{\pi \cdot 0}{4}\right) \cdot (0) - \cos\left(\frac{\pi \cdot 0}{4}\right) \cdot (0)$$

$$X_{a,real}(1) = (0) + \cos\left(\frac{\pi \cdot 1}{4}\right) \cdot (1.4142) - \sin\left(\frac{\pi \cdot 1}{4}\right) \cdot (1.4142)$$

$$X_{a,imag}(1) = (-1.9999) - \sin\left(\frac{\pi \cdot 1}{4}\right) \cdot (1.4142) - \cos\left(\frac{\pi \cdot 1}{4}\right) \cdot (1.4142)$$

$$X_{a,real}(2) = (1.4141) + \cos\left(\frac{\pi \cdot 2}{4}\right) \cdot (-1.4144) - \sin\left(\frac{\pi \cdot 2}{4}\right) \cdot (0)$$

$$X_{a,imag}(2) = (0) - \sin\left(\frac{\pi \cdot 2}{4}\right) \cdot (-1.4144) - \cos\left(\frac{\pi \cdot 2}{4}\right) \cdot (0)$$

$$X_{a,real}(3) = (0) + \cos\left(\frac{\pi \cdot 3}{4}\right) \cdot (1.4142) - \sin\left(\frac{\pi \cdot 3}{1}\right) \cdot (-1.4142)$$

$$X_{a,imag}(3) = (1.9999) - \sin\left(\frac{\pi \cdot 3}{4}\right) \cdot (1.4142) - \cos\left(\frac{\pi \cdot 3}{4}\right) \cdot (-1.4142) \quad (10\text{-}64)$$

Evaluating the sine and cosine terms in Eq. (10-64),

$$X_{a,real}(0) = (0) + (1) \cdot (0) - (0) \cdot (0) = 0,$$
$$X_{a,imag}(0) = (0) - (0) \cdot (0) - (1) \cdot (0) = 0,$$
$$X_{a,real}(1) = (0) + (0.7071) \cdot (1.4142) - (0.7071) \cdot (1.4142) = 0,$$
$$X_{a,imag}(1) = (-1.9999) - (0.7071) \cdot (1.4142) - (0.7071) \cdot (1.4142) = -3.9999,$$
$$X_{a,real}(2) = (1.4141) + (0) \cdot (-1.4144) - (1) \cdot (0) = 1.4141,$$
$$X_{a,imag}(2) = (0) - (1) \cdot (-1.4144) - (0) \cdot (0) = 1.4144,$$
$$X_{a,real}(3) = (0) + (-0.7071) \cdot (1.4142) - (0.7071) \cdot (-1.4142) = 0, \text{ and}$$
$$X_{a,imag}(3) = (1.9999) - (0.7071) \cdot (1.4142) - (-0.7071) \cdot (-1.4142) = 0 . (10\text{-}65)$$

Combining the results of the terms in Eq. (10-65), we have our final correct answer of

$$X_a(0) = X_{a,real}(0) + jX_{a,imag}(0) = 0 + j0 = 0 \angle 0°,$$
$$X_a(1) = X_{a,real}(1) + jX_{a,imag}(1) = 0 - j3.999 = 4 \angle -90°,$$
$$X_a(2) = X_{a,real}(2) + jX_{a,imag}(2) = 1.4141 + j1.4144 = 2 \angle 45°, \text{ and}$$
$$X_a(3) = X_{a,real}(3) + jX_{a,imag}(3) = 0 + j0 = 0 \angle 0°. \quad\quad (10\text{-}66)$$

After going through all the steps required by Eqs. (10-55) through (10-60), the reader might question the efficiency of this 2N-Point Real FFT algorithm. Using the same process as the above 2N-Point Real FFTs algorithm analysis, let's show that the 2N-Point Real FFT algorithm does provide some modest computational saving. First, we know that a single 2N-Point radix-2 FFT has $(2N/2) \cdot \log_2 2N = N \cdot (\log_2 N+1)$ butterflies and requires

2N-point complex FFT → $\quad 4N \cdot (\log_2 N+1)$ real multiplications $\qquad$ (10-67)

and

$$6N \cdot (\log_2 N+1) \text{ real additions.} \qquad (10\text{-}67')$$

If we add up the number of real multiplications and real additions required by the algorithm's N-point complex FFT, plus those required by Eqs. (10-55) through (10-58) and those required by Eqs. (10-59) and (10-60), the complete 2N-Point Real FFT algorithm requires

2N-Point Real FFT algorithm → $\ 2N \cdot \log_2 N + 8N$ real multiplications $\qquad$ (10-68)

and

$$3N \cdot \log_2 N + 8N \text{ real additions.} \qquad (10\text{-}68')$$

OK, using the same hardware considerations (multiplications only) we used to arrive at Eq. (10-52), the percentage gain in multiplication saving of the 2N-Point Real FFT algorithm relative to a 2N-point complex FFT is

$$\frac{4N \cdot (\log_2 N + 1) - (2N \cdot \log_2 N + 8N)}{4N \cdot (\log_2 N + 1)} \cdot 100\%$$

$$= \frac{2N \cdot \log_2 N + 2N - N \cdot \log_2 N - 4N}{2N \cdot \log_2 N + 2N} \cdot 100\%$$

$$= \frac{\log_2 N - 2}{2 \cdot \log_2 N + 2} \cdot 100\% \ . \qquad (10\text{-}69)$$

The computational (multiplications only) saving from Eq. (10-69) is plotted as the bottom curve of Figure 10-20. In terms of multiplications, the 2N-Point Real FFT algorithm provides a saving of >30% when N≥128 or whenever we transform input data sequences whose lengths are ≥256.

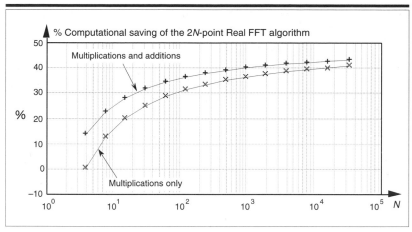

Figure 10-20 Computational saving of the 2*N*-Point Real FFT algorithm over that of a single 2*N*-point complex FFT. The top curve is the saving when both additions and multiplications are used in the comparison. The bottom curve indicates the saving when only multiplications are considered.

Again, for hardware using high-speed multipliers, we consider both multiplication and addition operations. The difference between those combined arithmetic operations in Eqs. (10-67) plus (10-67') and Eqs. (10-68) plus (10-68') is the appropriate comparison. In this case, the percentage gain in computational saving of our algorithm is

$$\frac{4N \cdot (\log_2 N + 1) + 6N \cdot (\log_2 N + 1) - (2N \cdot \log_2 N + 8N + 3N \cdot \log_2 N + 8N)}{4N \cdot (\log_2 N + 1) + 6N \cdot (\log_2 N + 1)} \cdot 100\%$$

$$= \frac{10 \cdot (\log_2 N + 1) - 5 \cdot \log_2 N - 16}{10 \cdot (\log_2 N + 1)} \cdot 100\%$$

$$= \frac{5 \; \log_2 N \; \; 6}{10 \cdot (\log_2 N + 1)} \cdot 100\%. \qquad (10\text{-}70)$$

The full computational (multiplications and additions) saving from Eq. (10-70) is plotted as a function of *N* in the top curve of Figure 10-20.

10.6 Calculating the Inverse FFT Using the Forward FFT

There are many signal processing applications where the capability to perform the inverse FFT is necessary. This can be a problem if available hardware or software routines have the capability to perform only the *forward* FFT. Fortunately, there are two slick ways to perform the inverse FFT using the forward FFT algorithm.

10.6.1 First Inverse FFT Method

The first inverse FFT calculation scheme is implemented following the processes shown in Figure 10-21. To see how this works, consider the expressions for the forward and inverse DFTs:

Forward DFT →
$$X(m) = \sum_{n=0}^{N-1} x(n)e^{-j2\pi nm/N} \quad,$$
(10-71)

and

Inverse DFT →
$$x(n) = \frac{1}{N} \sum_{m=0}^{N-1} X(m)e^{j2\pi mn/N} \quad.$$
(10-72)

To reiterate our goal, we want to use the process in Eq. (10-71) to implement Eq. (10-72).

The first step of our approach is to use complex conjugation. Remember, conjugation (represented by the superscript * symbol) is the reversal of the sign of a complex number's imaginary exponent—if $x = e^{j\varnothing}$, then, $x^* = e^{-j\varnothing}$. So, as a first step, we take the complex conjugate of both sides of Eq. (10-72) to give us

$$x^*(n) = \frac{1}{N} \left[\sum_{m=0}^{N-1} X(m)e^{j2\pi mn/N} \right]^*$$
(10-73)

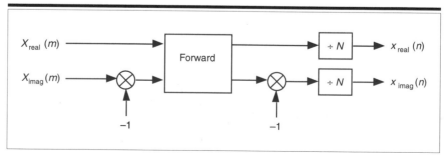

Figure 10-21 Processing diagram of first inverse FFT calculation method.

One of the properties of complex numbers, discussed in Appendix A, is that the conjugate of a product is equal to the product of the conjugates; that is, if $c = ab$, then $c^* = (ab)^* = a^*b^*$. Using this fact, we can show that the conjugate of the right side of Eq. (10-73) is given by

$$x^*(n) = \frac{1}{N} \sum_{m=0}^{N-1} X(m)^* \, (e^{j2\pi mn/N})$$

$$= \frac{1}{N} \sum_{m=0}^{N-1} X(m)^* \, e^{-j2\pi mn/N} \tag{10-74}$$

Hold on, we're almost there. Notice the similarity of Eq. (10-74) to our original forward DFT expression Eq. (10-71). If we perform a forward DFT on the conjugate of the $X(m)$ in Eq. (10-74) and divide the results by N, we get the conjugate of our desired time samples $x(n)$. Taking the conjugate of both sides of Eq. (10-74), we get a more straightforward expression for $x(n)$:

$$x(n) = \frac{1}{N} \left[\sum_{m=0}^{N-1} X(m)^* \, e^{-j2\pi mn/N} \right]^* \tag{10-75}$$

So, to get the inverse FFT of a sequence $X(m)$ using the first inverse FFT algorithm,

Step 1: Conjugate the $X(m)$ input sequence.

Step 2: Calculate the forward FFT of the conjugated sequence.

Step 3: Conjugate the forward FFT's results.

Step 4: Divide each term of the conjugated results by N to get $x(n)$.

10.6.2 Second Inverse FFT Method

The second inverse FFT calculation technique is implemented following the interesting data flow shown in Figure 10-22. In this clever inverse FFT scheme, we don't bother with conjugation. Instead, we merely swap the real and imaginary parts of sequences of complex data[22]. To see why this process works, let's look at the inverse DFT equation again while separating the input $X(m)$ term into its real and imaginary parts and remembering that $e^{j\varnothing} = \cos(\varnothing) + j\sin(\varnothing)$:

$$\text{Inverse DFT} \rightarrow \quad x(n) = \frac{1}{N} \sum_{m=0}^{N-1} X(m) e^{j2\pi mn/N}$$

$$= \frac{1}{N} \sum_{m=0}^{N-1} [X_{\text{real}}(m) + j X_{\text{imag}}(m)]$$

$$\cdot [\cos(2\pi mn/N) + j\sin(2\pi mn/N)] \ . \qquad (10\text{-}76)$$

Multiplying the complex terms in Eq. (10-76) gives us

$$x(n) = \frac{1}{N} \sum_{m=0}^{N-1} [X_{\text{real}}(m)\cos(2\pi mn/N) - X_{\text{imag}}(m)\sin(2\pi mn/N)]$$

$$+ j[X_{\text{real}}(m)\sin(2\pi mn/N) + X_{\text{imag}}(m)\cos(2\pi mn/N)] \ . \qquad (10\text{-}77)$$

Equation (10-77) is the general expression for the inverse DFT, and we'll now quickly show that the process in Figure 10-22 implements this equation. With $X(m) = X_{\text{real}}(m) + j X_{\text{imag}}(m)$ and, swapping these terms,

$$X_{\text{swap}}(m) = X_{\text{imag}}(m) + j X_{\text{real}}(m) \ . \qquad (10\text{-}78)$$

The forward DFT of our $X_{\text{swap}}(m)$ is

$$\text{Forward DFT} \rightarrow \quad \sum_{n=0}^{N-1} [X_{\text{imag}}(m) + j X_{\text{real}}(m)]$$

$$\cdot [\cos(2\pi mn/N) - j\sin(2\pi mn/N)] \ . \qquad (10\text{-}79)$$

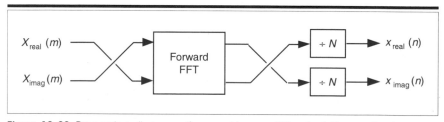

Figure 10-22 Processing diagram of second inverse FFT calculation method.

Multiplying the complex terms in Eq. (10-79) gives us

$$\text{Forward DFT} \rightarrow \sum_{n=0}^{N-1} [X_{\text{imag}}(m)\cos(2\pi mn/N) + X_{\text{real}}(m)\sin(2\pi mn/N)]$$

$$+ j[X_{\text{real}}(m)\cos(2\pi mn/N) - X_{\text{imag}}(m)\sin(2\pi mn/N)] \cdot \quad \text{(10-80)}$$

Swapping the real and imaginary parts of the results of this forward DFT gives us what we're after:

$$\text{Forward DFT}_{\text{swap}} \rightarrow \sum_{n=0}^{N-1} [X_{\text{real}}(m)\cos(2\pi mn/N) - X_{\text{imag}}(m)\sin(2\pi mn/N)]$$

$$+ j[X_{\text{imag}}(m)\cos(2\pi mn/N) + X_{\text{real}}(m)\sin(2\pi mn/N)] \quad \text{(10-81)}$$

If we divide Eq. (10-81) by N it would be equal to the inverse DFT expression in Eq. (10-77), and that's what we set out to show. To reiterate, we calculate the inverse FFT of a sequence $X(m)$ using this second inverse FFT algorithm in Figure 10-22:

Step 1: Swap the real and imaginary parts of the $X(m)$ input sequence.

Step 2: Calculate the forward FFT of the swapped sequence.

Step 3: Swap the real and imaginary parts of the forward FFT's results.

Step 4: Divide each term of the swapped sequence by N to get $x(n)$.

10.7 Fast FFT Averaging

Section 8.3 discussed the integration gain possible when averaging multiple FFT outputs to enhance signal-detection sensitivity. Well, there's a smart way to do this if we recall the linearity property of the DFT (which of course applies to the FFT) introduced in Section 3.3. If an input sequence $x_1(n)$ has an FFT of $X_1(m)$ and another input sequence $x_2(n)$ has an FFT of $X_2(m)$, then the FFT of the sum of these sequences $x_{\text{sum}}(n) = x_1(n) + x_2(n)$ is the sum of the individual FFTs, or

$$X_{sum}(m) = X_1(m) + X_2(m) \, . \tag{10-82}$$

So, if we want to average multiple FFT outputs, we can save considerable processing effort by averaging the individual FFT input sample sequences (frames), first, and then take a single FFT. Say, for example, that we wanted to average 20 FFTs to improve our FFT output signal-to-noise ratio. Instead of taking 20 FFTs of 20 frames of input signal data, we should average the 20 frames of input data, first, and then take a single FFT of that average. This avoids the number crunching necessary for 19 FFTs. By the way, for this technique to improve an FFT's signal-detection sensitivity, the original signal sampling must meet the criterion of coherent integration as described in Section 3.12.

10.8 Simplified FIR Filter Structure

If we need to implement an FIR digital filter using the standard structure in Figure 10-23(a), there's a way to simplify the necessary calculations when the filter has an odd number of taps. Let's look at the top of Figure 10-23(a) example where the 5-tap filter coefficients are $h(0)$ through $h(4)$ and the $y(n)$ output is given by

$$y(n) = h(4)x(n-4) + h(3)x(n-3) + h(2)x(n-2) + h(1)x(n-1) + h(0)x(n). \tag{10-83}$$

If the FIR filter's coefficients are symmetrical, we can reduce the number of necessary multipliers; that is, if $h(4) = h(0)$, and $h(3) = h(1)$, we can implement Eq. (10-83) by

$$y(n) = h(4)[x(n-4)+x(n)] + h(3)[x(n-3)+x(n-1)] + h(2) \cdot x(n-2) \, , \tag{10-84}$$

where only three multiplications are necessary, as shown at the bottom of Figure 10-23(b). In our 5-tap filter case, we've eliminated two multipliers at the expense of implementing two additional adders.

In the general case of symmetrical-coefficient FIR filters with S taps, we can trade $(S-1)/2$ multipliers for $(S-1)/2$ adders when S is an odd number. So, in the case of an odd number of taps, we need perform only $(S-1)/2 + 1$ multiplications for each filter output sample. For an even number of symmetrical taps as shown in Figure 10-23(b), the saving afforded by this technique reduces the necessary number of multiplications to $S/2$. For the half-band filters discussed in Section 5.7, with their alternating zero-valued coefficients, the simplified FIR structure in Figure 10-23(b) allows us to get away with only $(S+1)/4 + 1$ multiplications for

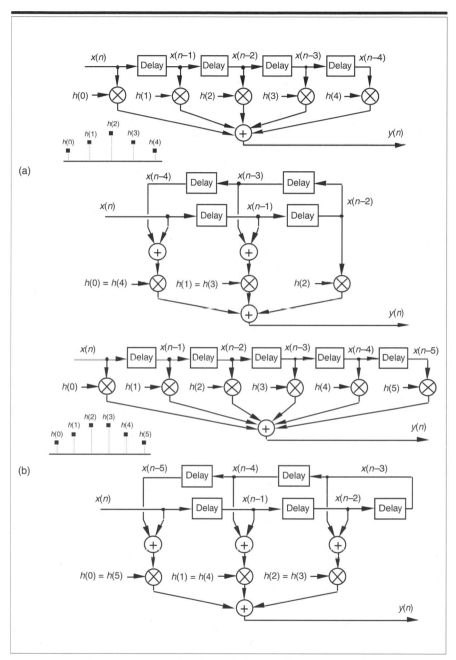

Figure 10-23 Simplified FIR filter implementations: (a) conventional and simplified, structures of an FIR filter with an odd number of taps; (b) conventional and simplified structures of an FIR filter with an even number of taps.

each filter output sample when S is odd and the first filter coefficient $h(0)$ is not zero.

We always benefit whenever we can exchange multipliers for adders. Because multiplication often takes a longer time to perform than addition, this symmetrical FIR filter simplification scheme may speed filter calculations performed in software. For a hardware FIR filter, this scheme can either reduce the number of necessary multiplier circuits or increase the effective number of taps for a given number of available hardware multipliers. Of course, whenever we increase the effective number of filter taps, we improve our filter performance for a given input signal sample rate.

10.9 Accurate A/D Converter Testing Technique

The manufacturers of A/D converters have recently begun to take advantage of digital signal processing techniques to facilitate the testing of their products. A traditional test method involves applying a sinusoidal analog voltage to an A/D converter and using the FFT to obtain spectrum of the digitized samples. Converter dynamic range, missing bits, harmonic distortion, and other nonlinearities can be characterized by analyzing the spectral content of the converter output. These nonlinearities are easy to recognize because they show up as spurious spectral components and increased background noise levels in the FFT spectra.

To enhance the accuracy of the spectral measurements, window functions were originally used on the time-domain converter output samples to reduce the spectral leakage inherent in the FFT. This was fine until the advent of 12- and 14-bit A/D converters. These converters have dynamic ranges so large that their small nonlinearities, evident in their spectra, were being swamped by the sidelobe levels of even the best window functions. (From Figure 9-4 we know that a 14-bit A/D converter can have an SNR ratio of well over 80 dB.) The clever technique that circumvents this problem is to use an analog sinusoidal input voltage whose frequency is an integral fraction of the A/D converter's sample frequency as shown in Figure 10-24(a). That frequency is mf_s/N where m is an integer, f_s is the sample frequency, and N is the FFT size. Figure 10-24(a) shows the $x(n)$ time-domain output of an ideal A/D converter under the condition that its analog input is a sinewave having exactly eight cycles over 128 output samples. In this case, the input frequency normalized to the sample rate f_s is $8f_s/128$ Hz. Recall, from Chapter 3, that the expression mf_s/N defined the analysis frequencies, or bin centers, of the DFT; and a DFT input whose frequency is at a bin center results in no leakage even without the use of a window function. Another way to look at this situation is to real-

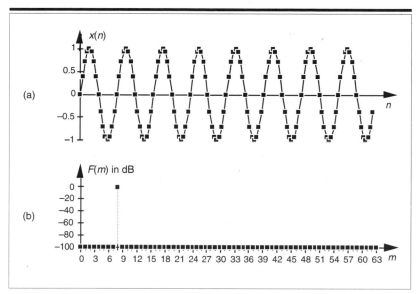

Figure 10-24 Ideal A/D converter output whose input is an analog $8f_s/128$ Hz sinusoid: (a) time-domain samples; (b) frequency-domain spectrum in dB.

ize that the analog mf_s/N frequency sinusoid will have exactly m complete cycles over the N FFT input samples, as indicated by Figure 3-7(b) in Chapter 3.

The first half of a 128-point FFT of $x(n)$ is shown in the logarithmic plot in Figure 10-24(b) where the input tone lies exactly at the $m = 8$ bin center, and DFT leakage has been avoided altogether. Specifically, if the sample rate were 1 MHz, then the A/D's input analog tone would have to be exactly $8 \cdot 10^6/128 = 62.5$ kHz. To implement this scheme, we need to ensure that the analog test generator be synchronized, exactly, with the A/D converter's clock frequency of f_s Hz. Achieving this synchronization is why this A/D converter testing procedure is referred to as *coherent sam pling*[23–25]. The analog signal generator and the A/D clock generator providing f_s must not drift in frequency relative to each other—they must remain coherent. (We must take care here from a semantic viewpoint because the quadrature sampling schemes described in Sections 7.1 and 7.2 are also sometimes called coherent sampling, but they are unrelated to this A/D converter testing procedure.)

As it turns out, some values of m are more advantageous than others. Notice in Figure 10-24(a), when $m = 8$, only nine different amplitude values are put out by the A/D converter. Those values are

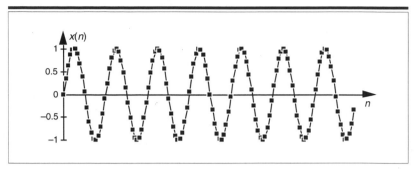

Figure 10-25 Seven-cycle sinusoidal A/D converter output.

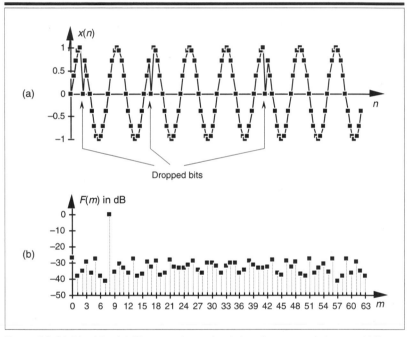

Figure 10-26 Nonideal A/D converter output showing several dropped bits:
(a) time-domain samples; (b) frequency-domain spectrum in dB.

repeated over and over. As shown in Figure 10-25, when $m = 7$, we exercise many more than nine different A/D output values. Because it's best to test as many A/D output binary words as possible, in practice, users of this A/D testing scheme have found that making m an odd prime number (3, 5, 7, 11, etc.) minimizes the number of redundant A/D output word values.

Figure 10-26(a) illustrates an extreme example of nonlinear A/D converter operation with several discrete output samples having dropped bits in the time domain $x(n)$ with $m = 8$. The FFT of this distorted $x(n)$ is shown in Figure 10-26(b) where we can see the greatly increased background noise level due to the A/D converter's nonlinearities compared to Figure 10-24(b).

To fully characterize the dynamic performance of an A/D converter, we'd need to perform this testing technique at many different input frequencies and amplitudes.[†] In addition, applying two analog tones to the A/D converter's input is often done to quantify the intermodulation distortion performance of a converter, which, in turn, characterizes the converter's dynamic range. In doing so, both input tones must comply with the mf_s/N restriction. The key issue here is that, when any input frequency is mf_s/N, we can take full advantage of the FFT's processing sensitivity while completely avoiding spectral leakage.

10.10 Fast FIR Filtering Using the FFT

While contemplating the convolution relationships in Eq. (5-31) and Figure 5-41, digital signal processing practitioners realized that convolution could sometimes be performed more efficiently using FFT algorithms than it could be using the direct convolution method[26,27]. This FFT-based convolution scheme, called *fast convolution*, is diagrammed in Figure 10-27. The standard convolution equation, for an M-tap FIR filter, given in Eq. (5-6) is repeated here for reference as

$$y(n) = \sum_{k=0}^{M-1} h(k)x(n-k) = h(k) * x(n) \ . \tag{10-85}$$

where $h(k)$ is the impulse response sequence (coefficients) of the FIR filter and the "$*$" symbol indicates convolution. It has been shown that, when the final $y(n)$ output sequence has a length greater than 30, the process in Figure 10-27 requires fewer multiplications than implementing the convolution expression in Eq. (10-85) directly. Consequently, this fast convolution technique is a very powerful signal processing tool, particularly, when used for

---

[†] The analog sinewave applied to an A/D converter must, of course, be as *pure* as possible. Any distortion inherent in the analog signal will show up in the final FFT output and could be mistaken for A/D nonlinearity.

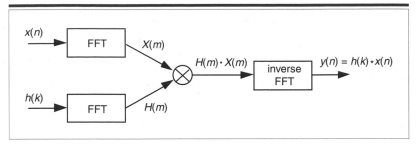

Figure 10-27 Processing diagram of fast convolution.

digital filtering. Very efficient FIR filters can be designed using this technique because, if their impulse response $h(k)$ is constant, then we don't have to bother recalculating $H(m)$ each time a new $x(n)$ sequence is filtered. In this case, the $H(m)$ sequence can be precalculated and stored in memory.

The necessary forward and inverse FFT sizes must, of course, be equal and are dependent on the length of the original $h(k)$ and $x(n)$ sequences. Recall from Eq. (5-29) that, if $h(k)$ is of length P and $x(n)$ is of length Q, the length of the final $y(n)$ sequence will be $(P+Q-1)$. For valid results from this fast convolution technique, the forward and inverse FFT sizes must be equal and greater than $(P+Q-1)$. This means that $h(k)$ and $x(n)$ must both be padded (or stuffed) with zero-valued samples at the end of their respective sequences, to make their lengths identical and greater than $(P+Q-1)$. This zero padding will not invalidate the fast convolution results. So, to use fast convolution, we must choose an N-point FFT size such that $N \geq (P+Q-1)$ and zero pad $h(k)$ and $x(n)$, so that they have new lengths equal to N.

An interesting aspect of fast convolution, from a hardware standpoint, is that the FFT indexing bit-reversal problem discussed in Sections 4.5 and 4.6 is not an issue here. If the identical FFT structures used in Figure 10-27 result in $X(m)$ and $H(m)$ having bit-reversed indices, the multiplication can still be performed directly on the scrambled $H(m)$ and $X(m)$ sequences. Then, an appropriate inverse FFT structure can be used that expects bit-reversed input data. That inverse FFT then provides an output $y(n)$ whose data index is in the correct order!

10.11 Calculation of Sines and Cosines of Consecutive Angles

There are times in digital signal processing when we need our software to calculate lots of sine and cosine values, particularly in implementing certain FFT algorithms[28,29]. Because trigonometric calculations are time consuming to perform, a clever idea has been used to calculate the sines and cosines of consecutive angles without having to actually call upon standard trigonometric functions.[†] Illustrating this scheme by way of example, let's say we want to calculate the sines and cosines of all angles from 0° to 90° in 1 degree increments. Instead of performing those 91 sine and 91 cosine trigonometric operations, we can use the following identities found in our trusty math reference book,

$$\sin(A+B) = \sin(A) \cdot \cos(B) + \cos(A) \cdot \sin(B) \qquad (10\text{-}86)$$

and

$$\cos(A+B) = \cos(A) \cdot \cos(B) - \sin(A) \cdot \sin(B) , \qquad (10\text{-}87)$$

to reduce our computational burden. To see how, lets make $A = \alpha$ and $B = n\alpha$, where $\alpha = 1°$ and n is an integral index $0 \le n \le 89$. Equations (10-86) and (10-87) now become

$$\sin(\alpha + n\alpha) = \sin(\alpha [1+n]) = \sin(\alpha) \cdot \cos(n\alpha) + \cos(\alpha) \cdot \sin(n\alpha) , \qquad (10\text{-}88)$$

and

$$\cos(\alpha + n\alpha) = \cos(\alpha [1+n]) = \cos(\alpha) \cdot \cos(n\alpha) - \sin(\alpha \cdot \sin(n\alpha) . \qquad (10\text{-}89)$$

OK, here's how we calculate the sines and cosines. First, we know the sine and cosine for the first angle of 0°; i.e., $\sin(0°) = 0$, and cosine$(0°) = 1$. Next we need to use a standard trigonometric function call to calculate and store, for later use, the sine and cosine of 1°; that is, $\sin(1°) = 0.017452$ and $\cos(1°) = 0.999848$. Now we're ready to calculate the sine and cosine of 2° by substituting $n = 1$ and $\alpha = 1°$ into Eqs. (10-88) and (10-89) giving us

$$\sin(1°[1+1]) = \sin(1°) \cdot \cos(1 \cdot 1°) + \cos(1°) \cdot \sin(1 \cdot 1°),$$

or

$$\sin(2°) = (0.017452) \cdot (0.999848) + (0.999848) \cdot (0.017452) = 0.034899, (10\text{-}90)$$

and

$$\cos(1°[1+1]) = \cos(1°) \cdot \cos(1 \cdot 1°) - \sin(1°) \cdot \sin(1 \cdot 1°),$$

or

$$\cos(2°) = (0.999848) \cdot (0.999848) + (0.017452) \cdot (0.017452) = 0.999391. (10\text{-}91)$$

[†] We're assuming here that there's insufficient memory space available to store all the required sine and cosine values for later recall.

Because we've already calculated sin(1°) and cos(1°), Eqs. (10-90) and (10-91) each required only two multiplies and an add. No trigonometric function needed to be called by software.

Next we're able to calculate the sine and cosine of 3° by substituting $n = 2$ and $\alpha = 1°$ in Eqs. (10-88) and (10-89); that is,

$$\sin(1°[1+2]) = \sin(1°) \cdot \cos(2 \cdot 1°) + \cos(1°) \cdot \sin(2 \cdot 1°),$$

or

$$\sin(3°) = (0.017452) \cdot (0.999391) + (0.999848) \cdot (0.034899) = 0.052336, (10\text{-}92)$$

and

$$\cos(1°[1+2]) = \cos(1°) \cdot \cos(2 \cdot 1°) - \sin(1°) \cdot \sin(2 \cdot 1°),$$

or

$$\cos(3°) = (0.999848) \cdot (0.999391) + (0.017452) \cdot (0.034899) = 0.998629. \ (10\text{-}93)$$

Again, because we previously calculated sin(2°) and cos(2°), Eqs. (10-92) and (10-93) only require us to perform four multiplications and two additions. The pattern of our calculations is clear now. For successive angles, we merely use the sine and cosine of 1° and the sine and cosine values obtained during the previous angle calculation. In our example, the angle increment was $\alpha = 1°$, but it's good to know that Eqs. (10-88) and (10-89) apply for any fixed angle increment.

10.12 Generating Normally Distributed Random Data

Section D.4 in Appendix D discusses the normal distribution curve as it relates to random data. A problem we may encounter is how to actually generate random data samples whose distribution follows that normal curve. There's a slick way to solve this problem using any software package that can generate uniformly distributed random data, as most of them do[30]. Figure 10-28 shows our situation pictorially where we require random data that's distributed normally with a mean (average) of μ' and a standard deviation of σ', as in Figure 10-28(a), and all we have available is a software routine that generates random data that's uniformly distributed between zero and one as in Figure 10-28(b). As it turns out, there's a principle in advanced probability theory, known as the *Central Limit Theorem*, that says, when random data from an arbitrary distribution is summed over M samples, the probability distribu-

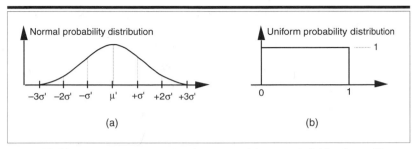

Figure 10-28 Probability distribution functions: (a) normal distribution with mean = μ', and standard deviation σ'; (b) uniform distribution between zero and one.

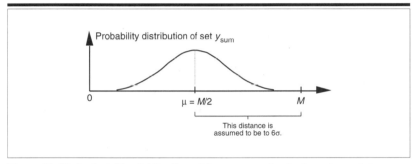

Figure 10-29 Probability distribution of the summed set of random data derived from uniformly distributed data.

tion of the sum begins to approach a normal distribution, as M increases[31,32]. In other words, if we generate a set of N random samples that are uniformly distributed between zero and one, we can begin adding other sets of N samples to the first set. As we continue summing additional sets, the distribution of the N-element set of sums becomes more and more *normal*. We can sound impressive and state that "the sum becomes asymptotically normal." Experience has shown that, for practical purposes, if we sum $M \geq 30$ times, the summed data distribution is essentially normal. With this rule in mind, we're halfway to solving our problem.

After summing M sets of uniformly distributed samples, the summed set y_{sum} will have a distribution like that shown in Figure 10-29. Because we've summed M sets, the mean of y_{sum} is $\mu = M/2$. To determine y_{sum}'s standard deviation σ, we assume that the six sigma point is equal to $M-\mu$; that is

$$6\sigma = M - \mu .$$
(10-94)

That assumption is valid because we know that the probability of an element in y_{sum} being greater than M is zero, and the probability of having a normal data sample at six sigma is one change in 6 billion, or essentially zero. Because $\mu = M/2$, then from Eq. (10-94), y_{sum}'s standard deviation is set to

$$\sigma = \frac{M - \mu}{6} = \frac{M - M/2}{6} = M/12 \; .\tag{10-95}$$

To convert the y_{sum} data set to our desired data set having a mean of μ' and a standard deviation of σ',

- subtract $M/2$ from each element of y_{sum} to shift its mean to zero,

- next, ensure that $6\sigma'$ is equal to $M/2$ by multiplying each element in the shifted data set by $12\sigma'/M$, and

- finally, center the new data set about the desired μ' by adding μ' to each element of the new data.

The steps in our algorithm are shown in Figure 10-30. If we call our desired normally distributed random data set $y_{desired}$, then the nth element of that set is described mathematically as

$$y_{desired}(n) = \frac{12\sigma'}{M} \cdot \left[\left(\sum_{k=1}^{M} x_k(n) \right) - \frac{M}{2} \right] + \mu' \; .\tag{10-96}$$

Our discussion thus far has had a decidedly software algorithmic flavor, but hardware designers also occasionally need to generate normally distributed (Gaussian) random data at high speeds in their designs. For your hardware designers, reference [33] presents an efficient hardware design technique to generate normally distributed random data using fixed-point arithmetic integrated circuits.

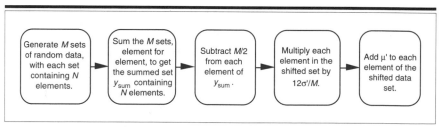

Figure 10-30 Processing steps required to generate normally distributed random data from uniformly distributed data.

References

[1] Freeny, S. "TDM/FDM Translation as an Application of Digital Signal Processing," *IEEE Communications Magazine*, January 1980.

[2] Considine, V. "Digital Complex Sampling," *Electronics Letters*, 19, 4 August 1983.

[3] Harris Semiconductor Corp. "A Digital, 16-Bit, 52 Msps Halfband Filter," *Microwave Journal*, September 1993.

[4] Hack, T. "IQ Sampling Yields Flexible Demodulators," *RF Design Magazine*, April 1991.

[5] Pellon, L. E. "A Double Nyquist Digital Product detector for Quadrature Sampling," *IEEE Trans. on Signal Processing*, Vol. 40, No. 7, July 1992.

[6] Waters, W. M., and Jarrett, B. R. "Bandpass Signal Sampling and Coherent Detection," *IEEE Trans. on Aerospace and Electronic Systems*, Vol. AES-18, No. 4, November 1982.

[7] Palacherls, A. "DSP-mP Routine Computes Magnitude," *EDN*, October 26, 1989.

[8] Mikami, N., Kobayashi, M., and Yokoyama, Y. "A New DSP-Oriented Algorithm for Calculation of the Square Root Using a Nonlinear Digital Filter," *IEEE Trans. on Signal Processing*, Vol. 40, No. 7, July 1992.

[9] Lyons, R. G. "Turbocharge Your Graphics Algorithm," *ESD: The Electronic System Design Magazine*, October 1988.

[10] Adams, W. T., and Brady, J. "Magnitude Approximations for Microprocessor Implementation," *IEEE Micro*, Vol. 3, No. 5, October 1983.

[11] Eldon, J. "Digital Correlator Defends Signal Integrity with Multibit Precision," *Electronic Design*, May 17, 1984.

[12] Smith, W. W. "DSP Adds Performance to Pulse Compression Radar," *DSP Applications*, October 1993.

[13] Harris Semiconductor Corp. HSP50110 Digital Quadrature Tuner Data Sheet, File Number 3651, February 1994.

[14] Bingham, C., Godfrey, M., and Tukey, J. "Modern Techniques for Power Spectrum Estimation," *IEEE Trans. on Audio and Electroacoust.*, Vol. AU-15, No. 2, June 1967.

[15] Harris, F. J. "On the Use of Windows for Harmonic Analysis with the Discrete Fourier Transform," *Proceedings of the IEEE*, Vol. 66, No. 1, January 1978.

[16] Nuttall, A. H. "Some Windows with Very Good Sidelobe Behavior," *IEEE Trans. on Acoust. Speech, and Signal Proc.*, Vol. ASSP-29, No. 1, February 1981.

[17] Cox, R. "Complex-Multiply Code Saves Clocks Cycles," *EDN*, June 25, 1987.

[18] Rabiner, L. R., and Gold, B. *Theory and Application of Digital Signal Processing*, Prentice-Hall, Englewood Cliffs, New Jersey, 1975, pp. 356.

[19] Sorenson, H. V., Jones, D. L., Heideman, M. T., and Burrus, C. S. "Real-Valued Fast Fourier Transform Algorithms," *IEEE Trans. on Acoust. Speech, and Signal Proc.*, Vol. ASSP-35, No. 6, June 1987.

[20] Cooley, J. W., Lewis, P. A., and Welch, P. D. "The Fast Fourier Transform Algorithm: Programming Considerations in the Calculation of Sine, Cosine and Laplace Transforms," *Journal Sound Vib.*, Vol. 12, July 1970.

[21] Brigham, E. O. *The Fast Fourier Transform and Its Applications*, Prentice-Hall, Englewood Cliffs, New Jersey, 1974, pp. 167.

[22] Burrus, C. S., et al. *Computer-Based Exercises for Signal Processing*, Prentice-Hall, Englewood Cliffs, New Jersey, 1994, pp. 53.

[23] Coleman, B., Meehan, P., Reidy, J., and Weeks, P. "Coherent Sampling Helps When Specifying DSP A/D Converters," *EDN*, October 1987.

[24] Ushani, R. "Classical Tests Are Inadequate for Modern High-Speed Converters," *EDN Magazine*, May 9, 1991.

[25] Meehan, P., and Reidy, J. "FFT Techniques Give Birth to Digital Spectrum Analyzer," *Electronic Design*, August 11, 1988, pp. 120.

[26] Stockham, T. G. "High-speed Convolution and Correlation," in *Digital Signal Processing*, Ed. by L. Rabiner and C. Rader, IEEE Press, New Jersey, 1972, pp. 330.

[27] Stockham, T. G. "High-Speed Convolution and Correlation with Applications to Digital Filtering," Chapter 7 in *Digital Processing of Signals*, by B. Gold et al., McGraw-Hill, New York, 1969, pp. 203.

[28] Dobbe, J. G. G. "Faster FFTs," *Dr. Dobb's Journal*, February 1995, pp. 125. *(Be careful here. The last equation in Example 5 is incorrect on page 133 of this reference, so be sure and use our Eq. (10-86) above.)*

[29] Crenshaw, J. W. "All About Fourier Analysis," *Embedded Systems Programming*, October 1994, pp. 70.

[30] Beadle, E. "Algorithm Converts Random Variables to Normal," *EDN Magazine*, May 11, 1995.

[31] Spiegel, M. R. *Theory and Problems of Statistics*, Shaum's Outline Series, McGraw-Hill Book Co., New York, 1961, pp. 142.

[32] Davenport, W. B., Jr., and Root, W. L. *Random Signals and Noise*, McGraw-Hill Book Co., New York, 1958, pp. 81.

[33] Salibrici, B. "Fixed-point DSP Chip Can Generate Real-time Random Noise," *EDN Magazine*, April 29, 1993.

The Arithmetic of
Complex Numbers

To understand digital signal processing, we have to get comfortable using complex numbers. The first step toward this goal is learning to manipulate complex numbers arithmetically. Fortunately, we can take advantage of our knowledge of real numbers to make this job easier. Although the physical significance of complex numbers is discussed in Appendix C, the following discussion provides the arithmetic rules governing complex numbers.

A.1 Graphical Representation of Real and Complex Numbers

To get started, real numbers are those positive or negative numbers we're used to thinking about in our daily lives. Examples of real numbers are 0.3, –2.2, 5.1, etc. Keeping this in mind, we see how a real number can be represented by a point on a one-dimensional axis, called the *real* axis, as shown in Figure A-1.

We can, in fact, consider that all real numbers correspond to all of the points on the real axis line on a one-to-one basis.

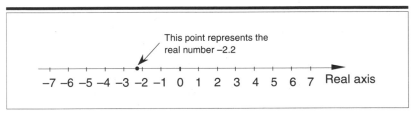

Figure A-1 The representation of a real number as a point on the one-dimensional real axis.

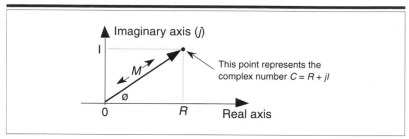

Figure A-2 The phasor representation of the complex number $C = R + jI$ on the complex plane.

A complex number, unlike a real number, has two parts: a real part and an imaginary part. Just as a real number can be considered to be a point on the one-dimensional real axis, a complex number can be treated as a point on a complex plane as shown in Figure A-2. We'll use this geometrical concept to help us understand the arithmetic of complex numbers.[†]

A.2 Arithmetic Representation of Complex Numbers

A complex number C is represented in a number of different ways in the literature, such as

Rectangular form: → $\qquad C = R + jI ,$ $\hfill$ (A-1)

Trigonometric form: → $\qquad C = M[\cos(\emptyset) + j\sin(\emptyset)] ,$ $\hfill$ (A-1')

Exponential form: → $\qquad C = M e^{j\emptyset} ,$ $\hfill$ (A-1'')

Magnitude and angle form: → $\quad C = M \angle \emptyset .$ $\hfill$ (A-1''')

Equations (A-1'') and (A-1''') remind us that the complex number C can also be considered the tip of a phasor on the complex plane, with magnitude M, in the direction of $\emptyset$ degrees relative to the positive real axis as shown in Figure A-2. (We'll avoid calling phasor M a *vector* because the term *vector* means different things in different contexts. In linear algebra, *vector* is the term used to signify a one-dimensional matrix. On the other hand, in mechanical engineering and field theory, vectors are used to signify magnitudes and directions, but there are vector operations (*scalar* or

---

[†] The complex plane representation of a complex number is sometimes called an Argand diagram—named after the French mathematician Jean Robert Argand (1768–1825).

dot product, and *vector* or *cross-product*) that don't apply to our definition of a phasor.) The relationships between the variables in this figure follow the standard trigonometry of right triangles. Keep in mind that C is a complex number, and the variables R, I, M, and ø are all real numbers. The magnitude of C, sometimes called the *modulus* of C, is

$$M = |C| = \sqrt{R^2 + I^2} , \tag{A-2}$$

and, by definition, the phase angle, or *argument*, of C is the arctangent of I/R, or

$$ø = \tan^{-1}\left(\frac{I}{R}\right) . \tag{A-3}$$

The variable ø in Eq. (A-3) is a general angle term. It can have dimensions of degrees or radians. Of course, we can convert back and forth between degrees and radians using π radians $= 180°$. So, if $ø_r$ is in radians and $ø_d$ is in degrees, then we can convert $ø_r$ to degrees by the expression

$$ø_d = \frac{180 ø_r}{\pi} . \tag{A-4}$$

Likewise, we can convert $ø_d$ to radians by the expression

$$ø_r = \frac{\pi ø_d}{180} . \tag{A-5}$$

The exponential form of a complex number has an interesting characteristic that we need to keep in mind. Whereas only a single expression in rectangular form can describe a single complex number, an infinite number of exponential expressions can describe a single complex number; that is, while, in the exponential form, a complex number C can be represented by $C = Me^{jø}$, it can also be represented by

$$C = Me^{jø} = Me^{j(ø + 2\pi n)} , \tag{A-6}$$

where $n = \pm 1, \pm 2, \pm 3, \ldots$ and ø is in radians. When ø is in degrees, Eq. (A-6) is in the form

$$C = Me^{jø} = Me^{j(ø + n360°)} . \tag{A-7}$$

Equations (A-6) and (A-7) are *almost* self-explanatory. They indicate that the point on the complex plane represented by the tip of the phasor C remains unchanged if we rotate the phasor some integral multiple of 2π radians or an integral multiple of $360°$. So, for example, if $C = Me^{j(20°)}$, then

$$C = Me^{j(20°)} = Me^{j(380°)} = Me^{j(740°)} . \tag{A-8}$$

The variable ø, the angle of the phasor in Figure A-2, need not be constant. We'll often encounter expressions containing a complex sinusoid that takes the form

$$C = Me^{j\omega t} . \tag{A-9}$$

Equation (A-9) represents a phasor of magnitude M whose angle in Figure A-2 is increasing linearly with time at a rate of ω radians each second. If $\omega = 2\pi$, the phasor described by Eq. (A-9) is rotating counterclockwise at a rate of 2π radians per second—one revolution per second—and that's why ω is called the radian frequency. In terms of frequency, Eq. (A-9)'s phasor is rotating counterclockwise at $\omega = 2\pi f$ radians per second, where f is the cyclic frequency in cycles per second (Hz). If the cyclic frequency is $f = 10$ Hz, the phasor is rotating 20π radians per second. Likewise, the expression

$$C = Me^{-j\omega t} \tag{A-9'}$$

represents a phasor of magnitude M that rotates in a clockwise direction about the origin of the complex plane at a negative radian frequency of $-\omega$ radians per second.

A.3 Arithmetic Operations of Complex Numbers

A.3.1 Addition and Subtraction of Complex Numbers

Which of the above forms for C in Eq. (A-1) is the best to use? It depends on the arithmetic operation we want to perform. For example, if we're adding two complex numbers, the rectangular form in Eq. (A-1) is the easiest to use. The addition of two complex numbers, $C_1 = R_1 + jI_1$ and $C_2 = R_2 + jI_2$, is merely the sum of the real parts plus j times the sum of the imaginary parts as

$$C_1 + C_2 = R_1 + jI_1 + R_2 + jI_2 = R_1 + R_2 + j(I_1 + I_2) . \tag{A-10}$$

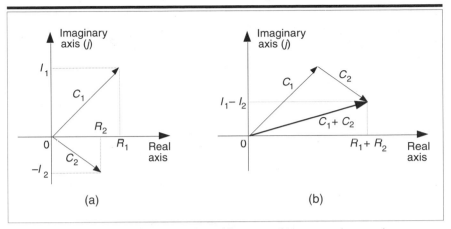

Figure A-3 Geometrical representation of the sum of two complex numbers.

Figure A-3 is a graphical depiction of the sum of two complex numbers using the concept of phasors. Here the sum phasor $C_1 + C_2$ in Figure A-3(a) is the new phasor from the beginning of phasor C_1 to the end of phasor C_2 in Figure A-3(b). Remember, the Rs and the Is can be either positive or negative numbers. Subtracting one complex number from the other is straightforward as long as we find the differences between the two real parts and the two imaginary parts separately. Thus

$$C_1 - C_2 = (R_1 + jI_1) - (R_2 + jI_2) = R_1 - R_2 + j(I_1 - I_2) \ . \qquad \text{(A-11)}$$

An example of complex number addition is discussed in Section 8.3 where we covered the topic of averaging fast Fourier transform outputs.

A.3.2 Multiplication of Complex Numbers

We can use the rectangular form to multiply two complex numbers as

$$C_1 C_2 = (R_1 + jI_1)(R_2 + jI_2) = (R_1 R_2 - I_1 I_2) + j(R_1 I_2 + R_2 I_1) \ . \qquad \text{(A-12)}$$

However, if we represent the two complex numbers in exponential form, their product takes the simpler form

$$C_1 C_2 = M_1 e^{j\varnothing_1} M_2 e^{j\varnothing_2} = M_1 M_2 e^{j(\varnothing_1 + \varnothing_2)} \qquad \text{(A-13)}$$

because multiplication results in the addition of the exponents.

As a special case of multiplication of two complex numbers, scaling is multiplying a complex number by another complex number whose imaginary part is zero. We can use the rectangular or exponential forms with equal ease as follows:

$$kC = k(R + jI) = kR + jkI ,$$ (A-14)

or in exponential form,

$$kC = k(Me^{j\varnothing}) = kMe^{j\varnothing} .$$ (A-15)

A.3.3 Conjugation of a Complex Number

The complex conjugate of a complex number is obtained by merely changing the sign of the number's imaginary part. So, if we denote C^* as the complex conjugate of the number $C = R + jI = Me^{j\varnothing}$, then, C^* is expressed as

$$C^* = R - jI = Me^{-j\varnothing} .$$ (A-16)

There are two characteristics of conjugates that occasionally come in handy. First, the conjugate of a product is equal to the product of the conjugates. That is, if $C = C_1C_2$, then from Eq. (A-13),

$$C^* = (C_1C_2)^* = (M_1M_2e^{j(\varnothing_1 + \varnothing_2)})^* = M_1M_2e^{-j(\varnothing_1 + \varnothing_2)}$$

$$= M_1e^{-j\varnothing_1}M_2e^{-j\varnothing_2} = C_1^*C_2^* .$$ (A-17)

Second, the product of a complex number and its conjugate is the complex number's magnitude squared. It's easy to show this in exponential form as

$$CC^* = Me^{j\varnothing} \cdot Me^{-j\varnothing} = M^2e^{j0} = M^2 .$$ (A-18)

(This property is often used in digital signal processing to determine the relative power of a complex sinusoidal phasor represented by $Me^{j\omega t}$.)

A.3.4 Division of Complex Numbers

The division of two complex numbers is also convenient using the exponential and magnitude and angle forms, such as

$$\frac{C_1}{C_2} = \frac{M_1 e^{j\emptyset_1}}{M_2 e^{j\emptyset_2}} = \frac{M_1}{M_2} e^{j(\emptyset_1 - \emptyset_2)} , \qquad \text{(A-19)}$$

and

$$\frac{C_1}{C_2} = \frac{M_1}{M_2} \angle \emptyset_1 - \emptyset_2 . \qquad \text{(A-19')}$$

Although not nearly so handy, we can perform complex division in rectangular notation by multiplying the numerator and the denominator by the complex conjugate of the denominator as

$$\frac{C_1}{C_2} = \frac{R_1 + jI_1}{R_2 + jI_2}$$

$$= \frac{R_1 + jI_1}{R_2 + jI_2} \cdot \frac{R_2 - jI_2}{R_2 - jI_2}$$

$$= \frac{(R_1 R_2 + jI_1 I_2) + j(R_2 I_1 - R_1 I_2)}{R_2^2 + I_2^2} . \qquad \text{(A-20)}$$

A.3.5 Inverse of a Complex Number

A special form of division is the inverse, or reciprocal, of a complex number. If $C = Me^{j\emptyset}$, its inverse is given by

$$\frac{1}{C} = \frac{1}{Me^{j\emptyset}} = \frac{1}{M} e^{-j\emptyset} . \qquad \text{(A-21)}$$

In rectangular form, the inverse of $C = R + jI$ is given by

$$\frac{1}{C} = \frac{1}{R + jI} = \frac{R - jI}{R^2 + I^2} . \qquad \text{(A-22)}$$

We get Eq. (A-22) by substituting $R_1 = 1$, $I_1 = 0$, $R_2 = R$, and $I_2 = I$ in Eq. (A-20).

A.3.6 Complex Numbers Raised to a Power

Raising a complex number to some power is easily done in the exponential form. If $C = Me^{j\varnothing}$, then

$$C^k = M^k (e^{j\varnothing})^k = M^k e^{jk\varnothing} .$$

(A-23)

For example, if $C = 3e^{j125°}$, then C cubed is

$$(C)^3 = 3^3(e^{j3 \cdot 125°}) = 27e^{j375°} = 27e^{j15°} .$$

(A-24)

   We conclude this appendix with four final complex arithmetic operations that are not very common in digital signal processing—but you may need them sometime.

A.3.7 Roots of a Complex Number

The kth root of a complex number C is the number that, multiplied by itself k times, results in C. The exponential form of C is the best way to explore this process. When a complex number is represented by $C = Me^{j\varnothing}$, remember that it can also be represented by

$$C = Me^{j(\varnothing + n360°)} .$$

(A-25)

In this case, the variable $\varnothing$ in Eq. (A-25) is in degrees. There are k distinct roots when we're finding the kth root of C. By distinct, we mean roots whose exponents are less than $360°$. We find those roots by using the following:

$$\sqrt[k]{C} = \sqrt[k]{Me^{j(\varnothing+n360°)}} = \sqrt[k]{M}e^{j(\varnothing+n360°)/k} .$$

(A-26)

Next, we assign the values 0, 1, 2, 3, . . ., $k–1$ to n in Eq. (A-26) to get the k roots of C. OK, we need an example here! Let's say we're looking for the cube (3rd) root of $C = 125e^{j(75°)}$. We proceed as follows:

$$\sqrt[3]{C} = \sqrt[3]{125e^{j(75°)}} = \sqrt[3]{125e^{j(75° +n360°)}} = \sqrt[3]{125}e^{j(75° +n360°)/3} .$$

(A-27)

Next we assign the values $n = 0$, $n = 1$, and $n = 2$ to Eq. (A-27) to get the three roots of C. So the three distinct roots are

1st root → $\quad \sqrt[3]{C} = 5e^{j(75° + 0·360°)/3} = 5e^{j(25°)}$;

2nd root → $\quad \sqrt[3]{C} = 5e^{j(75° + 1·360°)/3} = 5e^{j(435°)/3} = 5e^{j(145°)}$;

and

3rd root → $\quad \sqrt[3]{C} = 5e^{j(75° + 2·360°)/3} = 5e^{j(795°)/3} = 5e^{j(265°)}$.

A.3.8 Natural Logarithms of a Complex Number

Taking the natural logarithm of a complex number $C = Me^{j\varnothing}$ is straightforward using exponential notation; that is

$$\ln C = \ln(Me^{j\varnothing}) = \ln M + \ln(e^{j\varnothing}) = \ln M + j\varnothing , \qquad \text{(A-28)}$$

where $0 \le \varnothing < 2\pi$. By way of example, if $C = 12e^{j\pi/4}$, the natural logarithm of C is

$$\ln C = \ln(12e^{j\pi/4}) = \ln(12) + j\pi/4 = 2.485 + j0.785 . \qquad \text{(A-29)}$$

This means that $e^{(2.485 + j0.785)} = e^{2.485} \cdot e^{j0.785} = 12e^{j\pi/4}$.

A.3.9 Logarithm to the Base 10 of a Complex Number

We can calculate the base 10 logarithm of the complex number $C = Me^{j\varnothing}$ using

$$\log_{10}C = \log_{10}(Me^{j\varnothing}) = \log_{10}M + \log_{10}(e^{j\varnothing}) = \log_{10}M + j\varnothing \cdot \log_{10}(e) .^\dagger \quad \text{(A-30)}$$

Of course e is the irrational number, approximately equal to 2.71828, whose log to the base 10 is approximately 0.43429. Keeping this in mind, we can simplify Eq. (A-30) as

$$\log_{10}C \approx \log_{10}M + j(0.43429 \cdot \varnothing) . \qquad \text{(A-31)}$$

Repeating the above example with $C = 12e^{j\pi/4}$ and using the Eq. (A-31) approximation, the base 10 logarithm of C is

---

† For the second term of the result in Eq. (A-30) we used $\log_a(x^n) = n \cdot \log_a x$ according to the law of logarithms.

$$\log_{10} C = \log_{10}(12e^{j\pi/4}) = \log_{10}(12) + j(0.43429 \cdot \pi/4)$$

$$= 1.079 + j(0.43429 \cdot 0.785) = 1.079 + j0.341 \ . \qquad \text{(A-32)}$$

The result from Eq. (A-32) means that

$$10^{(1.079 + j0.341)} = 10^{1.079} \cdot 10^{j0.341} = 12 \cdot (e^{2.302})^{j0.341}$$

$$= 12e^{j(2.302 \cdot 0.341)} = 12e^{j0.785} = 12e^{j\pi/4} \ . \qquad \text{(A-33)}$$

A.3.10 Log to the Base 10 of a Complex Number Using Natural Logarithms

Unfortunately, some software mathematics packages have no base 10 logarithmic function and can calculate only natural logarithms. In this situation, we just use

$$\log_{10}(x) = \frac{\ln(x)}{\ln(10)} \qquad \text{(A-34)}$$

to calculate the base 10 logarithm of x. Using this *change of base* formula, we can find the base 10 logarithm of a complex number $C = Me^{j\o}$; that is,

$$\log_{10} C = \frac{\ln C}{\ln 10} = (\log_{10} e)(\ln C). \qquad \text{(A-35)}$$

Because $\log_{10}(e)$ is approximately equal to 0.43429, we use Eq. (A-35) to state that

$$\log_{10} C \approx 0.43429 \cdot (\ln C) = 0.43429 \cdot (\ln M + j\o) \ . \qquad \text{(A-36)}$$

Repeating, again, the example above of $C = 12e^{j\pi/4}$, the Eq. (A-36) approximation allows us to take the base 10 logarithm of C using natural logs as

$$\log_{10} C = 0.43429 \cdot (\ln(12) + \pi/4)$$

$$= 0.43429 \cdot (2.485 + j0.785) = 1.079 + j0.341 \ , \qquad \text{(A-37)}$$

giving us the same result as Eq. (A-32).

A.4 Some Practical Implications of Using Complex Numbers

At the beginning of Section A.3, we said that the choice of using the rectangular versus the polar form of representing complex numbers depends on the type of arithmetic operations we intend to perform. It's interesting to note that the rectangular form has a practical advantage over the polar form when we consider how numbers are represented in a computer. For example, let's say we must represent our complex numbers using a four-bit sign-magnitude binary number format. This means that we can have integral numbers ranging from –7 to +7, and our range of complex numbers covers a square on the complex plane as shown in Figure A-4(a) when we use the rectangular form. On the other hand, if we used 4-bit numbers to represent the magnitude of a complex number in polar form, those numbers must reside on or within a circle whose radius is 7 as shown in Figure A-4(b). Notice how the four shaded corners in Figure A-4(b) represent locations of valid complex values using the rectangular form, but are *out of bounds* if we use the polar form. Put another way, a complex number calculation, yielding an acceptable result in rectangular form, could result in an overflow error if we used polar notation in our computer. We could accommodate the complex value $7 + j7$ in rectangular form but not its polar equivalent because the magnitude of that polar number is greater than 7.

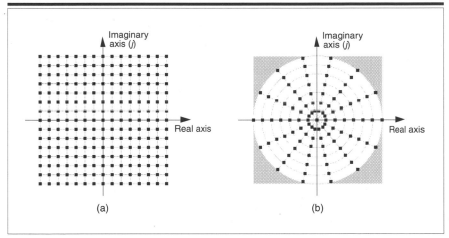

Figure A-4 Complex integral numbers represented as points on the complex plane using a four-bit sign-magnitude data format: (a) using rectangular notation; (b) using polar notation.

Although we avoid any further discussion here of the practical implications of performing complex arithmetic using standard digital data formats, it is an intricate and interesting subject. To explore this topic further, the inquisitive reader is encouraged to start with the references.

References

[1] Plauger, P. J. "Complex Math Functions," *Embedded Systems Programming*, August 1994.

[2] Kahan, W. "Branch Cuts for Complex Elementary Functions or Much Ado About Nothing's Sign Bit," *Proceedings of the Joint IMA/SIAM Conference on the State of the Art in Numerical Analysis*, Clarendon Press, 1987.

[3] Plauger, P. J. "Complex Made Simple," *Embedded Systems Programming*, July 1994.

Closed Form of a Geometric Series

In the literature of digital signal processing, we often encounter geometric series expressions like

$$\sum_{n=p}^{N-1} r^n = \frac{r^p - r^N}{1-r} \ ,$$

(B-1)

or

$$\sum_{n=0}^{N-1} e^{-j2\pi nm/N} = \frac{1 - e^{-j2\pi m}}{1 - e^{-j2\pi m/N}} \ .$$

(B-2)

Unfortunately, many authors make a statement like "and we know that," and drop Eqs. (B-1) or (B-2) on the unsuspecting reader who's expected to accept these expressions on faith. Assuming that you don't have a Ph.D. in mathematics, you may wonder exactly what arithmetic sleight of hand allows us to arrive at Eqs. (B-1) or (B-2). To answer this question, let's consider a general expression for a geometric series such as

$$S = \sum_{n=p}^{N-1} ar^n = ar^p + ar^{p+1} + ar^{p+2} + \ldots + ar^{N-1} \ ,$$

(B-3)

where n, N, and p are integers and a and r are any constants. Multiplying Eq. (B-3) by r, gives us

$$Sr = \sum_{n=p}^{N-1} ar^{n+1} = ar^{p+1} + ar^{p+2} + \ldots + ar^{N-1} + ar^N \ .$$

(B-4)

Subtracting Eq. (B-4) from Eq. (B-3) gives the expression

$$S - Sr = S(1 - r) = ar^p - ar^N ,$$

or

$$S = a \cdot \frac{r^p - r^N}{1 - r} . \tag{B-5}$$

So here's what we're after. The *closed form* of the series is

Closed form of a general geometric series: →
$$\sum_{n=p}^{N-1} ar^n = a \cdot \frac{r^p - r^N}{1 - r} . \tag{B-6}$$

(By closed form, we mean taking an infinite series and converting it to a simpler mathematical form without the summation.) When $a = 1$, Eq. (B-6) validates Eq. (B-1). We can quickly verify Eq. (B-6) with an example. Letting $N = 5$, $p = 0$, $a = 2$, and $r = 3$, for example, we can create the following list:

n	$ar^n = 2 \cdot 3^n$
0	$2 \cdot 3^0 = 2$
1	$2 \cdot 3^1 = 6$
2	$2 \cdot 3^2 = 18$
3	$2 \cdot 3^3 = 54$
4	$2 \cdot 3^4 = 162$
	The sum of this column is $$\sum_{n=0}^{4} 2 \cdot 3^n = 242 .$$

Plugging our example N, p, a, and r values into Eq. (B-6),

$$\sum_{n=p}^{N-1} ar^n = a \cdot \frac{r^p - r^N}{1 - r} = 2 \cdot \frac{3^0 - 3^5}{1 - 3} = 2 \cdot \frac{1 - 243}{-2} = 242 , \tag{B-7}$$

which equals the sum of the rightmost column in the list above.

As a final step, the terms of our earlier Eq. (B-2) are in the form of Eq. (B-6) as $p = 0$, $a = 1$, and $r = e^{-j2\pi m/N}$.[†] So plugging those terms from Eq. (B-2) into Eq. (B-6) gives us

$$\sum_{n=0}^{N-1} e^{-j2\pi nm/N} = 1 \cdot \frac{e^{-j2\pi m0/N} - e^{-j2\pi mN/N}}{1 - e^{-j2\pi m/N}} = \frac{1 - e^{-j2\pi m}}{1 - e^{-j2\pi m/N}} , \qquad \text{(B-8)}$$

confirming Eq. (B-2).

[†] From the math identity $a^{xy} = (a^x)^y$, we can say $e^{-j2\pi nm/N} = (e^{-j2\pi m/N})^n$, so $r = e^{-j2\pi m/N}$.

Complex Signals and Negative Frequency

Complex numbers are used in just about every field of science and engineering.[†] For us, complex signal notation is necessary to describe the effects of quadrature and Fourier processing. It may be fair to say that no other topic in digital signal processing causes more initial discomfort for the beginner than *complex* numbers with their *real* and *imaginary* parts. Why, even the terminology is bizarre. So here's where we review the fundamentals of complex numbers, get comfortable with how they're used to represent complex signals, and examine the notion of negative frequency as it relates to complex signal notation.

To start this discussion, it's important for us to be comfortable with the concept of negative frequency because it's essential in understanding the spectral replication effects of periodic sampling, discrete Fourier transforms, and the various quadrature signal processing techniques discussed in Chapter 7. The convention of negative frequency serves as a consistent and powerful mathematical tool in our analysis of signals. In fact, the use of negative frequency is mandatory when we represent *real signals*, such as a sine or cosine wave, in complex notation.

The difficulty in grasping the idea of negative frequency may be, for some, similar to the consternation felt in the parlors of mathematicians in the Middle Ages when they first encountered negative numbers. Until the thirteenth century, negative numbers were considered *fictitious* because numbers were normally used for counting and measuring. So up to that time, negative numbers just didn't make sense [1,2]. In those days, it was valid to ask, "How can you hold something in your hand that is less than nothing?" The idea of subtracting six from four must have seemed meaningless. Math historians suggest that negative num-

---

[†] That's because complex sinusoids are solutions to the second-order linear differential equations used to describe so much of nature.

bers were first analyzed in Italy. As the story goes, around the year 1200, the Italian mathematician Leonardo da Pisa (known as Fibonacci) was working on a financial problem whose only valid solution involved a negative number. Undaunted, Leo wrote, "This problem, I have shown to be insoluble unless it is conceded that the first man had a debt." Thus negative numbers arrived on the mathematics scene, never again to be disregarded.

Modern men and women can now appreciate that negative numbers have a direction associated with them. The direction is backward from zero in the context that positive numbers point forward from zero. For example, negative numbers can represent temperatures measured in degrees below zero, minutes before the present if the present is considered as zero time, or money we owe the tax collector when our income is considered positive dollars. So the notion of negative quantities is

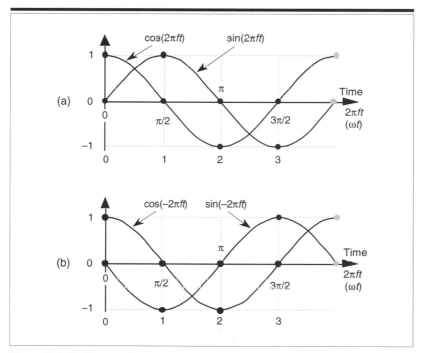

Figure C-1 Positive and negative frequency sinusoids: (a) sine and cosine waves at a positive radian frequency ($2\pi ft$); (b) sine and cosine waves whose arguments are a negative frequency ($-2\pi ft$).

[†] Shortly, we'll represent negative frequency sinusoids graphically using phasor notation and see that the negative frequency phasors are merely rotating in the direction opposite from an arbitrary positive reference direction.

perfectly valid if we just define it properly. As comfortable as we now are with negative numbers, negative frequency remains a troublesome and controversial concept for many engineers [1,2]. This author once encountered a paper in a technical journal which stated: "since negative frequencies cannot exist—." Well, like negative numbers, negative frequency is a perfectly valid concept as long as we define it properly relative to what we're used to thinking of as positive frequency. In radians, a negative frequency sinusoid is merely one whose angular argument is negative, as shown in Figure C-1(b).[†]

C.1 Development of Imaginary Numbers

Before we go further in our discussion of negative frequency by way of complex notation, we must first take a deep breath and enter the Twilight Zone of the "j" operator and imaginary numbers. You've seen the definition $j = \sqrt{-1}$ before. Stated in words, we say that j represents a number which, when multiplied by itself, results in a negative one. Well, this definition causes initial difficulty for the beginner because we know that any number multiplied by itself always results in a positive number. (Unfortunately digital signal processing textbooks typically define j and, then, with justified haste, swiftly carry on with all the ways that the j operator can be used to analyze sinusoidal signals. Readers soon forget about the question: What does $j = \sqrt{-1}$ actually mean? Well, $\sqrt{-1}$ had been on the mathematical scene for some time but wasn't taken seriously until it had to be used in the sixteenth century to solve cubic equations [3, 4]. Mathematicians reluctantly began to accept the abstract concept of $\sqrt{-1}$ without having to visualize it, as long as its mathematical properties were consistent with the arithmetic of *normal* real numbers. It was Euler's introduction of the complex plane, used to solve second-order differential equations, that further legitimized the notion of $\sqrt{-1}$ to Europe's mathematicians in the eighteenth century.[†] Euler showed how complex numbers using the $\sqrt{-1}$ operator had a clean, consistent relationship to the well-known trigonometric functions of sines and cosines [3].

We can get more comfortable with the complex plane representation of imaginary numbers by examining the mathematical properties of the $j = \sqrt{-1}$ operator as shown in Figure C-2(a). Multiplying any number on

---

[†] Leonhard Euler, born in Switzerland in 1707, is considered by many historians as the world's greatest mathematician. By the way, in case you don't speak German, the name *Euler* is pronounced as "oiler."

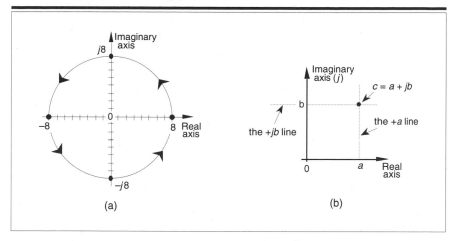

Figure C-2 Numbers on the complex plane: (a) the 90° rotations when a number is multiplied by *j*; (b) a complex number, *c* = *a+jb*, having both real and imaginary parts on the complex plane.

the real axis by *j* results in an imaginary product that lies on the imaginary axis. The example in Figure C-2(a) shows that if +8 is represented by the dot lying on the positive real axis, multiplying +8 by *j* results in an imaginary number, *+j8*, whose position has been rotated 90° counterclockwise from +8, putting it on the positive imaginary axis. Similarly, multiplying *+j8* by *j* results in another 90° rotation yielding the –8 lying on the negative real axis because $j^2 = -1$. Multiplying –8 by *j* results in a further 90° rotation giving the *–j8* lying on the negative imaginary axis. Whenever any number represented by a dot is multiplied by *j*, the result is a counterclockwise rotation of 90° on the complex plane. (Conversely, multiplication by –*j* results in a clockwise rotation of –90° on the complex plane.) The complex plane enables us to represent complex numbers having both real and imaginary parts. For example in Figure C-2(b), the complex number *c* = *a+jb* is a point lying on the complex plane on neither the real nor the imaginary axis. We locate point *c* by going +*a* units along the real axis and up +*b* units along the imaginary axis. (Just to refresh the reader's memory, Appendix A provides a discussion of the arithmetic of complex numbers.) We can, if we care to, think of all numbers as being complex. Thus, real numbers are complex numbers with no imaginary part, and imaginary numbers are complex numbers with no real part.

Let's pause for a moment here to catch our breath. Don't worry if the ideas of imaginary numbers and the complex plane seem a little mysterious. It's that way for everyone at first—you'll get comfortable with them the more you use them. (Remember, the $\sqrt{-1}$ operator puzzled Europe's heavyweight mathematicians for hundreds of years.) Not only is the mathematics of complex numbers a bit strange at first, but the terminology is unfortunate. Although the term *imaginary* is an unfortunate one to use, the term *complex* is downright weird. When first encountered, the phrase *complex numbers* makes one think "complicated numbers." This is regrettable because the concept of complex numbers is not really all that complicated.[†]

C.2 Representing Real Signals Using Complex Phasors

OK, now we can turn our attention back to the original discussion of understanding negative frequency when using the mathematics of complex numbers. We start by representing a real sinusoid in the context of the complex plane. To see how this is done, consider the waveform in Figure C-3. Think of a cosine wave defined by $2\cos(\omega t)$ oscillating back and forth on the real axis as time passes. (Here the ω term is frequency in radians/s, and it corresponds to a frequency of $2\pi f$ cycles/s where f is a single cycle/s or 1 Hz.) Now, consider two complex phasors rotating in opposite directions about the time axis, one rotating counterclockwise with a positive angular change ωt and the other phasor rotating with a negative angle, $-\omega t$. The angles, of course, change as a function of time making the $e^{j\omega t}$ and $e^{-j\omega t}$ phasors rotate. Figure C-4 attempts to show, as time passes, one complete rotation of the $e^{j\omega t}$ phasor, for example. That phasor's tip follows a corkscrew path spiraling along and centered about the time axis.

Let's show how the cosine wave is the phasor sum of the two complex phasors. We can visualize the phasors' ωt and $-\omega t$ angles better if we orient the three-dimensional view in Figure C-3 so that we're looking right straight down the time axis in the negative time direction as shown in Figure C-5. The time arrow is coming straight out of the page toward the reader and the two complex phasors, frozen in time, are represented by the bold arrows. Figure C-5 looks reasonably simple, but it contains a wealth of information. Let's first consider the $e^{j\omega t}$ phasor. Using the real and imaginary axis, we can see why the $e^{j\omega t}$ phasor is defined by

---

[†] The brilliant American engineer Charles P. Steinmetz, who pioneered the use of real and imaginary numbers in electrical circuit analysis in the early twentieth century, refrained from using the term *complex numbers*—he called them "general numbers."

Euler's equation: →
$$e^{j\omega t} = \cos(\omega t) + j\sin(\omega t) .$$
(C–1)

Named in his honor, Eq. (C-1) is now called *Euler's equation*. This expression defines the position of the tip of the $e^{j\omega t}$ arrow in the complex plane. The $\cos(\omega t)$ term in Eq. (C-1) is the phasor's component along the real axis, and the $\sin(\omega t)$ term is the phasor's component along the imaginary, or j, axis. It's very important to realize that the phasor's component projected on the imaginary axis is the real number $\sin(\omega t)$, and *not* $j\sin(\omega t)$. The $+j$ in the second term of Eq. (C-1) merely means that the phasor has a component along

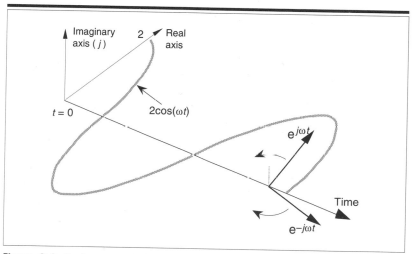

Figure C-3 Cosine represented by two rotating complex phasors.

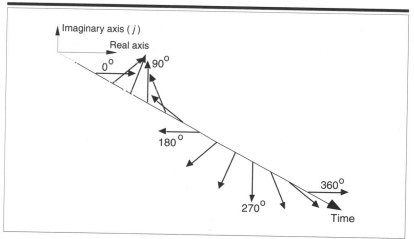

Figure C-4 The rotation of the $e^{j\omega t}$ phasor about the time axis.

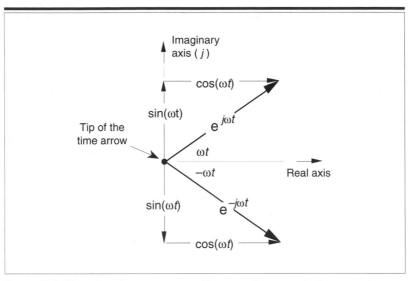

Figure C-5 Complex phasors representing a cosine wave in the complex plane.

the positive j axis whose magnitude is $\sin(\omega t)$. Likewise, the $e^{-j\omega t}$ phasor has a component along the negative j axis whose magnitude is $\sin(\omega t)$. We describe the $e^{-j\omega t}$ phasor with the expression

Another version of Euler's equation: $\rightarrow$ $\qquad e^{-j\omega t} = \cos(\omega t) - j\sin(\omega t)$.† $\qquad$ (C-2)

The positive angle ωt is (arbitrarily) defined as the counterclockwise angle between the real axis and the phasor. As time t increases, angle ωt increases and the $e^{j\omega t}$ phasor rotates counterclockwise. The $e^{-j\omega t}$ phasor rotates in the clockwise direction at the same angular frequency as the $e^{j\omega t}$ phasor; that is, when ωt is zero radians, both phasors are at 3:00 o'clock, along the real axis. As times passes, when the $e^{j\omega t}$ phasor is at twelve o'clock, the $e^{j\omega t}$ phasor is at six o'clock. As the phasors continue rotating, they pass each other, again, exactly at nine o'clock.

Let's look further and see how the sum of the $e^{j\omega t}$ and $e^{-j\omega t}$ phasors represents a real cosine wave. In Figure C-6(a), the phasor angle is $\omega t = \pi/4$ (45 degrees). The bold arrow representing the real cosine wave, $2\cos(\omega t)$, is the phasor sum of the $e^{j\omega t}$ and $e^{-j\omega t}$ phasors. Notice how the cosine wave's arrow

---

† Other versions of Euler's equation, which we can derive from Eqs. (C-1) and (C-2), are $\sin(\omega t) = (e^{j\omega t} - e^{-j\omega t})/2j$, and $\cos(\omega t) = (e^{j\omega t} + e^{-j\omega t})/2$.

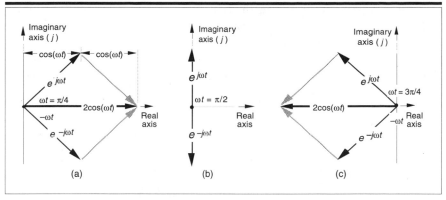

Figure C-6 Complex phasors representing a cosine wave: (a) when $\omega t = \pi/4$; (b) when $\omega t = \pi/2$; (c) when $\omega t = 3\pi/4$.

tip is along the real axis. At a later time, when $\omega t = \pi/2$ radians, the $e^{j\omega t}$ phasor has rotated counterclockwise, the $e^{-j\omega t}$ phasor has rotated clockwise, and the $2\cos(\omega t)$ wave's arrow tip has moved to the left and sits exactly at the zero point on the real axis as shown in Figure C-6(b). We know this is correct because $2\cos(\pi/2) = 0$. As time passes and $\omega t = 3\pi/4$ radians, the $e^{j\omega t}$ and $e^{-j\omega t}$ phasors and their resulting $2\cos(\omega t)$ phasor are shown in Figure C-6(c), which is real and negative. As the complex phasors continue rotating, the $2\cos(\omega t)$ phasor tip merely slides back and forth always remaining on the real axis. Figure C-3 attempts to illustrate this by showing the $2\cos(\omega t)$ wave always lying in the plane of the real axis in our three-dimensional depiction.

Thinking about these phasors, it's clear, now, why the cosine wave can be equated to the sum of two complex phasors by

$$\cos(\omega t) = \frac{e^{j\omega t} + e^{-j\omega t}}{2} \, , \tag{C-3}$$

which is half the sum of Eqs. (C-1) and (C-2). (The example in Figure C-3 used a cosine wave with a peak amplitude of 2 merely to avoid cluttering Figures C-3 through C-6 with a denominator of 2 below each of the $e^{j\omega t}$ and $e^{-j\omega t}$ terms.) If Eq. (C-3) describes a real cosine wave in the complex plane, how do we represent a real sinewave? Well, we can combine Eqs. (C-1) and (C-2) and solve for $\sin(\omega t)$ to get the following standard expression for a sinewave in complex notation:

$$\sin(\omega t) = \frac{e^{j\omega t} - e^{-j\omega t}}{2j} \, .$$
$$\tag{C-4}$$

To investigate the meaning of this expression, we can multiply both sides of Eq. (C-4) by 2 to get the following alternate expression for a sinewave in terms of complex exponentials:

$$2\sin(\omega t) = \frac{e^{j\omega t} - e^{-j\omega t}}{j} = -j(e^{j\omega t} - e^{-j\omega t}) \ . \tag{C-5}$$

Equation (C-5) looks a little strange, doesn't it? It states that a real sinewave is equal to $-j$ times the sum of two complex numbers! Well, we can decipher Eq. (C-5) by graphically illustrating the complex exponentials as phasors just as we did in Figure C-5. A depiction of just the $(e^{j\omega t} - e^{-j\omega t})$ factor from Eq. (C-5) is the phasors in Figure C-7(a). To account for the $-j$ term in Eq. (C-5), we rotate the phasors in Figure C-7(a) by $\pi/2$, or 90°, in the clockwise direction. So the full $-j(e^{j\omega t} - e^{-j\omega t})$ expression on the right side of Eq. (C-5) is shown in Figure C-7(b). Because the $-e^{-j\omega t}$ phasor rotates clockwise and the $e^{\omega t}$ phasor rotates counterclockwise, their phasor sum representing $2\sin(\omega t)$ will always lie on the real axis. We know this is true because the $2\sin(\omega t)$ sinewave in Eq. (C-5) is real—like a cosine wave, it has no imaginary part.

To belabor this point a bit further, consider Figure C-8. When the angle $\omega t = 0$, both the $je^{-j\omega t}$ and $-je^{j\omega t}$ phasors lie on the imaginary axis, and their phasor sum sits exactly at the zero point of both axes, as shown in Figure C-8(a). We know this is correct because $2\sin(0) = 0$. At a later time when $\omega t = \pi/4$ radians, the $je^{-j\omega t}$ phasor has rotated clockwise, the $-je^{j\omega t}$ phasor

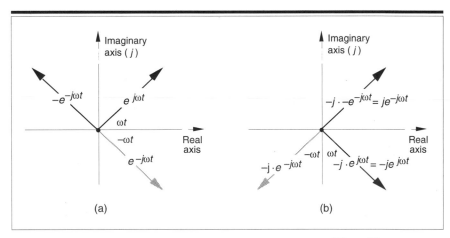

Figure C-7 Complex phasors: (a) phasor representation of $(e^{j\omega t} - e^{-j\omega t})$; (b) phasor representation of $-j(e^{j\omega t} - e^{-j\omega t})$, or $2\sin(\omega t) = -je^{j\omega t} + je^{-j\omega t}$.

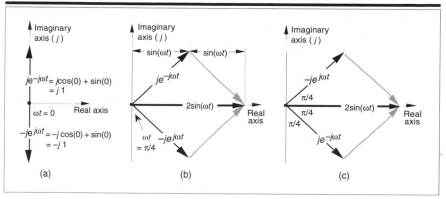

Figure C-8 Complex phasors representing a $2\sin(\omega t)$ sinewave: (a) when $\omega t = 0$; (b) when $\omega t = \pi/4$; (c) when $\omega t = 3\pi/4$.

has rotated counterclockwise, and the $2\sin(\omega t)$ wave's arrow tip has moved to the right to a positive point on the real axis, as shown in Figure C-8(b). As time passes and $\omega t = 3\pi/4$ radians, the $-je^{j\omega t}$ phasor is now in the northeast quadrant, and $je^{-j\omega t}$ phasor is directed southeast. They and their $2\sin(\omega t)$ phasor sum are shown in Figure C-8(c). As the complex phasors continue rotating, the $2\sin(\omega t)$ phasor oscillates back and forth, always remaining on the real axis.

To keep the reader's mind from spinning like our complex phasors, please realize that the sole purpose of Figures C-6 through C-8 is to validate the complex expressions of the cosine and sinewave given in Eqs. (C-3) and (C-4).

C.3 Representing Real Signals Using Negative Frequencies

OK, let's keep in mind one of the goals of this discussion—why the concept of negative frequency is valid when real signals are represented in complex notation. We can do this by comparing the relationship between a $\cos(\omega_o t)$ wave and a $\sin(\omega_o t)$ wave, using Eqs. (C-3) and (C-4) and combining them graphically on a three-dimensional complex frequency axis as shown in Figure C-9. If we say that the angular frequency is ω_o radians/second = $2\pi f_o$ Hz, the time-varying composite angle is $\omega_o t$ radians = $2\pi f_o t$ cycles at time t. Look at Figure C-9 very carefully—it's a snapshot of the complex phasors that comprise a real cosine wave (solid arrows) from Eq. (C-3) and a real sinewave (shaded arrows) from Eq. (C-4) when time equals zero, or $\omega_o t = 0$.

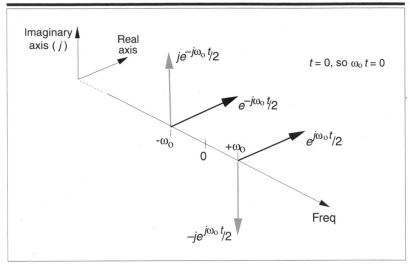

Figure C-9 Relationship of the complex representations of the $\cos(\omega_o t) = (e^{j\omega_o t} + e^{-j\omega_o t})/2$ (solid arrows) and the $\sin(\omega_o t) = (e^{j\omega_o t} - e^{-\omega_o t})/j2 = (-je^{j\omega_o t} + je^{-\omega_o t})/2$ (shaded arrows), when $\omega_o t = 0$.

This figure shows that, when we use complex notation, $e^{j\omega t}$ and $e^{-j\omega t}$ are the fundamental constituents of a sinusoid, not $\sin(\omega t)$ or $\cos(\omega t)$ because both $\sin(\omega t)$ and $\cos(\omega t)$ are made up of $e^{j\omega t}$ and $e^{-j\omega t}$ components. On the frequency axis, the notion of negative frequency is seen as those phasors located at $-\omega_0$ radians on the frequency axis. (If we were to take the discrete Fourier transform of a cosine wave and plot the complex results, we'd get exactly those bold phasors in Figure C-9. Likewise, the discrete Fourier transform of a sinewave results in the shaded phasors in Figure C-9.) Figure C-9 also reiterates the fact that a cosine wave is merely a sinewave shifted in time. If we let the phasors in Figure C-9 rotate 90° ($\pi/2$ radians, or a quarter of a cycle), they're oriented as shown in Figure C-10. So the sinewave's shaded phasors, in Figure C-10, are oriented exactly as the cosine wave's phasors in Figure C-9. The relationship between Figures C-9 and C-10 illustrates the two following trigonometric identities:

$$\sin(\varnothing) = \cos(\varnothing - \pi/2) , \tag{C-6}$$

and

$$\cos(\varnothing) = -\sin(\varnothing - \pi/2) . \tag{C-7}$$

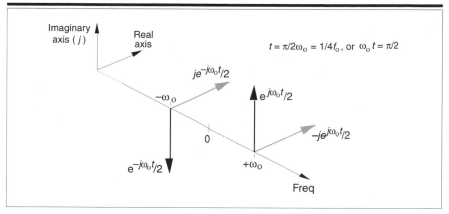

Figure C-10 Relationship of the complex representations of $\cos(\omega_o t)$ (solid arrows) and $\sin(\omega_o t)$ (shaded arrows) when $\omega_o t = \pi/2$.

Putting Eqs. (C-6) and (C-7) into words relative to the two figures above, we can say that the sine phasors in Figure C-10 are equal to the cosine phasors in Figure C-9, and the cosine phasors in Figure C-10 are equal to the negative of the sine phasors in Figure C-9. This tells us that a sinusoidal wave can be described as a sinewave or a cosine wave depending on the time we start looking at it; that is, a delay in the time domain manifests itself as a phase shift in the frequency domain. This effect is very important in digital signal processing, so remember it.[†]

If we use the same axis as before, we can illustrate a sinusoidal wave at an arbitrary phase angle ø, $A\cos(\omega_o t + ø)$, as shown by the bold phasors in Figure C-11(a). The magnitude of the two bold phasors is, of course, $A/2$, so they'll sum to A when they're aligned along the real axis. Figure C-11(a) shows that the phasors are the *phasor sum* of their real and imaginary parts. Figure C-11(b) shows the real and imaginary components aligned with the real and imaginary axis. Why bother showing the real and imaginary parts as in Figure C-11(b)? Because this signal representation leads us to the graphical form used so often in the digital signal processing literature of quadrature signal processing.

Think for a moment of a real signal comprising seven sinusoids over a bandwidth B centered about a carrier frequency of ω_o. We could represent that signal's spectrum by the bold phasors in the complex frequency domain of Figure C-12(a). If we had another signal with an unlimited number of sinusoids over the bandwidth B, we could forego drawing

---

[†] For additional examples of this shifting property, check out Section 3.6.

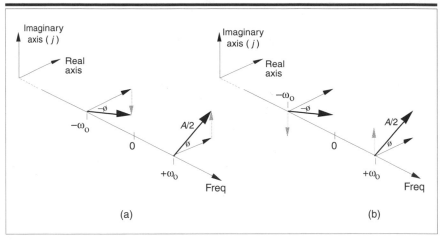

Figure C-11 Two depictions of the complex representation of $A\cos(\omega_o t + \varnothing)$ indicated by the bold phasors. The thin solid arrows are the real parts, and the thin shaded arrows are the imaginary parts.

individual phasors and merely show the spectrum by joining the tips of the phasors with the bold lines in Figure C-12(b). This signal has a continuous spectral envelope over the bandwidth B, and that's the typical signal spectrum representation used in the literature of quadrature processing. Figure C-12(b) shows the projection of the signal's constant-amplitude rectangular spectrum on both the real and the imaginary axis.

In quadrature processing, by convention, the real part of the spectrum is called the *in-phase* component, and the imaginary part of the spectrum is called the *quadrature*-phase component. The signal in Figure C-12(b) is real, and, in the time domain, it can be represented digitally by a series of amplitude values that have nonzero real parts and zero-valued imaginary parts. We're not forced to use complex notation to represent it in the time domain—the signal is real. (If it was a continuous physical signal, it could be transmitted over a single conductor.) Real signals always have positive and negative frequency spectral components. For any real signal, the positive and negative frequency components of its in-phase (real) spectrum always have even symmetry about the zero frequency point; that is, the in-phase part's positive and negative frequency components are mirror images of each other. Conversely, the positive and negative frequency components of its quadrature (imaginary) spectrum are always complex conjugates of each other [6]. This means that the phase angle of any given positive quadrature frequency component is the negative of the phase angle of the correspond-

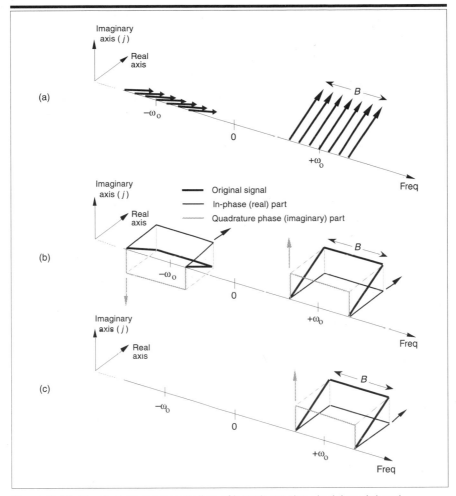

Figure C-12 Quadrature representation of bandpass signals: (a) real signal containing seven sinusoids over bandwidth *B*; (b) real signal containing an infinite number of sinusoids over bandwidth *B*; (c) complex signal of bandwidth *B*.

ing quadrature negative frequency component, as shown by the dashed arrows in Figure C-11. This is the invariant nature of real signals when their spectra are represented using complex notation.

C.4 Complex Signals and Quadrature Mixing

Complex signals, on the other hand, are a different story. Figure C-12(c) shows the spectrum of an arbitrary complex signal of bandwidth *B*. In the time domain, it can only be represented by amplitude values that have nonzero real and nonzero imaginary parts. We're forced to use complex notation to represent it in the time domain—the signal is complex. (As a continuous physical signal, a complex signal could only be transmitted using two conductors.) Complex signals aren't forced to have the symmetric real and the conjugate symmetric imaginary spectral components characteristic of real signals—complex signals can have just positive (or just negative) frequency components, as shown in Figure C-12(c).

We can summarize all that we've covered in this chapter by considering an interesting topic—the spectrum translation effects of quadrature mixing [7]. Figure C-13(a) shows the triangular spectrum of an arbitrary real signal centered about ω_o radians/s. In quadrature mixing, we translate the original signal's spectrum down to the zero frequency point before digitization. This is typically done by multiplying the original signal by a cosine wave of ω_o radians/s, in the time domain, to get the in-phase spectral component of the input signal as shown in Figure C-13(b). It's important to see that Figure C-13(b) is the product of Figure C-13(a) and the solid cosine phasors in Figure C-9. The $e^{-j\omega_o t}/2$ complex cosine phasor from Figure C-9 translates both Figure C-13(a)'s original input signal spectral bandwidths toward the negative frequency direction, as shown in Figure C-13(b). Likewise, the $e^{j\omega_o t}/2$ complex cosine phasor from Figure C-9 translates both original input signal spectral bandwidths toward the positive frequency direction, also shown in Figure C-13(b). Although not obvious in the figure, the magnitude in Figure C-13(b) is half the magnitude of that in Figure C-13(a) because of the $1/2$ scale factors in the cosine expressions in Eq. (C-3) and Figure C-9.

To get the quadrature-phase spectral components of our input signal, we multiply the input signal by a sinewave of ω_o radians/s to get the spectrum shown in Figure C-13(c). Again, Figure C-13(c) is the product of Figure C-13(a) and the dashed sine phasors in Figure C-9. The $je^{-j\omega_o t}/2$ dashed complex sine phasor from Figure C-9 translates both Figure C-13(a)'s original input signal spectral bandwidths toward the negative frequency direction, but aligned with the imaginary $+j$ axis as shown in Figure C-13(c). The $-je^{j\omega_o t}/2$ dashed complex sine phasor from Figure C–9 translates both original input signal spectral bandwidths toward the positive frequency direction aligned with the imaginary $-j$ axis, also shown in Figure C-13(c). Figure C-13(d) shows both the in-phase and quadrature spectral bands combined on the complex frequency plane axis. (Examples of this quadrature mixing scheme are discussed in Sections 7.1 and 7.2.)

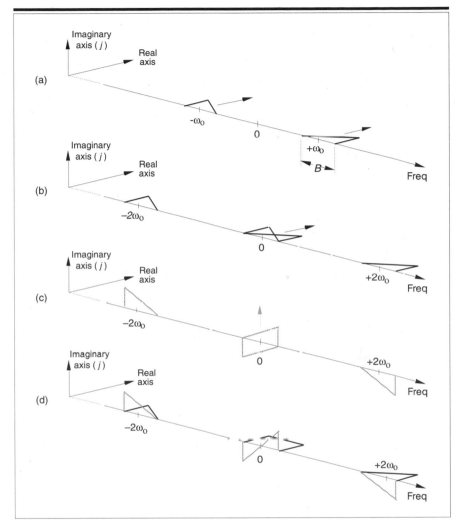

Figure C-13 Quadrature mixing of a real bandpass signal with even symmetric spectral components: (a) original real input signal of bandwidth B; (b) in-phase spectral components of the input signal; (c) quadrature-phase spectral components of the input signal; (d) combined in-phase and quadrature-phase components.

Providing another example to demonstrate the frequency translation that takes place in complex signal mixing, Figure C-14 shows the spectral results of quadrature mixing when the input signal is real and has conjugate positive and negative frequency components. Section 10.1, with its discussion of digital mixing sequences, provides several practical examples of

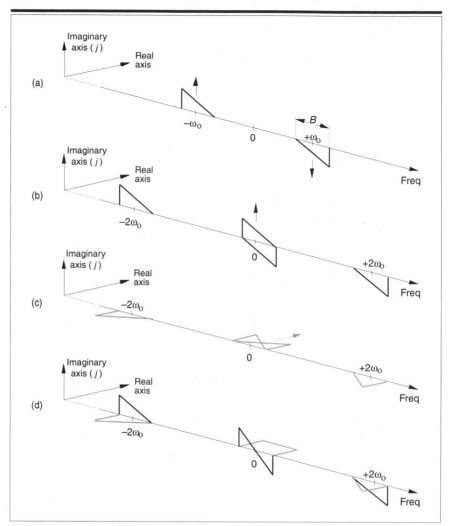

Figure C-14 Quadrature mixing of a real bandpass signal with complex conjugate symmetric spectral components: (a) original real input signal of bandwidth B; (b) in-phase spectral components of the input signal; (c) quadrature-phase spectral components of the input signal; (d) combined in-phase and quadrature-phase components.

the spectral translation effects of quadrature mixing. That concludes our introduction to complex signals and negative frequency. For further material on these topics, the reader is encouraged to review reference [8], a terrific series of papers detailing how complex signals and negative frequencies relate to modern-day digital communications signals.

References

[1] Lewis, L. J., et al. *Linear Systems Analysis*, McGraw-Hill Inc., New York, 1969, pp. 193.

[2] Schwartz, M. *Information, Transmission, Modulation, and Noise*, McGraw-Hill Inc., New York, 1970, pp. 35.

[3] Struik, D. *A Concise History of Mathematics*, Dover Publications Inc., New York, 1967, pp. 92.

[4] Bergamini, D. *Mathematics*, Life Science Library, Time Inc., New York, 1963, pp. 152.

[5] Hsu, H. *Fourier Analysis*, Simon and Schuster, New York, 1970, pp. 75.

[6] Chester, D. B., and Phillips, G. "Single Chip Digital Down Converter Simplifies RF DSP Applications," *RF Design*, November 1992.

[7] Boutin, N., "Complex Signals," *RF Design*, December 1989.

Mean, Variance, and Standard Deviation

In our studies, we're often forced to consider noise functions. These are descriptions of noise signals that we cannot explicitly describe with a time-domain equation. Noise functions can be quantified, however, in a worthwhile way using the statistical measures of mean, variance, and standard deviation. Although we only touch, here, on the very broad and important field of statistics, we will describe why, how, and when to use these statistical indicators, so that we can add them to our collection of signal analysis tools. First we'll determine how to calculate these statistical values for a series of discrete data samples, cover an example using a continuous analytical function, and conclude this appendix with a discussion of the probability density functions of several random variables that are common in the field of digital signal processing. So let's proceed by sticking our toes in the chilly waters of the mathematics of statistics to obtain a few definitions.

D.1 Statistical Measures

Consider a continuous sinusoid having a frequency of f_o Hz with a peak amplitude of A_p expressed by the equation

$$x(t) = A_p\sin(2\pi f_o t) \ . \tag{D-1}$$

Equation (D-1) completely specifies $x(t)$—that is, we can determine $x(t)$'s exact value at any given instant. For example, when time $t = 1/4f_o$, we know that $x(t)$'s amplitude will be A_p and, at the later time $t = 1/2f_o$, $x(t)$'s amplitude will be zero. On the other hand, we have no definite way to express the successive values of a random function or of random noise.[†] There's no equation like Eq. (D-1) available to predict future noise-amplitude values for example. (That's why they call it random noise.) Statisticians have, however, developed powerful mathematical tools to characterize several properties of random functions. The most important of these properties have been given the names *mean*, *variance*, and *standard deviation*.

Mathematically, the *mean*, or *average*, of N separate values of a sequence x, denoted x_{ave}, is defined as [1]

$$x_{ave} = \frac{1}{N} \sum_{n=1}^{N} x(n) = \frac{x(1) + x(2) + x(3) + \ldots + x(N)}{N} \, . \tag{D-2}$$

Equation (D-2), already familiar to most people, merely states that the average of a sequence of N numbers is the sum of those numbers divided by N. Graphically, the average can be depicted as that value about which a series of sample values cluster, or congregate, as shown in Figure D-1. If the eight values depicted by the dots in Figure D-1 represent some measured quantity and we applied those values to Eq. (D-2), the average of the series is 5.17, as shown by the dotted line.

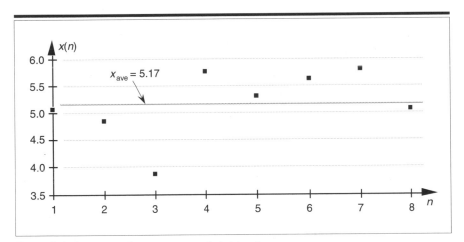

Figure D-1 Average of a sequence of eight values.

[†] We define *random noise* to be unwanted, unpredictable, disturbances contaminating a signal or a data sequence of interest.

Now that we've defined *average*, another key definition is the variance of a sequence, σ^2, defined as

$$\sigma^2 = \frac{1}{N} \sum_{n=1}^{N} [x(n) - x_{ave}]^2$$

$$= \frac{[x(1) - x_{ave}]^2 + [x(2) - x_{ave}]^2 + [x(3) - x_{ave}]^2 + \dots + [x(N) - x_{ave}]^2}{N}. \quad \text{(D-3)}$$

Sometimes, in the literature, we'll see σ^2 defined with a $1/(N-1)$ factor before the summation instead of the $1/N$ factor in Eq. (D-3). There are subtle statistical reasons why the $1/(N-1)$ factor sometimes gives more accurate results [2]. However, when N is greater than, say 20, as it will be for our purposes, the difference between the two factors will have no practical significance.

Variance is a very important concept because it's the yardstick with which we measure, for example, the effect of quantization errors and the usefulness of signal-averaging algorithms. It gives us an idea how the aggregate values in a sequence fluctuate about the sequence's average and provides us with a well-defined quantitative measure of those fluctuations. (Because the positive square root of the variance, the standard deviation, is typically denoted as σ in the literature, we'll use the conventional notation of σ^2 for the variance.) Equation (D-3) looks a bit perplexing if you haven't seen it before. Its meaning becomes clear if we examine it carefully. The $x(1) - x_{ave}$ value in the bracket, for example, is the difference between the $x(1)$ value and the sequence average x_{ave}. For any sequence value $x(n)$, the $x(n) - x_{ave}$ difference, which we denote as $\Delta(n)$, can be either positive or negative, as shown in Figure D-2. Specifically, the differences $\Delta(1)$, $\Delta(2)$, $\Delta(3)$, and $\Delta(8)$ are negative because their corresponding sequence values are below the sequence average shown by the dotted line. If we replace the $x(n) - x_{ave}$ difference terms in Eq. (D-3) with $\Delta(n)$ terms, the variance can be expressed as

$$\sigma^2 = \frac{1}{N} \sum_{n=1}^{N} [\Delta(n)]^2, \text{ where } \Delta(n) = x(n) - x_{ave} . \quad \text{(D-4)}$$

The reader might wonder why the squares of the differences are summed, instead of just the differences themselves. If we just add the differences, some of the negative $\Delta(n)$s will cancel some of the positive $\Delta(n)$s

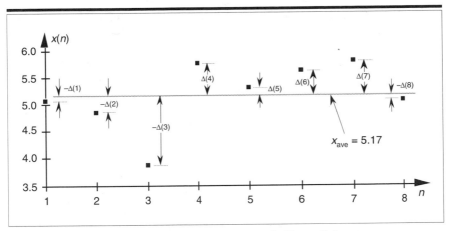

Figure D-2 Difference values $\Delta(n)$ of the sequence in Figure D-1.

resulting in a sum that may be too small. For example, if we add the $\Delta(n)$s in Figure D-2, the positive $\Delta(6)$ and $\Delta(7)$ values and the negative $\Delta(3)$ value will just about cancel each other out and we don't want that. Because we need an unsigned measure of each difference, we use the difference-squared terms as indicated by Eq. (D-4). In that way, individual $\Delta(n)$ difference terms will contribute to the overall variance regardless of whether the difference is positive or negative. Plugging the $\Delta(n)$ values from the example sequence in Figure D-2 into Eq. (D-4), we get a variance value of 0.34. Another useful measure of a signal sequence is the square root of the variance known as the *standard deviation*. Taking the square root of Eq. (D-3) to get the standard deviation σ,

$$\sigma = \sqrt{\sigma^2} = \sqrt{\frac{1}{N}\sum_{n=1}^{N}[x(n) - x_{\text{ave}}]^2} \ . \tag{D-5}$$

So far, we have three measurements to use in evaluating a sequence of values: the average x_{ave}, the variance σ^2, and the standard deviation σ. Where x_{ave} indicates about what constant level the individual sequence values vary, σ^2 is a measure of the magnitude of the noise fluctuations about the average x_{ave}. If the sequence represents a series of random signal samples, we can say that x_{ave} specifies the average, or constant, value of the signal. The variance σ^2 is the magnitude squared, or power, of the fluctuating component of the signal. The standard deviation, then, is an indication of the magnitude of the fluctuating component of the signal.

D.2 Standard Deviation, or RMS, of a Continuous Sinewave

For sinewaves, electrical engineers have taken the square root of Eq. (D-3), with $x_{ave} = 0$, and defined a useful parameter, called the *rms* value, that's equal to the standard deviation. For discrete samples, that parameter, $x(n)_{rms}$, is defined as

$$x(n)_{rms} = \sqrt{\frac{1}{N} \sum_{n=1}^{N} x(n)^2} \;.$$ (D-6)

The $x(n)_{rms}$ in Eq. (D-6), obtained by setting $x_{ave} = 0$ in Eq. (D-5), is the square *root* of the *mean* (average) of the *squares* of the sequence $x(n)$. For a continuous sinusoid $x(t) = A_p \sin(2\pi f t) = A_p \sin(\omega t)$ whose average value is zero, $x(t)_{rms}$ is $x(t)_{rms\text{-}sinewave}$ defined as

$$x(t)_{rms-sinewave} = \sqrt{\frac{1}{2\pi} \int_0^{2\pi} [A_p \sin(\omega t)]^2 d(\omega t)}$$

$$= \sqrt{\frac{A_p^2}{2\pi} \int_0^{2\pi} \frac{1}{2} [1 - \cos(\omega t)] d(\omega t)}$$

$$= \sqrt{\frac{A_p^2}{2\pi} \cdot \left[\frac{\omega t}{2} - \frac{1}{2}(\sin(\omega t))\right]_0^{2\pi}} = \sqrt{\frac{A_p^2}{2\pi} \cdot \left[\frac{2\pi}{2}\right]}$$

$$= \frac{A_p}{\sqrt{2}} \;.$$ (D-7)

This $x(t)_{rms\text{-}sinewave} = A_p / \sqrt{2}$ expression is a lot easier to use for calculating average power dissipation in circuit elements than taking the integral of more complicated expressions for instantaneous power dissipation. The variance of a sinewave is, of course, the square of Eq. (D-7), or $A_p^2 / 2$.

We've provided the equations for the mean (average) and variance of a sequence of discrete values, introduced an expression for the standard deviation or rms of a sequence, and given an expression for the rms of a continuous sinewave. The next question is "How can we characterize random functions for which there are no equations to predict their values and we have no discrete sample values with which to work?" The answer is that we must use probability density functions.

D.3 The Mean and Variance of Random Functions

To determine the mean or variance of a random function, we use what's called the *probability density function*. The probability density function (PDF) is a measure of the likelihood of a particular value occurring in some function. We can explain this concept with simple examples of flipping a coin or throwing dice as illustrated in Figure D-3(a) and (b). The result of flipping a coin can only be one of two possibilities: heads or tails. Figure D-3(a) indicates this PDF and shows that the probability (likelihood) is equal to one-half for both heads and tails. That is, we have an equal chance that the coin side facing up will be heads or tails. The sum of those two probability values is one, meaning that there's a 100% probability that either a head or a tail will occur.

Figure D-3(b) shows the probability of a particular sum of the upper faces when we throw a pair of dice. This probability function is not uniform because, for example, we're six times more likely to have the die faces add to seven than add to two (snake eyes). We can say that after tossing the dice a large number of times, we should expect that 6/36 = 16.7 percent of those tosses would result in sevens, and 1/36 = 2.8 percent of the time we'll get snake eyes. The sum of those eleven probability values

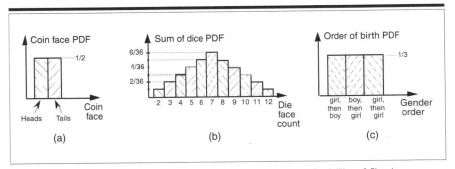

Figure D-3 Simple probability density functions: (a) the probability of flipping a single coin; (b) the probability of a particular sum of the upper faces of two die; (c) the probability of the order of birth of the girl and her sibling.

in Figure D-3(b) is also one, telling us that this PDF accounts for all (100%) of the possible outcomes of throwing the dice.

The fact that PDFs must account for all possible results is emphasized in an interesting way in Figure D-3(c). If a woman says, "Of my two children, one is a girl. What's the probability that she has a sister?" Be careful now—curiously enough, the answer to this controversial question is not a 50–50 chance. There are more possibilities to consider than just the girl having a brother or a sister. We can think of all the possible combinations of birth order of two children such that one child is a girl. Because we don't know the gender of the first-born child, there are three gender order possibilities: girl, then boy; boy, then girl; and girl, then girl as shown in Figure D-3(c). So the possibility of the daughter having a sister is 1/3 instead of 1/2! (Believe it.) Again, the sum of those three 1/3rd probability values is one.

Two important features of PDFs are illustrated by the examples in Figure D-3: PDFs are always positive, and the areas under their *curves* must be equal to unity. The very concept of PDFs make them a positive *likelihood* that a particular result will occur, and the fact that some result must occur is equivalent to saying that there's a probability of one (100% chance) that we'll have a result. For continuous probability density functions, $p(f)$, we indicate these two characteristics by

PDF values are never negative: $\rightarrow$ $\qquad\qquad p(f) \geq 0 ,$ $\qquad\qquad\qquad$ (D-8)

and

The sum of all the PDF values is one: $\rightarrow$ $\qquad\qquad \int\limits_{-\infty}^{\infty} p(f)df = 1 .$ $\qquad\qquad$ (D-8')

In Section D.1 we illustrated how to calculate the average (mean) and variance of discrete samples. We can also determine these statistical measures for a random function if we know the PDF of the function. Using μ_f to denote the average of a random function of f, then, μ_f is defined as

$$\mu_f = \int\limits_{-\infty}^{\infty} f \cdot p(f)df ,$$
$\qquad\qquad\qquad\qquad\qquad\qquad\qquad\qquad$ (D-9)

and the variance of f is defined as [3]

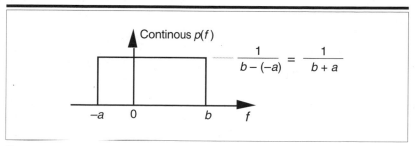

Figure D-4 Continuous, uniform probability density function.

$$\sigma_f^2 = \int_{-\infty}^{\infty}(f-\mu_f)^2 \cdot p(f)df = \int_{-\infty}^{\infty}f^2 \cdot p(f)df - \mu_f^2 \ . \tag{D-10}$$

In digital signal processing, we'll encounter continuous probability density functions that are uniform in value similar to the examples in Figure D-3. In these cases, it's easy to use Eqs. (D-9) and (D-10) to determine their average and variance. Figure D-4 illustrates a uniform, continuous PDF indicating a random function whose values have an equal probability of being anywhere in the range from $-a$ to b. From Eq. (D-8), we know that the area under the curve must be unity (i.e., the probability is 100% that the value will be somewhere under the curve). So the amplitude of $p(f)$ must be the area divided by the width, or $p(f) = 1/(b + a)$. From Eq. (D-9), the average of this $p(f)$ is given by

$$\mu_f = \int_{-\infty}^{\infty}f \cdot p(f)df = \int_{-\infty}^{\infty}f \cdot \frac{1}{b+a}df$$

$$= \frac{1}{b+a}\int_{-a}^{b}f\,df = \frac{1}{b+a} \cdot \left[\frac{f^2}{2}\right]_{-a}^{b} = \frac{b^2-a^2}{2(b+a)}$$

$$= \frac{(b+a)(b-a)}{2(b+a)} = \frac{b-a}{2} \tag{D-11}$$

which happens to be the midpoint in the range from $-a$ to b. The variance of the PDF in Figure D-4 is given by

$$\sigma_f^2 = \int_{-\infty}^{\infty} f^2 \cdot p(f) df - \mu_f^2 = \int_{-a}^{b} f^2 \cdot \frac{1}{b+a} df - \frac{(b-a)^2}{4}$$

$$= \frac{1}{b+a} \cdot \left[\frac{f^3}{3} \right]_{-a}^{b} - \frac{(b+a)^2}{4} = \frac{1}{3(b+a)} \cdot (b^3 + a^3) - \frac{(b-a)^2}{4}$$

$$= \frac{(b+a)(b^2 - ab + a^2)}{3(b+a)} - \frac{b^2 - 2ab + a^2}{4}$$

$$= \frac{b^2 + 2ab + a^2}{12} = \frac{(b+a)^2}{12} \ . \tag{D-12}$$

We use the results of Eqs. (D-11) and (D-12) in Chapter 9 to analyze the errors induced by quantization from analog-to-digital converters and the effects of finite word lengths of hardware registers.

D.4 The Normal Probability Density Function

A probability density function that's so often encountered in nature deserves our attention. This function is so common that it's actually called the *normal* probability density function.† This function, whose shape is shown in Figure D-5, is important because random data having this distribution is very useful in testing both software algorithms and hardware processors. The normal PDF is defined mathematically by

$$p(x) = \frac{1}{\sigma\sqrt{2\pi}} e^{-(x-x_{\text{ave}})^2 / 2\sigma^2} \ . \tag{D-13}$$

The area under the curve equals one, and the percentages at the bottom of Figure D-5 tell us that, for random functions having a normal distribution, there's a 68.27 percent chance that any particular value of x will differ from the mean by $\leq\sigma$. Likewise, 99.73 percent of all the x data values will be within 3σ of the mean x_{ave}.

---

† The *normal probability density function* is sometimes called the Gaussian function. A scheme for generating data to fit this function is discussed in Section 10.12.

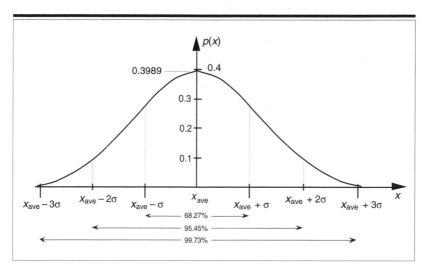

Figure D-5 A normal PDF with mean = x_{ave} and standard deviation = σ.

References

[1] Papoulis, A. *Probability Random Variables, and Stochastic Processes*, McGraw-Hill, New York, 1965, pp. 266–68.

[2] Miller, I., and Freund, J. *Probability and Statistics for Engineers*, 2nd Ed., Prentice-Hall, Englewood Cliffs, New Jersey, 1977, pp. 118.

[3] Bendat, J., and Piersol, A. *Measurement and Analysis of Random Data*, John Wiley and Sons, New York, 1972, pp. 61.

Decibels
(dB and dBm)

This appendix introduces the logarithmic function used to improve the magnitude resolution of frequency-domain plots when we evaluate signal spectrums, digital filter magnitude responses, and window function magnitude responses. When we use a logarithmic function to plot signal levels in the frequency domain, the vertical axis unit of measure is *decibels*.

E.1 Using Logarithms to Determine Relative Signal Power

In discussing decibels, it's interesting to see how this unit of measure evolved. When comparing continuous (analog) signal levels, early specialists in electronic communications found it useful to define a measure of the difference in powers of two signals. If that difference was treated as the logarithm of a ratio of powers, it could be used as a simple additive measure to determine the overall gain or loss of cascaded electronic circuits. The positive logarithms associated with system components having gain could be added to the negative logarithms of those components having loss to quickly determine the overall gain or loss of the system. With this in mind, the difference between two signal power levels (P_1 and P_2), measured in bels, was defined as the base 10 logarithm of the ratio of those powers, or

$$\text{Power difference} = \log_{10}\left(\frac{P_1}{P_2}\right) \text{ bels.}^\dagger \qquad \text{(E-1)}$$

---

† The dimensionless unit of measure *bel* was named in honor of Alexander Graham Bell.

486

The use of Eq. (E-1) led to another evolutionary step because the unit of bel was soon found to be inconveniently large. For example, it was discovered that the human ear could detect audio power level differences of one-tenth of a bel. Measured power differences smaller than one bel were so common that it led to the use of the decibel (bel/10), effectively making the unit of bel obsolete. The decibel (dB), then, is a unit of measure of the relative power difference of two signals defined as

$$\text{Power difference} = 10 \cdot \log_{10}\left(\frac{P_1}{P_2}\right) \text{dB} \ . \tag{E-2}$$

The logarithmic function $10 \cdot \log_{10}$, plotted in Figure E-1, doesn't seem too beneficial at first glance, but its application turns out to be very useful. Notice the large change in the function's value, when the power ratio (P_1/P_2) is small, and the gradual change when the ratio is large. The effect of this nonlinearity is to provide greater resolution when the ratio P_1/P_2 is small, giving us a good way to recognize very small differences in the power levels of signal spectrums, digital filter responses, and window function frequency responses.

Let's demonstrate the utility of the logarithmic function's variable resolution. First, remember that the power of any frequency-domain sequence representing signal magnitude $|X(m)|$ is proportional to

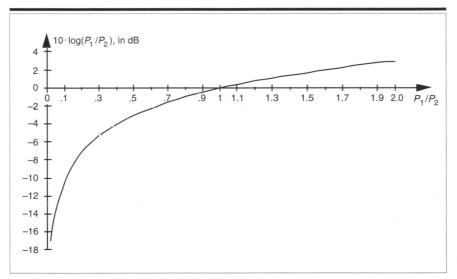

Figure E-1 Logarithmic decibel function of Eq. (E-2).

$|X(m)|$ squared. For convenience, the proportionality constant is assumed to be one, so we say the power of $|X(m)|$ is

$$\text{discrete power spectrum of } X(m) = |X(m)|^2. \qquad \text{(E-3)}$$

Although equation (E-3) may not actually represent power (in watts) in the classical sense, it's the squaring operation that's important here, because it's analogous to the traditional magnitude squaring operation used to determine the power of continuous signals. (Of course, if $X(m)$ is complex, we can calculate the power spectrum sequence using $|X(m)|^2 = X_{real}(m)^2 + X_{imag}(m)^2$.) Taking 10 times the log of Eq. (E-3) allows us to express a power spectrum sequence $X_{dB}(m)$ in dB as

$$X_{dB}(m) = 10 \cdot \log_{10}(|X(m)|^2) \text{ dB}. \qquad \text{(E-4)}$$

Because $\log(x^2) = \log(x) + \log(x) = 2\log(x)$, we can eliminate the squaring operation in Eq. (E-4) by doubling the factor of 10 and represent the power spectrum sequence by the expression

$$X_{dB}(m) = 20 \cdot \log_{10}(|X(m)|) \text{ dB}. \qquad \text{(E-5)}$$

Without the need for the squaring operation, Eq. (E-5) is a more convenient way, than Eq. (E-4), to calculate the $X_{dB}(m)$ power spectrum sequence from the $X(m)$ sequence.

Equations (E-4) and (E-5), then, are the expressions used to convert a linear magnitude axis to a logarithmic magnitude-squared, or power, axis measured in dB. What we most often see in the literature are normalized log magnitude spectral plots where each value of $|X(m)|^2$ is divided by the first $|X(0)|^2$ power value (for $m = 0$) as

$$\text{normalized } X_{dB}(m) = 10 \cdot \log_{10}\left(\frac{|X(m)|^2}{|X(0)|^2}\right) = 20 \cdot \log_{10}\left(\frac{|X(m)|}{|X(0)|}\right) \text{dB}. \qquad \text{(E-6)}$$

The division by the $|X(0)|^2$ or $|X(0)|$ value always forces the first value in the normalized log magnitude sequence $X_{dB}(m)$ equal to 0 dB.[†] This makes it easy for us to compare multiple log magnitude spectral plots. To illustrate, let's look at the frequency-domain representations of the Hanning

---

[†] That's because $\log_{10}(|X(0)|/|X(0)|) = \log_{10}(1) = 0$.

and triangular window functions. The magnitudes of those frequency-domain functions are plotted on a linear scale in Figure E-2(a) where we've arbitrarily assigned their peak values to be two. Comparing the two linear scale magnitude sequences, $W_{\text{Hanning}}(m)$ and $W_{\text{triangular}}(m)$, we can see some minor differences between their magnitude values. If we're interested in the power associated with the two window functions, we square the two magnitude functions and plot them on a linear scale as in Figure E-2(b). The difference between the two window function's power sequences is impossible to see above the frequency of, say, $m = 8$ in Figure E-2(b). Here's where the dB scale helps us out. If we plot the normalized log magnitude versions of the two magnitude-squared sequences on a logarithmic dB scale using Eq. (E-6), the difference between the two functions will become obvious.

Normalization, in the case of the Hanning window, amounts to calculating the log magnitude sequence normalized over $|W_{\text{Hanning}}(0)|$ as

$$W_H(m) = 10 \cdot \log_{10}\left(\frac{|W_{\text{Hanning}}(m)|^2}{|W_{\text{Hanning}}(0)|^2}\right) = 20 \cdot \log_{10}\left(\frac{|W_{\text{Hanning}}(m)|}{|W_{\text{Hanning}}(0)|}\right) \text{dB} . \quad \text{(E-7)}$$

The normalized log magnitude sequences are plotted in Figure E-2(c). We can now clearly see the difference in the magnitude-squared window functions in Figure E-2(c) as compared to the linear plots in Figure E-2(b). Notice how normalization forced the peak values for both log magnitude functions in Figure E-2(c) to be zero dB. (The dots in Figure E-2 are connected by lines to emphasize the sidelobe features of the two log magnitude sequences.)

Although we've shown the utility of dB plots using window function frequency responses as examples, the dB scale is equally useful when we're plotting signal-power spectrums or digital filter frequency responses. We can further demonstrate the dB scale using a simple digital filter example. Let's say we're designing an 11-tap highpass FIR filter whose coefficients are shown in Figure E-3(a). If the center coefficient $h(5)$ is -0.48, the filter's frequency magnitude response $|H_{-0.48}(m)|$ can be plotted as the white dots on the linear scale in Figure E-3(b). Should we change $h(5)$ from -0.48 to -0.5, the new frequency magnitude response $|H_{-0.5}(m)|$ would be the black dots in Figure E-3(b). It's difficult to see much of a difference between $|H_{-0.48}(m)|$ and $|H_{-0.5}(m)|$ on a linear scale. If we used Eq. (E-6) to calculate two normalized log magnitude sequences, they could be plotted as shown in Figure E-3(c) where the filter sidelobe effects of changing $h(5)$ from -0.48 to -0.5 are now easy to see.

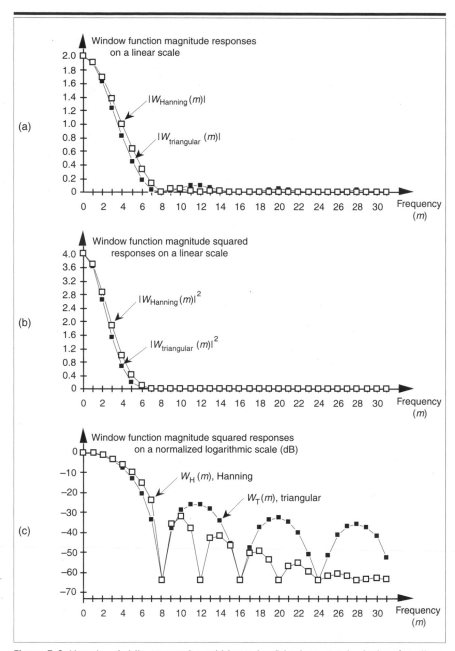

Figure E-2 Hanning (white squares) and triangular (black squares) window functions in the frequency domain: (a) magnitude responses using a linear scale; (b) magnitude-squared responses using a linear scale; (c) log magnitude responses using a normalized dB scale.

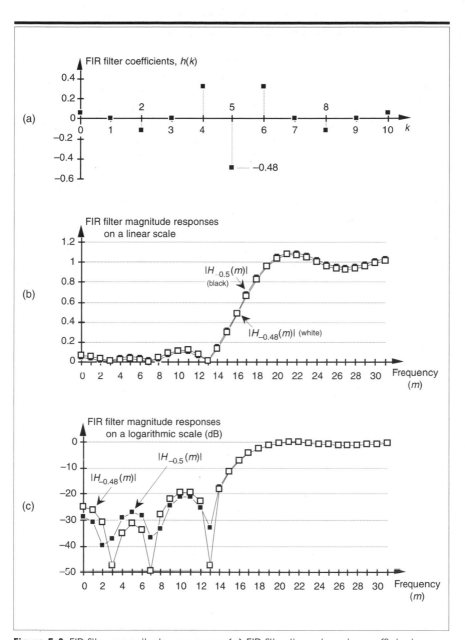

Figure E-3 FIR filter magnitude responses: (a) FIR filter time-domain coefficients; (b) magnitude responses using a linear scale. $|H_{-0.48}(m)|$ is shown by the white squares and $|H_{-0.5}(m)|$ is shown by the black squares; (c) log magnitude responses using the dB scale.

E.2 Some Useful Decibel Numbers

If the reader finds him- or herself using dB scales on a regular basis, there are a few constants worth committing to memory. A power difference of 3 dB corresponds to a power factor of 2; that is, if the magnitude-squared ratio of two different frequency components is 2, then from Eq. (E-2),

$$\text{power difference} = 10 \cdot \log_{10}\left(\frac{2}{1}\right) = 10 \cdot \log_{10}(2) = 3.01 \approx 3 \text{ dB} . \qquad \text{(E-8)}$$

Likewise, if the magnitude-squared ratio of two different frequency components is 1/2, then the relative power difference is –3 dB because

$$\text{power difference} = 10 \cdot \log_{10}\left(\frac{1}{2}\right) = 10 \cdot \log_{10}(0.5) = -3.01 \approx -3 \text{ dB} . \qquad \text{(E-9)}$$

Table E-1 lists several magnitude and power ratios vs. dB values worth remembering. Keep in mind that decibels indicate only relative power

Table E-1 Some Useful Logarithmic Relationships

Magnitude Ratio	Magnitude-Squared Power (P_1/P_2) Ratio	Relative dB (approximate)	
$10^{-1/2}$	10^{-1}	-10	$\leftarrow P_1$ is one-tenth P_2
2^{-1}	$2^{-2} = 1/4$	-6	$\leftarrow P_1$ is one-fourth P_2
$2^{-1/2}$	$2^{-1} = 1/2$	-3	$\leftarrow P_1$ is one-half P_2
2^0	$2^0 = 1$	0	$\leftarrow P_1$ is equal to P_2
$2^{1/2}$	$2^1 = 2$	3	$\leftarrow P_1$ is twice P_2
2^1	$2^2 = 4$	6	$\leftarrow P_1$ is four times P_2
$10^{1/2}$	$10^1 = 10$	10	$\leftarrow P_1$ is ten times P_2
10^1	$10^2 = 100$	20	$\leftarrow P_1$ is one hundred times P_2
$10^{3/2}$	$10^3 = 1000$	30	$\leftarrow P_1$ is one thousand times P_2

relationships. For example, if we're told that signal A is 6 dB above signal B, we know that the power of signal A is four times that of signal B, and that the magnitude of signal A is twice the magnitude of signal B. We may not know the absolute power of signals A and B in watts, but we do know that the power ratio is $P_A/P_B = 4$.

E.3 Absolute Power Using Decibels

Let's discuss another use of decibels that the reader may encounter in the literature. It's convenient for practitioners in the electronic communications field to measure continuous signal-power levels referenced to a specific absolute power level. In this way, they can speak of absolute power levels in watts while taking advantage of the convenience of decibels. The most common absolute power reference level used is the milliwatt. For example, if P_2 in Eq. (E-2) is a reference power level of one milliwatt, then

$$\text{absolute power of } P_1 = 10 \cdot \log_{10}\left(\frac{P_1}{P_2}\right) = 10 \cdot \log_{10}\left(\frac{P_1 \text{ in watts}}{1 \text{ milliwatt}}\right) \text{dBm. (E-10)}$$

The dBm unit of measure in Eq. (E-10) is read as "dB relative to a milliwatt." Thus, if a continuous signal is specified as having a power of 3 dBm, we know that the signal's absolute power level is 2 times one milliwatt, or 2 milliwatts. Likewise, a –10 dBm signal has an absolute power of 0.1 milliwatts.[†]

The reader should take care and not inadvertently use dB and dBm interchangeably. They mean very different things. Again, dB is a relative power level relationship, and dBm is an absolute power level in milliwatts.

[†] Other absolute reference power levels can be used. People involved with high-power transmitters sometimes use a single watt as their reference power level. Their unit of power using decibels is the dBW, read as "dB relative to a watt." In this case, for example, 3 dBW is equal to a 2 watt power level.

Digital Filter Terminology

The first step in becoming familiar with digital filters is to learn to speak the language used in the filter business. Fortunately, the vocabulary of digital filters corresponds very well with the mother tongue used for continuous (analog) filters—so we don't have to unlearn anything that we already know. This appendix is an introduction to the terminology of digital filters.

Allpass filter — an IIR filter whose magnitude response is unity over its entire frequency range, but whose phase response is variable. Allpass filters are typically appended in a cascade arrangement following a standard IIR filter, $H_1(z)$, as shown in Figure F-1.

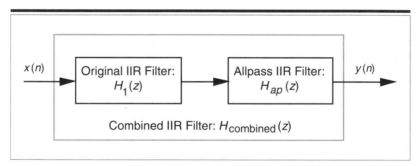

Figure F-1 Typical use of an allpass filter.

An allpass filter, $H_{ap}(z)$, can be designed so that its phase response compensates for, or *equalizes*, the nonlinear phase response of an original IIR filter [1–3]. Thus, the phase response of the combined filter, $H_{combined}(z)$, is more linear than the original $H_1(z)$, and this is particularly desirable in communications systems. In this context, an allpass filter is sometimes called a *phase equalizer*.

Attenuation — an amplitude loss, usually measured in dB, incurred by a signal after passing through a digital filter. Filter attenuation is the ratio, at a given frequency, of the signal amplitude at the output of the filter divided by the signal amplitude at the input of the filter, defined as

$$\text{attenuation} = 20 \cdot \log_{10}\left(\frac{a_{\text{out}}}{a_{\text{in}}}\right) \text{dB}. \qquad \text{(F-1)}$$

For a given frequency, if the output amplitude of the filter is smaller than the input amplitude, the ratio in Eq. (F-1) is less than one, and the attenuation is a negative number.

Band Reject Filter — a filter that rejects (attenuates) one frequency band and passes both a lower and a higher frequency band. Figure F-2(a) depicts the frequency response of an ideal band reject filter. This filter type is sometimes called a *notch filter*.

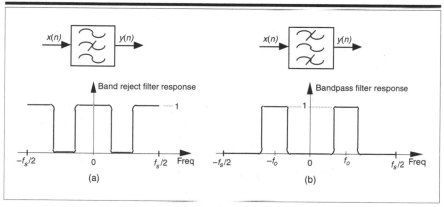

Figure F-2 Filter symbols and frequency responses: (a) band reject filter; (b) bandpass filter.

Bandpass Filter — a filter, as shown in Figure F-2(b), that passes one frequency band and attenuates frequencies above and below that band.

Bandwidth — the frequency width of the passband of a filter. For a lowpass filter, the bandwidth is equal to the cutoff frequency. For a bandpass filter, the bandwidth is typically defined as the frequency difference between the upper and lower 3 dB points.

Bessel function — a mathematical function used to produce the most linear phase response of all IIR filters with no consideration of the frequency magnitude response. Specifically, filter designs based on Bessel functions have maximally constant group delay.

Butterworth function — a mathematical function used to produce maximally flat filter magnitude responses with no consideration of phase linearity or group delay variations. Filter designs based on a Butterworth function have no amplitude ripple in either the passband or the stopband. Unfortunately, for a given filter order, Butterworth designs have the widest transition region of the most popular filter design functions.

Cascaded filters — a filtering *system* where multiple individual filters are connected in series; that is, the output of one filter drives the input of the following filter as illustrated in Figures F-1 and 6-38(a).

Center Frequency (f_o) — the frequency lying at the midpoint of a bandpass filter. Figure F-2(b) shows the f_o center frequency of a bandpass filter.

Chebyshev function — a mathematical function used to produce passband or stopband ripples constrained within fixed bounds. There are families of Chebyshev functions based on the amount of ripple, such as 1 dB, 2 dB, and 3 dB of ripple. Chebyshev filters can be designed to have a frequency response with ripples in the passband and flat passbands (Chebyshev Type I), or flat passbands and ripples in the stopband (Chebyshev Type II). Chebyshev filters cannot have ripples in both the passband and the stopband. Digital filters based upon Chebyshev functions have steeper transition region roll-off but more nonlinear-phase response characteristics than, say, Butterworth filters.

Cutoff Frequency — the upper passband frequency for low-pass filters, and the lower passband frequency for highpass filters. A cutoff frequency is determined by the −3 dB point of a filter magnitude response relative to a peak passband value. Figure F-3 illustrates the f_c cutoff frequency of a low-pass filter.

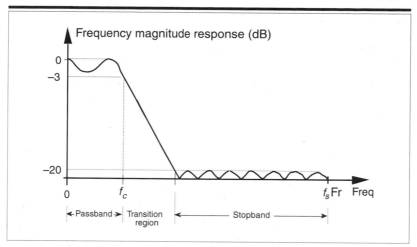

Figure F-3 A low-pass digital filter frequency response. The stopband relative amplitude is –20 dB.

Decibels (dB) — a unit of attenuation, or gain, used to express the relative voltage or power between two signals. For filters, we use decibels to indicate cutoff frequencies (–3 dB) and stopband signal levels (–20 dB) as illustrated in Figure F-3. Appendix E discusses decibels in more detail.

Decimation filter — a low-pass digital FIR filter where the output sample rate is less than the filter's input sample rate. As discussed in Section 7.3, to avoid aliasing problems, the output sample rate must not violate the Nyquist criterion.

Digital filter — computational process, or algorithm, transforming a discrete sequence of numbers (the input) into another discrete sequence of numbers (the output) having a modified frequency-domain spectrum. Digital filtering can be in the form of a software routine operating on data stored in computer memory or can be implemented with dedicated hardware.

Elliptic function — a mathematical function used to produce the sharpest roll-off for a given number of filter taps. However, filters designed by using elliptic functions, also called *Cauer filters*, have the poorest phase linearity of the most common IIR filter design functions. The ripple in the passband and stopband are equal with elliptic filters.

Envelope delay — *see* group delay.

Filter coefficients — the set of constants, also called *tap weights*, used to multiply against delayed signal sample values within a digital filter structure. Digital filter design is an exercise in determining the filter coefficients that will yield the desired filter frequency response. For an FIR filter, by definition, the filter coefficients are the impulse response of the filter.

Filter order — a number describing the highest exponent in the numerator or denominator of the z-domain transfer function of a digital filter. For FIR filters, there is no denominator in the transfer function, and the filter order is merely the number of taps used in the filter structure. For IIR filters, the filter order is equal to the number of delay elements in the filter structure. Generally, the larger the filter order, the better the frequency magnitude response performance of the filter.

Finite impulse response (FIR) filter — a class of digital filters that has only zeros on the z-plane. The key implications of this are that FIR filters are always stable and have linear phase responses (as long as the filter's coefficients are symmetrical). For a given filter order, FIR filters have a much more gradual transition region roll-off than digital IIR filters.

Frequency magnitude response — a frequency-domain description of how a filter interacts with input signals. The frequency magnitude response in Figure F-3 is a curve of filter attenuation (in dB) vs. frequency. Associated with a filter's magnitude response is a phase response.

Group delay — the derivative of a filter's phase with respect to frequency, $G = \Delta\phi/\Delta f$, or the slope of a filter's $H_\phi(m)$ phase response curve. The concept of group delay deserves additional explanation beyond a simple definition. For an ideal filter, the phase will be linear and the group delay would be constant. Group delay, whose unit of measure is time in seconds, can also be thought of as the propagation time delay of the envelope of an amplitude-modulated signal as it passes through a digital filter. (In this context, group delay is often called *envelope delay*.) Group delay distortion occurs when signals at different frequencies take different amounts of time to pass through a filter. If the group delay is denoted G, then the relationship between group delay, a $\Delta\phi$ increment of phase, and a Δf increment of frequency is

$$G = \frac{\Delta\varnothing_{\text{degrees}} / 360}{\Delta f} = \frac{\Delta\varnothing_{\text{radians}} / 2\pi}{\Delta f} \text{ seconds.} \qquad \text{(F-2)}$$

If we know a linear phase filter's phase shift ($\Delta\varnothing$) in degrees/Hz, or radians/Hz, we can determine the group delay in seconds using

$$G \cdot \Delta f = G \cdot 1 = G = \frac{\Delta\varnothing_{\text{degrees/Hz}}}{360} = \frac{\Delta\varnothing_{\text{radians/Hz}}}{2\pi} \text{ seconds.} \qquad \text{(F-3)}$$

To demonstrate Eq. (F-3) and illustrate the effect of a nonlinear phase filter, let's assume that we've digitized a continuous waveform comprising four frequency components defined by

$$x(t) = \sin(2\pi \cdot 1 \cdot t) + \sin(2\pi \cdot 3 \cdot t)/3 + \sin(2\pi \cdot 5 \cdot t)/5 + \sin(2\pi \cdot 7 \cdot t)/7. \qquad \text{(F-4)}$$

The $x(t)$ input comprises the sum of 1-Hz, 3-Hz, 5-Hz, and 7-Hz sinewaves, and its discrete representation is shown in Figure F-4(a). If we applied the discrete sequence representing $x(t)$ to the input of an ideal 4-tap linear-phase low-pass digital FIR filter with a cutoff frequency of greater than 7 Hz, and whose phase shift is 0.25 radians/Hz, the filter's output sequence would be that shown in Figure F-4(b).

Because the filter's phase shift is 0.25 radians/Hz, Eq. (F-3) tells us that the filter's constant group delay G in seconds is

$$G = \frac{\Delta\varnothing_{\text{radians/Hz}}}{2\pi} = \frac{0.25}{2\pi} = 0.04 \text{ seconds.} \qquad \text{(F-5)}$$

With a constant group delay of 0.04 seconds, the 1-Hz input sinewave is delayed at the filter output by 0.25 radians, the 3-Hz sinewave is delayed by 0.75 radians, the 5-Hz sinewave by 1.25 radians, and the 7-Hz sinewave by 1.75 radians. Notice how a linear-phase (relative to frequency) filter results in an output that's merely a time shifted version of the input as seen in Figure F-4(b). The amount of time shift is the group delay of 0.04 seconds. Figure F-4(c), on the other hand, shows the distorted output waveform if the filter's phase was nonlinear, for whatever reason, such that the phase shift was 3.5 radians instead of the ideal 1.75 radians at 7 Hz. Notice the distortion of the beginning of the output waveform envelope in Figure F-4(c) compared to Figure F-4(b). The point here is that, if the desired informa-

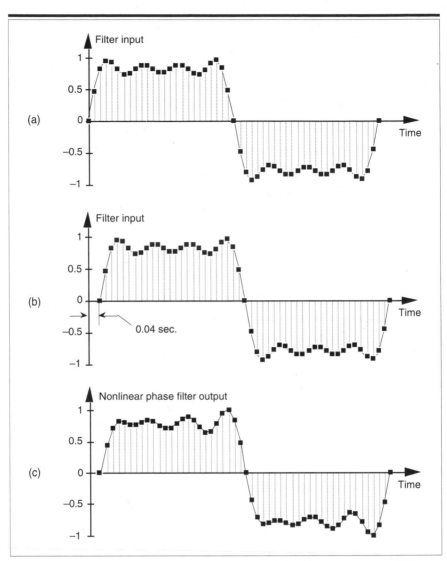

Figure F-4 Filter time-domain response examples: (a) filter input sequence; (b) linear-phase filter output sequence that's time shifted by 0.04 seconds, duplicating the input sequence; (c) distorted output sequence due to a filter with a nonlinear phase.

tion is contained in the envelope of a signal that we're passing through a filter, we'd like that filter's passband phase to be as linear as possible with respect to frequency. In other words, we'd prefer the filter's group delay to vary as little as possible in the passband. (Additional aspects of nonlinear-phase filters are discussed in Section 5.8.)

Half-Band filter — a type of FIR filter whose transition region is centered at one quarter of the sampling rate, or $f_s/4$. Specifically, the end of the passband and the beginning of the stopband are equally spaced about $f_s/4$. Due to their frequency-domain symmetry, half-band filters are often used in decimation filtering schemes because half of their time-domain coefficients are zero. This reduces the number of necessary filter multiplications, as described in Section 5.7.

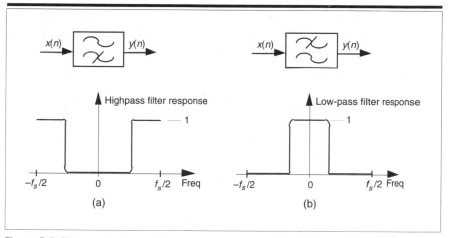

Figure F-5 Filter symbols and frequency responses: (a) highpass filter; (b) low-pass filter.

Highpass filter — a filter that passes high frequencies and attenuates low frequencies, as shown in Figure F-5(a). We've all experienced a kind of highpass filtering in our living rooms. Notice what happens when we turn up the treble control (or turn down the bass control) on our home stereo systems. The audio amplifier's normally flat frequency response changes to a kind of analog highpass filter giving us that sharp and *tinny* sound as the high-frequency components of the music are being accentuated.

Impulse response — a digital filter's time-domain output sequence when the input is a single unity-valued sample (impulse) preceded and followed by zero-valued samples. A digital filter's frequency-domain response can be calculated by taking the discrete Fourier transform of the filter's time-domain impulse response [4].

Infinite impulse response (IIR) filter — a class of digital filters that may have both zeros and poles on the z-plane. As such, IIR filters are not guaranteed to be stable and almost always have nonlinear phase responses. For a given filter order (number of IIR feedback taps), IIR filters have a much steeper transition region roll-off than digital FIR filters.

Linear-phase filter — a filter that exhibits a constant change in phase angle (degrees) as a function of frequency. The resultant filter phase plot vs. frequency is a straight line. As such, a linear-phase filter's group delay is a constant. To preserve the integrity of their information-carrying signals, linear phase is an important criteria for filters used in communication systems.

Low-pass filter — a filter that passes low frequencies and attenuates high frequencies as shown in Figure F-5(b). By way of example, we experience low-pass filtering when we turn up the bass control (or turn down the treble control) on our home stereo systems giving us that dull, muffled sound as the low-frequency components of the music are being intensified.

Notch filter — *see* band reject filter.

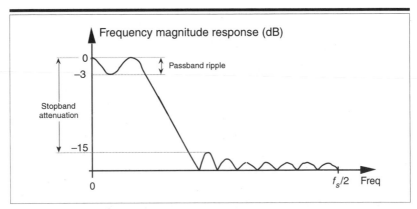

Figure F-6 Low-pass digital filter frequency response showing passband ripple and stopband attenuation.

Passband — the frequency range over which a filter passes signal energy. Usually defined as the frequency range where the filter's frequency response is equal to or greater than –3 dB, as depicted in Figure F-3.

Passband ripple — fluctuations, or variations, in the frequency magnitude response within the passband of a filter. Passband ripple, measured in dB, is illustrated in Figure F-6.

Phase response — the difference in phase, at a particular frequency, between an input sinewave and the output sinewave at that frequency. The phase response, sometimes called *phase delay*, is usually depicted by a curve showing the filter's phase shift vs. frequency. Section 5.8 discusses digital filter phase response in more detail.

Phase wrapping — an artifact of arctangent software routines, used to calculate phase angles, that causes apparent phase discontinuities. When a true phase angle is in the range of –180° to –360°, some software routines automatically convert those angles to their equivalent positive angles in the range of 0° to +180°. Section 5.8 illustrates an example of phase wrapping when the phase of an FIR filter is calculated.

Quadrature filter — a dual-path digital filter operating on complex signals, as shown in Figure F-7. One filter operates on the in-phase $i(n)$ data, and the other filter processes the quadrature-phase $q(n)$ signal data. Quadrature filtering is normally performed with low-pass filters.

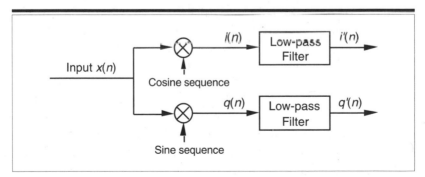

Figure F-7 Two low-pass filters used to implement quadrature filtering.

Relative attenuation — attenuation measured relative to the largest magnitude value. The largest signal level (minimum attenuation) is typically assigned the reference level of zero dB, as depicted in Figures F-2 and F-5, making all other magnitude points on a frequency-response curve negative dB values.

Ripple — refers to fluctuations (measured in dB) in the passband, or stopband, of a filter's frequency-response curve. Elliptic and Chebyshev-based filters have equiripple characteristics in that their ripple is constant across their passbands. Bessel and Butterworth derived filters have no ripple in their passband responses. Ripples in the stopband response are sometimes called *out-of-band ripple*.

Roll-off — a term used to describe the steepness, or slope, of the filter response in the transition region from the passband to the stopband. A particular digital filter may be said to have a roll-off of 12 dB/octave, meaning that the second-octave frequency would be attenuated by 24 dB, and the third-octave frequency would be attenuated by 36 dB, and so on.

Shape factor — a term used to indicate the steepness of a filter's roll-off. Shape factor is normally defined as the ratio of a filter's passband width divided by the passband width plus the transition region width. The smaller the shape factor value, the steeper the filter's roll-off. For an ideal filter with a transition region of zero width, the shape factor is unity. The term *shape factor* is also used to describe analog filters.

Stopband — that band of frequencies attenuated by a digital filter. Figure F-3 shows the stopband of a low-pass filter. Although the stopband attenuation in Figure F-3 is –20 dB, not all filters have stopband lobes of equal amplitude. Figure F-6 shows that stopband attenuation is measured between the peak passband amplitude and the largest stopband lobe amplitude.

Structure — refers to the block diagram showing how a digital filter is implemented. A recursive filter structure is one in which feedback takes place and past filter output samples are used, along with past input samples, in calculating the present filter output. IIR filters are almost always implemented with recursive filter structures. A nonrecursive filter structure is one in which only past input samples are used in calculating the present filter output. FIR filters are almost always implemented with nonrecursive filter structures. *See* Chapter 6 for examples of various digital filter structures.

Tap weights — *see* filter coefficients.

Tchebyschev function — *see* Chebyshev.

Transfer function — a mathematical expression of the ratio of the output of a digital filter divided by the input of the filter. Given the transfer function, we can determine the filter's frequency magnitude and phase responses.

Transition region — the frequency range over which a filter transitions from the passband to the stopband. Figure F-3 illustrates the transition region of a low-pass filter. The transition region is sometimes called the *transition band*.

Transversal filter — in the field of digital filtering, *transversal filter* is another name for FIR filters implemented with the nonrecursive structures described in Chapter 5.

References

[1] Rabiner, L. R., and Gold, B. *The Theory and Application of Digital Signal Processing*, Prentice-Hall, Englewood Cliffs, New Jersey, 1975, pp. 206, 273, and 288.

[2] Oppenheim, A. V., and Schafer, R. W. *Discrete-Time Signal Processing*, Prentice-Hall, Englewood Cliffs, New Jersey, 1989, pp. 236 and 441.

[3] Laakso, T. I., et al. "Splitting the Unit Delay," *IEEE Signal Processing Magazine*, January 1972, pp. 46.

[4] Pickerd, J. "Impulse-Response Testing Lets a Single Test Do the Work of Thousands," *EDN*, April 27, 1995.

INDEX